D0688042

Genetic Explanations

GENETIC EXPLANATIONS

Sense and Nonsense

EDITED BY

SHELDON KRIMSKY

JEREMY GRUBER

Harvard University Press

Cambridge, Massachusetts, and London, England 2013

Copyright © 2013 by the President and Fellows of Harvard College
All rights reserved
Printed in the United States of America

Library of Congress Cataloging-in-Publication Data

Genetic explanations : sense and nonsense / edited by Sheldon Krimsky and Jeremy Gruber.
p. ; cm.
Includes bibliographical references and index.
ISBN 978-0-674-06446-1 (alk. paper)
I. Krimsky, Sheldon. II. Gruber, Jeremy.
[DNLM: 1. Genetic Phenomena. 2. Genetic Determinism. 3. Interdisciplinary Communication. QU 500]
616'.042—dc23 2012005988

This book is a project of the Council for Responsible Genetics (www.councilforresponsiblegenetics.org).

For Audrey, always and forever

JG

For my remarkable mother Rose, wife Carolyn, daughter Alyssa and her husband Will, son Eliot, and grandchildren Benjamin and Andrew.

SK

Contents

Foreword

We must never lose sight of the immensely powerful role played by metaphors in our understanding of the natural world. Nothing has influenced our explanation of the properties of living organisms more than Descartes's *bête machine* and the later inclusion of humans in the metaphor as the *homme machine* by La Mettrie. If living organisms, including humans, are to be understood as machines, then the nature of the materials of which they are composed and the exact way in which those materials are put together must provide us with an explanation of the organisms' properties. In particular, how are we to explain that, despite the fact that the fertilized eggs of a species all look very much like one another, some individuals, after development, may turn out to be very different?

The answer that dominates modern biology is that the way in which the organism develops is largely determined by its genes. But because the genes are composed of DNA, it is the similarities and differences in DNA, the "blueprints" for the machines' construction, that determine or at least have a powerful influence over the similarities and differences in the structure and function of the developed organism. The program for the investigation of the similarities and differences that we observe among human beings is then dictated: sequence their DNA. We expect "genes for" not only skin color and eye shape but also intelligence, disease, and musical ability.

Although we accept that organisms are material objects that ultimately owe their properties to their material nature, it is an error to suppose that the DNA sequence of an organism predicts its total nature and life history. The reader of this page could not have known, from a prior knowledge of the complete DNA sequence of its author, what its content would be, nor, more important, could she have predicted the various diseases and disabilities that its author would have experienced. The attempt to find in humans the genes "for" the leading causes of mortality has largely been a failure. Simple single-mutation causes of diseases like sickle-cell anemia are an exception. Although the mutation of a single gene that greatly increases the probability of breast cancer is known, the large majority of breast cancer patients do not carry the mutation, a situation that is typical of so-called genetic predispositions to disease.

The problem of finding the genes "for" a disease illustrates a deep issue in the understanding of the machine metaphor. The evidence for the "heritability" of a characteristic comes from the greater resemblance of relatives than of unrelated persons. Characteristics "run" in families, and individuals in one racial or ethnic group are more likely to resemble one another than they do members of other groups. This has given rise to a concern over the source of "missing heritability," the existence of similarity among relatives in some characteristic, say, the occurrence of a disease state, in the absence of any demonstration of particular genetic differences that can be associated with the disorder. Elaborate theories of gene interaction are offered as an explanation. One consequence of the usual application of the machine metaphor is to ascribe similar malfunctions of two copies of the machine to similar faults in their construction. But this form of explanation ignores the fact that the functioning of a machine depends not only on its details of construction but on its history of operation. Does one company provide its machines with a better grade of fuel than another enterprise, or more frequent servicing, or more skillful operators, or less constant operation of the machines at high speed? That is, the machine metaphor, when properly applied, includes environmental sources of similarities and differences in performance, differences that exist independently of any variation in machine design and manufacture, quite aside from variation in the blueprints of the mechanism.

Putting to one side simple cases of morphological differences like skin color or eye shape, or metabolic disorders that can be traced to single-gene mutations, how are we to know whether any particular difference between individuals or families or races is a consequence of genetic or environmental or chance idiosyncratic developmental variations ("developmental noise"), or of our cultural histories, or of powerful historical social and economic forces?

Unlike the ontological question of what the real causes of differences are, we are faced with the epistemological question of how to find out and of what the reigning conventions of investigation and explanation are at any time. We currently are in a situation where molecular investigations dominate biology, and where DNA is at the center of interest. Although we still understand and operate with “inheritance” as a cultural phenomenon as well as a biological one, the widespread general interest in DNA has seemingly validated the nineteenth-century view that everything about us as individuals is specified in our biological ancestry. Now it is said that “it’s all in our DNA.” This book is meant as a challenge to that convention and the current dominant metaphor of DNA or genetic reductionism that drives it.

Richard Lewontin
Alexander Agassiz Professor of Zoology in the
Museum of Comparative Zoology, Emeritus
Harvard University

Genetic Explanations

Introduction

Evolving Narratives of Genetic Explanation across Disciplines

SHELDON KRIMSKY

In the aftermath of every major breakthrough or revolution in the natural sciences, each newly adopted theory evolves into four narrative forms. First, there is the canonical form of the theory introduced by the person or persons identified with the discovery, usually in their published writings. These are the primary texts of discovery, which include Newton's *Principia* and Einstein's 1905 paper on relativity. Second, there are extensions to and extrapolations from the canonical theory by other scientists in the same field, usually over a period of years. Gaps are filled, corrections are made, interpretations are offered, and questions are raised that expand the research program, extend the theory, and generate new empirical observations. Third, social scientists exploit the science to inform or build a new foundation for their own fields. Thus behavioral scientists seek a more reductive explanation by adapting principles from chemistry, physics, or biology. These narratives sometimes introduce ethical concepts and explore the social and legal implications of the scientific findings. For example, is a person who is claimed to possess a genetic tendency toward aggression culpable for his actions? Fourth, popularizers of the theory, including science writers, scientists themselves, and the media, introduce analogies and metaphors for the purpose of communicating scientific ideas to the general public. These metaphors are sometimes used strategically to win public support for funding new areas of science or for policies based on claims that human cognitive or behavioral traits are determined by molecular events.

During the eighteenth century many Enlightenment intellectuals were influenced by Newton's discoveries. The economist Adam Smith had an interest in astronomy that made him receptive to Newton's methods of analysis and his conception of the universe. "Adam Smith took Newton's conception of nature as a law-bound system of matter in motion as his model when he represented society as a collection of individuals pursuing their self-interest in an economic order governed by laws of supply and demand."[1] Smith modeled his method on Newton's *Principia,* which, in the view of some historians, begins with induction from phenomena to framing principles and then deduces the phenomena from the principles.[2] After seeing the first edition of Newton's *Principia,* John Locke revised his initial idea in his *Essay Concerning Human Understanding* that it was impossible for a body to act on another body that it did not touch.[3] David Hume, deemed the "Newton of the moral sciences," embedded Newtonian terminology in his *Treatise on Human Nature.*[4]

Charles Darwin's theory of natural selection was embraced by social scientists such as Herbert Spencer who drew implications from it in sociology and ethics to explain human social development. The Pulitzer Prize–winning intellectual historian Richard Hofstadter introduced the concept of social Darwinism into American intellectual life in his 1944 classic *Social Darwinism in American Thought.* In his introduction to the revised edition Hofstadter wrote: "Periodically in modern history scientists have set forth new theories whose consequences go far beyond the internal development of science as a system of knowledge and beyond such practical applications as they may happen to have."[5] His book criticized those social theorists who misused Darwin's discoveries for ideological purposes.

When Niels Bohr proposed a solar-system-like model of the atom with a nucleus and electrons circulating in quantized circular orbits, it accounted for some spectral emission lines of the hydrogen atom but not more complex atomic structures. His model had to be refined and revised by Werner Heisenberg and Erwin Schrödinger in the new quantum theory. New theories are usually first-approximation idealizations of reality.

All the major scientific revolutions of the sixteenth to the twentieth centuries (Copernican, Newtonian, Darwinian, Einsteinian, Freudian, quantum, nuclear) witnessed aspects of the four narratives discussed above. By the mid-twentieth century the revolution in molecular biology began to take hold, and a similar recapitulation of the four narrative forms ensued. The molecular-genetics revolution has been reaching full force over a period of about fifty years. The milestones include discovery of the structure of the DNA molecule (1953);[6] the mechanism of protein synthesis from DNA (1958);[7] the discovery of the "operon" (regulation of genetic expression) (1961);[8] the discovery of the genetic code (1961);[9]

the central dogma: nucleic acid to protein but not the reverse (1970);[10] recombinant DNA (1973);[11] and sequencing the human genome (2000).[12]

During the half century in which molecular-genetic theory developed, many hypotheses and revisions to those hypotheses were made regarding the nature of bacterial, viral, animal, and plant genomes; the relationship between genes and phenotype (such as disease, behavior, and physical traits); gene-environment interactions; the process by which cells decode the information stored on DNA; and the estimated number of genes in the human genome. Even the concept of the gene has been revised. The early concept of the genome likened it to a Lego structure, composed of segments of DNA linked together and differentiated by the sequence of four nucleotides (the bases adenine, cytosine, guanine, and thymine). An early physical model of the DNA molecule that James Watson and Francis Crick fabricated was a metallic double helix "made of flat plates of galvanized metal with narrow brass tubes for bonds."[13] The static model was an early representation of the way genes were organized in the genome, which later underwent considerable revision. Initially, genes were viewed as the segments of DNA that held coding information about proteins.

Although some segments of the DNA in the human genome have a coding function, most of the 3 billion base pairs, until recently, were considered junk DNA with no coding or regulatory functions—the flotsam and jetsam of evolution. The remaining functional DNA (estimated at about 2 percent of the genome) was divided into discrete segments, each assigned to the coding for one of the unique 100,000 proteins estimated to be present/active in the human body. Recently the junk-DNA hypothesis, at least as interpreted as DNA that has no function in the organism, has been largely discounted as scientists have discovered that more and more of the noncoding DNA is transcribed into RNA with functions as yet uncharacterized. Also, new estimates put the number of human coding genes at between 20,000 and 30,000, signaling that some DNA segments contain the coding information for more than one protein.

Francis Crick first postulated the central dogma of molecular-genetics theory in 1957 at a meeting of the Society of Experimental Biology in a talk titled "On Protein Synthesis."[14] According to Crick, genetic information transfers from nucleic acid (DNA or RNA) to nucleic acid, or from nucleic acid to protein, but never from protein to nucleic acid. In other words, proteins do not contain the information for duplicating themselves. The central dogma has often been simplified as "DNA makes RNA makes protein." Early popular conceptions of the genetic mechanism gave the false impression that DNA is a self-replicating master molecule. In fact, proteins play a critical role in directing the orchestral process of protein synthesis.[15] "DNA may be a large complex molecule, but alone it does nothing. It does not have powers of self-replication, nor those to create new generations of life."[16] In a popular magazine article Barry Commoner gave a more realistic view of the role of proteins in all aspects of DNA replication and transcription:

> In the living cell the gene's nucleotide code can be replicated faithfully only because an array of specialized proteins intervenes to prevent most of the errors—which DNA by itself is prone to make—and to repair the few remaining ones. . . . Genetic information arises not from DNA alone but through its essential collaboration with protein enzymes—a contradiction of the central dogma's precept that inheritance is uniquely governed by the self-replication of the DNA double helix.[17]

Initially, molecular geneticists believed that the function of a gene was to control the production of a single protein. Then it was discovered that genes carried the code for forms of RNA that do not become proteins. From the late 1960s to the present the details of the central dogma have been filled in or revised with some variations in how information flows in viruses and retroviruses.

An example of replication without nucleic acid is given by prions, newly discovered proteins responsible for mad cow disease (kuru). Prions can replicate even though they do not contain nucleic acid. They alter normal brain proteins, which adapt to the prion's shape, thus, in a sense, replicating themselves. This phenomenon is well documented in yeast, where fungal prions can undergo a structural conversion and can become self-propagating and infectious.[18] These yeast prions exemplify an epigenetic phenomenon in which information is encoded not in the nuclear DNA but within the protein. If information can be transferred from protein to protein, the universality of the central dogma of molecular biology is called into question.

The Lego model of the genome seems as simplistic as the Bohr model of the atom. Rather than seeing genes as fixed entities in a static structure awaiting self-activation, the current conception views the genome as more characteristic of an ecosystem—more fluid, more dynamic, and more interactive than the Lego model implies.

> The assumption that identifiable bits of DNA sequence are even "genes" for particular proteins has turned out not to be generally true. Alternative splicing of fragments of particular sequences, alternative reading frames, and post-transcriptional editing—some of the things that happen between the transcription of DNA and the formatting of a final protein product—are

> among the processes the discovery of which had led to a radically different view of the genome.[19]

My first encounter with the static Lego-like model of the genome came in 1976 during the controversy over the safety of recombinant DNA research. Panels had been convened at the national and municipal levels to evaluate the risks of transplanting genetic material across species. While I was serving on the Cambridge Experimentation Review Board in Cambridge, Massachusetts, I had the opportunity to query molecular biologists about the possible consequences of moving segments of DNA across species.[20] At the time many believed that if functional DNA was moved from cells of one species to another, the new cells would either use the DNA to synthesize exactly what the gene coded for in its native species or produce nothing at all. The concept of emergent properties in molecular genetics was missing from the discourse. In the mid-1980s, during the peak of the debate over the potential risks of releasing genetically modified microorganisms into agricultural fields, one Berkeley scientist likened the plant genome to the keys of a piano. In a videotaped message to the public, Steven Lindow stated that removing one gene is like removing one key from a piano; all the other keys will function the same.[21]

Within a decade scientists began to acknowledge that such a view was far too simplistic, and the complexity of the genome began to reveal itself. There are gene-gene interactions, DNA expressing different products when it is situated in different parts of the chromosome (the position effect), and segments of DNA that could be read differently in different organisms or in the same organism because of alternative splicing, which may result in different reading frames.[22] By 2001 scientists at the Food and Drug Administration recognized this complexity when they were reviewing food-safety issues arising from genetically modified crops. An agency document included the following statement:

> It is also possible with bioengineering that the newly introduced genetic material may be inserted into the chromosome of a food plant in a location that causes the food derived from the plant to have higher levels of toxins than normal, or lower levels of a significant nutrient. In the former case the food may not be safe to eat, or may require special preparation to reduce or eliminate the toxic substance. In the latter case the food may require special labeling, so that consumers would know that they were not receiving the level of nutrients they would ordinarily expect from consuming a comparable food.[23]

This model of the plant genome is far afield from the piano metaphor where the added or subtracted key does not interact with the other keys or affect the system as a whole other than adding a new protein. Also, there are other reasons to question a simple relationship among DNA, RNA, and proteins. A study published in 2001 reported that information in DNA is not always faithfully transferred to RNA in transcription—RNA bases were found that did not match the corresponding DNA sequence.[24]

While the genetics paradigm was undergoing significant revision after the discovery of the structure of DNA, social scientists had begun the search for DNA sequence variations within the human genome that correlated with variations in human phenotype. Molecular genetics had provided the grist for new hereditarian theories in social science. The fields of anthropology, psychology, sociology, and political science began to turn to genetics as a way to give natural scientific grounding to the so-called softer sciences. New research findings in behavioral genetics, cognition, and psychosocial issues were reported widely in the media, including the following: genes may cause bed-wetting; genes provide the master switch for right-left symmetry; genes explain longevity; and genes are linked to thrill seeking, generosity, homosexuality, childhood obesity, voter turnout, sports ability, food choices, religiosity, congeniality, marital discord, attraction to pornography, and gang membership. Despite the burgeoning research in gene-environment interactions, the discovery of multiple levels of developmental structure that affect the relationship between DNA and medical illness, and new findings in epigenetics, stories in the media highlighted discoveries of genes touted as being genes "for" breast, prostate, pancreatic, skin, and colon cancers, as well as the gene to protect against cancer and the genes for deafness, dyslexia, blindness, Alzheimer's disease, bed-wetting, infidelity, and kidney disease. Because the methodology or validity of such claims was never part of the public discussion, gene-centric thinking became the centerpiece of popular discourse. Let us consider, as an example, how spurious claims about genetic explanation can enter into important societal decisions.

After Gary Cossey pleaded guilty to possession of child pornography literature, the trial judge issued a six-and-a-half-year sentence to the defendant because he believed that Cossey had a "pornography gene," which, the judge believed, would ensure recidivism. Some scientific studies concluded that pedophilia is familial, fixed, and immutable.[25] A federal appeals court overturned Casey's conviction on the basis that the trial judge used the existence of "an undiscovered gene" as the grounds for the sentence.[26] Let us assume for a moment that there are familial links in pedophilia. How does that justify the conclusion that a single gene is responsible for this kind of behavior? Or let us suppose that pedophiles produce higher levels of a hormone than nonpedophiles, or even that they show different brain scans of their hypothalamus.[27] There are so many levels of structure and development and so many possible interactions between the production of the hormone or the structure of a brain region and any behavior that, without irrefutable causal evidence, it would take an act of pure imagination to make the leap that there is a pedophile gene. In a gene-centric framework there is a tendency to draw the simplest reductionist explanation for a behavior or trait and neglect the complexities of multigenetic, gene-environment, and epigenetic interactions or some complex combination involving multiple causation. As Martin Richards noted: "Molecular genetics often has the feel of greedy reductionism, trying to explain too much, too fast, under-estimating the complexity and skipping over whole levels of process in the rush to link everything to the foundations of DNA."[28]

Within the social sciences, psychology, anthropology, and sociology have thus far been most influenced by the genetics revolution. Cognitive and behavioral psychologists have been drawn to genetics to explain differences in human intelligence among population groups and to provide a genetic cause for sociopathic behavior. Anthropologists have embraced genetics as a tool for understanding migration patterns of ancient populations. Increasingly, sociologists have applied findings in genetics to understand why people associate in certain groups or adopt certain ideologies. Lately, political scientists have used genetics to explain differences in political behavior—a subfield called molecular politics.

One of the most controversial of these studies appeared in the prestigious *American Political Science Review* in 2008. The authors correlated data from public records of voter turnout in Los Angeles with data from a twin registry. They used a twin model and a statistical linear regression, which assumes the additivity of three factors that could account for the behavior of identical or fraternal twins: genetic makeup, unshared environment, or shared environment. Their model did not account for the influences of gene-environment interactions. The authors reached the conclusion that a "significant proportion of the variation in voting turnout can be accounted for by genes."[29]

At one level, there is an intuitive dimension to their findings. If one studies the behavior of monozygote (identical) and dizygote (fraternal) twins and discovers a closer correlation in behaviors and choices of the former (identical) than in behaviors and choices of the latter (fraternal) twins, at first glance it seems reasonable to conclude that the behavior in question has a genetic component, only if one ignores the contested assumption of the equal environment hypothesis in classical twin studies.[30]

But at another level, these results—like all twin-study results—are counterintuitive. Genetic differences in individuals may result in variations in the regulation of genes and the expression of proteins. But what is this assumed imaginary pathway from proteins to voting behavior, and in what sense can such a pathway be deterministic, given the many other levels of biological and social organization, gene-environment interactions, and the incalculable number of potential social influences on whether a person goes to the polls? There may be some behaviors that are highly correlated between identical twins even when they are reared in separate environments, behaviors that can be more easily understood as physiological—such as food choices or response to stress—but political choices have no intuitive grounding in genetic causation. The modern concept of the genome as "ecosystem-like" considers genes and environment in constant interaction throughout development, and thus they are not in a state of static partition.[31]

James Fowler and Christopher Dawes of the University of California at San Diego have reduced political behavior to a mechanistic level by linking specific genes and biochemical pathways to individual choice. "We hypothesize that genes may influence voting and political participation because they influence a generalized tendency to engage in prosocial behavior via their functional role in neurochemical processes."[32] The two genes cited are *MAOA* and *5HTT;* both are associated with the metabolism of serotonin in the brain. Serotonin is a chemical (a neurotransmitter) that transmits messages between brain cells. Levels of serotonin affect the regulation of sleep, appetite, and mood, among other functions. Serotonin deficiency has been linked to mental health problems, including depression and anxiety. The authors build on studies that show that polymorphisms (alleles) of the two genes are linked to antisocial behavior; they next argue that it is reasonable to hypothesize that such polymorphisms will also correlate with lower motivation to vote.

Few would have imagined that political scientists would turn to genetic explanations to account for people's political choices when so many other social variables are at work. Notwithstanding its counterintuitive aspects, the studies on the genetic basis of political choices have gained considerable support from journal editors and represent a new stage in the geneticization of the social sciences, somewhat like the growth of interest in the 1960s and 1970s in the search for a genetic basis of intelligence. Jay Joseph introduced a thought experiment to highlight the counterintuitive aspects of molecular politics:

> Suppose that one male MZA twin [monozygote-identical reared apart] is placed at birth in an aristocratic Japanese family in 1802. The other male MZA co-twin is placed at birth in a poor peasant family living in the highlands of El Salvador in 1960. . . . Would we expect a study of genetically identical pairs of this type to find sizable correlations for political behavior and social attitude?[33]

With the sequencing of the human genome and the proliferation of genetic databases, social scientists have embarked on a new frontier for investigating links between genotype and cognitive, social, and cultural traits using the same tools that are used in medical genetics to find correlations between mutations and disease. Ironically, it is very rare in medicine to find monogenic diseases. Despite its recurring failure, the ideology of genetic reductionism persists. Genetic explanation of social behavior implies that a "gene" is a persistent, single entity that is conserved across generations. But this idea is a vestige of earlier and discarded views in molecular genetics. As noted by Evelyn Fox Keller in *The Century of the Gene,* new data in molecular genetics "threaten to throw the very concept of 'the gene'—either as a unit of structure or as a unit of function—into blatant disarray."[34] Genetics will continue to play a central role in the life sciences, but the evidence suggests a far less reductionist role as scientists fully appreciate the dynamic nature of biological systems and the complex interactions among DNA, proteins, and the environment at all levels of organization within the system.

The idea for this book grew out of the work of Ruth Hubbard, professor emerita of biology at Harvard University. She has written, lectured, and advocated against genetic reductionism in science and medicine. In my earliest recollection, her views on genetic reductionism were expressed at a 1977 meeting of the National Academy of Sciences on recombinant DNA research and its applications. Hubbard stated her disagreement with the idea that "the more and more we know about smaller and smaller [biological] units," the greater our knowledge will be for curing disease. According to Hubbard, the first essay she published that systematically discussed genetic reductionism appeared as a chapter titled "The Theory and Practice of Genetic Reductionism—From Mendel's Laws to Genetic Engineering" in a book edited by Steven Rose called *Towards a Liberatory Biology.*[35] In this chapter she wrote: "It is important to stress that genes reproduce (or DNA replicates) as part of the metabolic activities of living cells that involve enzymes, substrates, sources of energy, etc. They do not reproduce themselves, as it is often phrased. And even if we understood in detail the timing and control of DNA synthesis (replication) by cells—which we do not—this could not tell us how gene replication is translated into traits."[36] A year after her chapter on genetic reductionism appeared in print, Hubbard helped found the Council for Responsible Genetics, where, as a board member and editor of *GeneWatch,* she provided critical perspectives on society's growing "genophilia" to the exclusion of cells, the organism, and the environment.

Her 1993 book *Exploding the Gene Myth,* coauthored with Elijah Wald, gave scientists and nonscientists a framework for placing genes and DNA in their proper context while dethroning the "gene" as a "master molecule" or the "holy grail" of biology. *Exploding the Gene Myth* also brought attention to fallacies of "geneticization"—a process by which individual differences in health, behavior, and cognition are reduced to their DNA code.[37]

In Chapter 1, "The Mismeasure of the Gene," Hubbard begins the series of chapters in Part I ("New Understanding of Genetic Science") by focusing on how DNA and proteins actually function in living cells. She argues that the gene is both a material object and an ideology carrying forward a theme in her 1982 essay: "Genes and DNA, as they are often conceptualized, are the reductionist self-fulfillment of hereditarianism, which is the social impulse behind genetics."[38]

In Chapter 2, "Evolution Is Not Mainly a Matter of Genes," Stuart A. Newman, professor of cell biology and anatomy at New York Medical College, shows that evolutionary change can be accounted for by developmental mechanisms that include, but are not exclusively reducible to, genes. By drawing on examples from embryology, Newman demonstrates how genes and physical laws work in tandem to produce evolutionary change.

Evelyn Fox Keller, professor emerita of history and philosophy of science at the Massachusetts Institute of Technology, contributes Chapter 3, "Genes as Difference Makers." Largely adapted from her book *The Mirage of a Space between Nature and Nurture,* this chapter questions the conventional meaning of "genetic disease" and "disease-causing genes" and argues that a causal analysis of disease for the purpose of therapeutic intervention should focus more on metabolic pathways and less on the "particulate gene."

David S. Moore, professor of psychology at Pitzer College and Claremont Graduate University, contributes Chapter 4, "Big B, Little b: Myth #1 Is That Mendelian Genes Actually Exist." Moore argues that "the genes most of us envision inside us, calling the shots and determining our characteristics, are myths." In "The Myth of the Machine-Organism"

(Chapter 5) Stephen L. Talbott, senior researcher at the Nature Institute, argues that the influential mechanistic model of DNA and genes, which has been pervasive in molecular genetics, has run its course and must be replaced by organismic imagery without falling prey to a new vitalism.

Part II of this volume, "Medical Genetics," comprises six chapters on the genetics of disease, covering psychiatry, autism, cancer, disease causation, and the role of genes as predictors of disease. In "Some Problems with Genetic Horoscopes" (Chapter 6), Eva Jablonka, a professor and theoretical evolutionary biologist at Tel Aviv University's Cohn Institute for the History and Philosophy of Science and Ideas, discusses genetic plasticity and epigenetic inheritance. She argues that "organisms can respond to changing environments by changing their development and their phenotypes, and sometimes these plastic responses can be stabilized and inherited." Carlos Sonnenschein and Ana M. Soto are professors of anatomy and cellular biology at Tufts University School of Medicine. In Chapter 7, "Cancer Genes: The Vestigial Remains of a Fallen Theory," they offer a scientific critique of the "somatic mutation theory of cancer." In its place they introduce the "tissue organization field theory" of cancer, which supersedes the "epicycles" of the dominant theory. In Chapter 8, "The Fruitless Search for Genes in Psychiatry and Psychology: Time to Reexamine a Paradigm," Jay Joseph, licensed psychologist and author, and Carl Ratner, director of the Institute for Cultural Research and Education, team up to challenge the idea that the major psychiatric disorders have an underlying genetic basis. The authors recount the failed efforts to find genes for psychiatric disorders and argue that research attention should focus on the methodological problems and unsupported theoretical assumptions of previous studies of families, twins, and adoptees.

Carl F. Cranor, Distinguished Professor of Philosophy at the University of California, Riverside, contributes Chapter 9, "Assessing Genes as Causes of Human Disease in a Multicausal World." He explores the nuanced meanings of "gene X is the cause of disease Y" and compares how we currently understand genetic causation with how we should understand it.

Martha R. Herbert, a pediatric neurologist and assistant professor of neurology at Harvard Medical School, contributes Chapter 10, "Autism: From Static Genetic Brain Defect to Dynamic Gene-Environment-Modulated Pathophysiology." This chapter reveals the failed attempts to apply genetic reductionism—"DNA makes the rules; everything else obeys"—to autism spectrum disorders. Taking a systems-biology approach to autism spectrum, Herbert finds that neither a single gene nor a single gene-environment interaction explains the origins and progression of the disease; rather, she sees it as a complex chronic and dynamic pathophysiology consisting of multiple "combinatorial" pathways. In "The Prospects of Personalized Medicine" (Chapter 11), David Jones, A. Bernard Ackerman Professor of the Culture of Medicine at Harvard University, looks at the prospects of pharmacogenomics in medical care. His examples capture the promise but also the mixed legacy of personalized genomic medicine and looks at the factors that have contributed to this legacy.

In Part III of this volume, "Genetics in Human Behavior and Culture," the authors address how genetics has been portrayed in popular culture and how this portrayal influences social attitudes about human behavior. This part begins with Chapter 12, "The Persistent Influence of Failed Scientific Ideas," by Jonathan Beckwith, American Cancer Society Professor at Harvard Medical School. Through a series of cases, Beckwith shows how half-baked hypotheses, preliminary data, and speculative ideas in behavioral genetics, which are often accompanied by negative social consequences, are prematurely accepted as scientific fact. Susan Lindee, a historian and associate dean of the social sciences at the University of Pennsylvania, contributes Chapter 13, "Map Your Own Genes! The DNA Experience." Lindee shows how the mass marketing of genetic testing has shaped social expectations and questions whether the promises of genetic testing will outlive their proven utility.

In Chapter 14, "Creating a 'Better Baby': The Role of Genetics in Contemporary Reproductive Practices," Shirley Shalev, a faculty member of the Women, Gender and Health Concentration at Harvard School of Public Health, discusses the exaggerations and misconceptions in the argument that by undertaking more aggressive efforts of genetic selection one can find a more perfect mate and conceive a more perfect child. According to Shalev, with increased variability in the methods of procreation and the different forms of parental relations, genetic identity and genetic selection have to be reexamined.

William C. Thompson, professor in the Department of Criminology, Law and Society and the School of Law at the University of California, Irvine, takes on the popular myth that DNA forensic evidence is foolproof in Chapter 15, appropriately titled "Forensic DNA Evidence: The Myth of Infallibility." He uses case examples, statistical knowledge, law, and science to demonstrate that any combination of human error, dubious inferences, and misleading statistics can contribute to false claims about forensic DNA evidence. In Chapter 16, "Nurturing Nature: How

Parental Care Changes Genes," Mae-Won Ho, a geneticist and director/cofounder of the Institute of Science in Society, explains that "neither genetic nor environmental determinism rules." Discoveries in epigenetics have shown that environmental factors such as parental care or lack thereof can affect whether genes get expressed. In "Conclusion: The Unfulfilled Promise of Genomics," Jeremy Gruber, lawyer and president of the Council for Responsible Genetics, discusses the promotion of genomic-based medical research by the commercial sector, government, scientists, and the media. He looks back over the ten years since the human genome was sequenced, analyzes the state of its exaggerated contributions to human health, and suggests finding a new balance in research priorities.

The chapters in this volume provide a counterargument to exaggerated, erroneous, or overly simplified claims about the role that DNA and genes play in cells, organisms, evolution, human behavior, and culture. Decoding the human genome, once hailed as deciphering the "book of life," has grossly understated the complexity of biological processes. The role of genetic explanation must be tempered by the new scientific understanding of systems biology, proteomics, epigenetics, and gene-environment interactions.

I am grateful to Eva Jablonka, David Moore, Stuart Newman, George Smith, and Carlos Sonnenschein for their helpful suggestions on earlier drafts of this chapter.

PART ONE

New Understanding of Genetic Science

CHAPTER ONE

The Mismeasure of the Gene

RUTH HUBBARD

> Conscious fraud is probably rare in science. It is also not very interesting, for it tells us little about the nature of scientific activity. Liars, if discovered, are excommunicated; scientists declare that their profession has properly policed itself, and they return to work, mythology unimpaired and objectively vindicated. The prevalence of unconscious finagling, on the other hand, suggests a general conclusion about the context of science. For if scientists can be honestly self-deluded . . . then prior prejudice may be found anywhere, even in the basics of measuring bones and toting sums.
>
> —Stephen Jay Gould, *The Mismeasure of Man*

SCIENCE IS AN INTERPRETATION OF NATURE and, like other forms of interpretation, fits into the cultural framework of its time. I shall illustrate this fact by tracing some of the threads that, in the course of the twentieth century, have led to the notion that genes determine virtually all physical and social characteristics of humans and other animals. Currently, everything about us is "in the genes," and this view offers the hope that once we learn to read our "genetic blueprint," we will be able to change it and live happily ever after.

The most obvious place to begin this story is with the Austrian monk Gregor Mendel, who in the 1860s developed what have come to be known as Mendel's laws of inheritance. Using pea plants as his experimental objects, Mendel examined the transmission of flower color and of the shape and texture of the seeds to successive generations. He deliberately selected these traits because they are transmitted in an all-or-nothing fashion, unlike traits that vary continuously, such as weight or size. After performing large numbers of crosses between plants that had been shown to breed true, he was able to describe the numerical regularities in the way the traits were passed from parents to successive generations of offspring that have come to be known as Mendel's laws. However, he did not speculate about what mechanisms might account for the transmission of traits from one generation to the next and merely suggested that they probably involved "factors" within the plants.

That few scientists paid attention to Mendel's paper when it was published in 1865 presumably had to do with the fact that there was no larger context into which to put his observations. This situation had changed dramatically by 1900, when his paper was independently "rediscovered" in three laboratories. By that time biologists had observed well-defined structures within the cell's nucleus that took up chemical stains and were therefore called chromosomes. They had further noted that when cells divide, their chromosomes also divide, so that the two daughter cells end up with the same number of chromosomes as were present in the parent cell. The chromosomes, therefore, were generally accepted as the bearers of heredity, and the idea took hold that Mendel's "factors" bore some relationship to them.

In 1905 the Danish botanist Wilhelm Johannsen coined the word "gene" to lend more concrete reality to Mendel's "factors." At a time when invisible atoms, electrons, and quanta were being accepted into the world of chemistry and physics, biologists had little problem accepting that heredity also was mediated by invisible material particles. Soon a series of groundbreaking experiments, done mainly with fruit flies and corn (maize), led them to decide that the genes must lie along the chromosomes, like beads on a string, and that when the chromosomes were replicated during cell division, the genes also got copied.

During the first half of the twentieth century, biologists became increasingly interested in exploring the molecular constitution of cells and the ways in which molecules participate in the metabolism and growth of organisms. They came up with molecular explanations of human diseases known to have hereditary components, such as sickle-cell disease and phenylketonuria, and identified the specific molecules associated with such conditions.

Chemists and biochemists described various biologically important substances, including vitamins and hormones, and characterized their biological functions in chemical terms. In the process they identified a series of hitherto-unknown carbohydrates and fats and also very large and complex proteins, which had previously been thought to be ill-defined aggregates and not discrete molecules at all. It was an exciting period in which chemically oriented biologists spoke of bringing biology to the molecular level. At the same time, they also tried to understand how different chemical components are integrated into the way whole organisms function, writing books with such titles as *The Organism as a Whole, The Wisdom of the Body,* and *Dynamic Aspects of Biochemistry.*[1]

These kinds of explorations led biochemists to identify protein molecules that function as enzymes, others that mediate muscular contraction and relaxation, and yet others that transport oxygen and CO_2 around the body. As part of these kinds of explorations, biochemists came to realize that chromosomes contain both proteins and another type of very large molecule, called DNA, and this raised the question of the chemical nature of genes: are they made of proteins, DNA, or both?

Initially, many biologists favored the idea that DNA forms an inert chromosomal framework to which protein molecules attach themselves as genes. The reason was that although DNA is a very large molecule, it is made up of only six different components: a type of phosphate, a sugar, and the four so-called bases that are now familiar to us by the abbreviations A (adenine), G (guanine), C (cytosine), and T (thymine). The naturally occurring proteins, in contrast, contain twenty different subunits (called amino acids), strung together in many different combinations, and come in many different shapes and sizes. It therefore was easier to imagine that different proteins would be the ones to transmit the various traits for which genes are now assumed to be responsible.

In the late 1940s and early 1950s, however, experiments with bacteria and viruses showed that the hereditary material—the gene—consists of DNA. By then it had become clear that genes are involved in the synthesis of proteins, and biologists had concluded that DNA, in fact, specifies the composition of proteins, but the mechanism by which this happens was an open question. It is, however, crucial to realize that intriguing as this puzzle was, all this time, DNA was looked on as just one of the sorts of molecules that are important to the way cells and organisms function.

All this changed in April 1953 when James Watson and Francis Crick proposed their double-helix model of the structure of DNA.[2] Since then DNA has come to be considered the most important molecule in biology, and "molecular biology" has come to refer exclusively to the biological functions of DNA.

To understand this shift in outlook, it is important to consider the social and political dimensions of how DNA and the double helix came to be propelled into the center of biological interest. Watson has described the discovery of the structure of DNA, from his point of view, in his bestselling memoir *The Double Helix.*[3] Although it may be hazardous to do so, it is worth speculating how the story of DNA might have unfolded if one of the other two groups of scientists who were trying to elucidate its structure at that time had "won the race." I am referring to the great chemist Linus Pauling and his group at the California Institute of Technology in Pasadena and to Rosalind Franklin and Maurice Wilkins, two experts in X-ray diffraction analysis at King's College London.

For one thing, neither of these two groups was racing. They did not even know there was a race. Only Watson and Crick were racing. As for Pauling, he and his colleagues had recently elucidated the structure of the α-helix, a basic structural component of many of the proteins of biological importance. That was an enormous achievement for which Pauling was shortly awarded a Nobel Prize. Before turning to the structure of DNA, Pauling's group had already determined the three-dimensional structure of the bases that compose DNA. It therefore seems reasonable to assume that had Pauling been the first to describe the full structure of DNA, it would have been exciting, but it would have been just another of his many major accomplishments.

By all accounts, at the time Watson and Crick unveiled their DNA model, Rosalind Franklin was close to solving the structure herself. She had been working on it for about two years, and although no one (including Franklin) knew it at the time, Watson and Crick drew heavily on her X-ray measurements and on the structural information she derived from them to come up with the double helix.[4] Had Franklin been the one to solve the DNA structure, she would, of course, have published it, but she might not have announced it with great fanfare because that was not her style. The structure in itself was beautiful, and people would have been extremely interested, but it might well not have become the biology-shaking event of the century.

In contrast, from the moment Watson and Crick began to think about how to figure out the structure of DNA and long before they had bothered to find out what was known about its chemical composition, they thought of DNA as "the secret of life." Indeed, Watson writes in *The Double Helix* that even before they had quite clinched their model, Crick rushed into the pub they frequented to announce in a booming voice that they had "found the secret of life."[5] That is also how they communicated the news of the structure to their colleagues and mentors, although their note in *Nature* struck the proper objective tone.

What the Watson-Crick model showed (and most people nowadays can find out by reading the newspapers) is that DNA can be pictured as two spiral ribbons wound in parallel to form a double helix. The four bases—the As, Gs, Cs, and Ts—are attached to the ribbons at regular intervals and point toward the center of the helix, hence toward one another much like the teeth on a zipper, except that in DNA the teeth meet rather than overlap one another. What makes the Watson-Crick model so exciting is the fact that in order to get the bases to fit into the double helix, an A on one ribbon, or strand, must abut on a T on the other, and a C on one must abut on a G on the other.

This geometric arrangement means that to copy DNA—the "gene"—the double helix must merely begin to unwind (or, in this metaphor, become unzipped). Each strand can then serve as a template for the synthesis of its partner. As this synthesis progresses, the old strands and their newly formed partners simply zip up to form two identical copies of the original. In other words, the double-helix structure itself explains how DNA—the gene—can get copied. The simplicity of this model has had several ideological consequences. One is that the way DNA is copied has been called "self-replication," and DNA has come to be referred to as a "self-replicating" molecule. Of course, it is nothing of the sort. DNA does not replicate itself. Cells and, in real life, organisms copy their DNA using each strand of the double helix as the template for the synthesis of its partner. This process requires a whole series of physical and chemical conditions and reactions within the cell.

An important consequence of thinking of DNA as a "self-replicating" molecule, however, was that it sparked the imagination of a number of distinguished physicists and mathematicians who, until then, had shown little interest in biological and biochemical systems and, indeed, perhaps a temperamental aversion to their inherent messiness. At the end of World War II, after two atomic bombs had been dropped on two Japanese cities, many physicists had become disillusioned with physics (the harbinger of death) and were only too glad to turn their attention to biology (the harbinger of life). Following the lead of the German exile and Nobel Prize physicist Erwin Schrödinger, who in his short book *What Is Life?* had referred to the gene as a code and hailed it as the secret of life,[6] they got excited about DNA. Familiar with wartime uses of cybernetics and code breaking, they decided to try to crack the "genetic code" by devising formal solutions for the way different sequences of A, G, C, and T could get translated into the sequences of amino acids that constitute different proteins.

It is important to pay attention to the differences in the conceptual and physical tools the scientists attacking these questions used. As the messy biochemical work of grinding up tissues and isolating their cells and molecules yielded first place to the skills of code breaking, centrifuges, spattered lab coats, and dirty glassware were replaced by paper and pencil and soon by computers. In the process different sequences of

A, G, C, and T molecules became a "code," and the biological and chemical complexities of living organisms were reduced to abstractions about how to translate the linear "code" of DNA into the linear array of the amino acids that make up proteins.

In the process what was conceptually pushed aside was the fact that this "translation" ordinarily happens inside dividing and metabolizing cells of organisms, which live in complicated relationships with their environments. The complexities of such biological and social realities got erased as scientific interest focused on computations and codes rather than on the interrelationships of gooey cells and A molecules and, indeed, of the organisms and social structures among which life gets played out. And although in the end the messy biochemists were the first to work out correspondences between the base sequences in DNA and the composition of proteins (for which they duly got their Nobel Prizes), much of the intellectual drama went with the more theoretical aspects of "breaking the code."

Before moving on, it is important to remember that by itself, DNA is an inert, sticky glop. It takes organisms or, at least, the enzyme systems extracted from them, along with other essential molecules, to perform the synthetic processes within which DNA specifies either the composition of its own copies or the composition of proteins. As soon as we think of DNA as part of the living cells of living organisms, we realize that even a relatively simple trait, such as eye color, cannot possibly be "caused" by a single gene. Just the synthesis of the pigments that color the iris of our eyes involves the participation of several proteins, the composition of each of which is specified by a different DNA sequence (or "gene"). Further proteins are required to knit the base sequences of these genes together, these proteins require further genes for their synthesis, and so on. Up to this point, we have not even begun to consider how the pigment gets deposited in the proper location in the iris or how our eyes, including the iris, get formed during embryonic development.

We are dealing with a situation in which even the "simplest" inherited trait about which we speak as though it were transmitted by a single gene, such as sickle-cell disease or phenylketonuria, involves the participation of many proteins and therefore of many "genes" (DNA sequences). The synthesis of these genes, in turn, requires further proteins, and so on and on. The usual shorthand "the gene for" must not be taken literally. Yet this way of thinking about genes has turned DNA into the "master molecule," while proteins are said to fulfill "housekeeping" functions. (And one need not be a raving postmodernist to detect class, race, and gender biases in this way of describing the molecular relationships.)

Another level of complexity in the way DNA functions has to do with the fact that a "gene"—the piece of DNA that gets translated into a particular protein—often does not exist as a continuous base sequence on the chromosome. Presumably because of our long evolutionary history, a base sequence that specifies the composition of a given protein may be interrupted by sequences that were, until recently, thought to be meaningless gibberish. As a shorthand, molecular biologists sometimes call the coding (or "expressed") sequences—those that get translated into protein—exons, and the presumably meaningless sequences, introns. But so far, no one understands how cells know how to cobble appropriate exons together and to splice out the gibberish so as to produce the final sequence (or "message") that specifies the composition of a particular protein. To make things even more complicated, exons often overlap, and different parts of a given base sequence may function in different genes. In addition, pieces of expressed coding sequence can be buried inside what are thought to be meaningless introns. Such kinds of complexities have led many molecular biologists to stop using words like "gene," "exon," and "intron" and to speak only of coding or noncoding sequences.

These sorts of largely unanticipated complexities suggest that the base sequence of the human genome, which President Clinton hailed as "the language in which God created life" when it was announced with great fanfare in June 2000, is a very complex tongue indeed. Of the strings of bases that constitute the human genome, only some 3 percent are thought to be involved in specifying the composition of proteins. These 3 percent are by no means consecutive, and combinations of them can switch around or produce redundancies. Some of them, indeed, appear to get spliced into hundreds or even thousands of different "genes." How the remaining 97 percent of bases function, or whether they even have a function, is as yet unknown.

It will take a long time to identify all the coding sequences and figure out how they combine to specify the composition of the many proteins that function in the human body. It will also be no small task to understand how the relevant metabolic systems "decide" when, where, and at what rates different proteins are to be synthesized. The fact that the human genome turns out to harbor only about a third as many coding sequences as scientists had expected will make it all the more difficult to understand how coding sequences get cobbled together so as to perform all the functions attributed to them and to what extent they actually do so. The fact that the composition and number of the coding sequences of humans are quite similar to those found in mice (and even yeasts), despite the rather significant differences between us, will make it no easier to figure these relationships out.

In a sense, spelling out the sequences of As, Gs, Cs, and Ts that constitute the human genome does not put us conceptually that far ahead of where we were at the beginning of the twentieth century, when biologists first decided that chromosomes and their genes play a fundamental role in the way cells and organisms are replicated but had no idea how that might happen. At present, the translation of DNA into proteins seems straightforward only as long as we ignore the dynamic changes in which DNA, proteins, and our other body constituents participate from one moment to the next and in different locations in our bodies. There is no way even to imagine the extent and the ways DNA participates in the transformations our cells and bodies undergo in the course of our lives.

Unfortunately, these are not just interesting scientific or philosophical puzzles, because the contrast between the actual complexities and the conceptual simplifications scientists use when they try to explain them in terms of the dance of DNA has created a dangerous situation. Biotechnology—the industry of "genetic engineering"—is built on the pretense that scientists not only understand but also can anticipate and direct the functions of the DNA sequences they isolate from organisms or manufacture in the laboratory. The industry cheerfully promises that it can foresee the potential effects of transferring specific DNA sequences, wherever and however obtained, into bacteria, plants, or animals, including humans, and thus improve targeted characteristics.

In reality, such operations can have three possible outcomes: (1) in the inhospitable environment of the cells of the host species, inserted DNA sequences do not succeed in specifying the intended proteins, so nothing new happens; (2) the inserted sequence mediates the synthesis of the desired protein product in the right amounts and at the right time and location; and (3) unpredicted and unintended consequences follow because the inserted DNA gets spliced into the wrong place in the genome of the host organism and disrupts or adversely alters one or more of its vital functions. The first alternative wastes time and money, the second is the hope, and the third spells danger. Yet which of them happens cannot be predicted a priori, or from one genetic manipulation to another, because the conditions within and around the host organisms are likely to change over time.

Clearly, the model underlying the promise of genetic engineering is overly simplistic. But what makes the situation even more problematic is that DNA sequences, once isolated or synthesized, as well as the cells, organs, or organisms into which they are inserted, can be patented and thereby become forms of intellectual property. The science and the business of genetic engineering have become one, and efforts at basic understanding compete with the pursuit of profits. The usual professional rivalries are enhanced by major financial rivalries, and the complete interlinking of government, universities, and industry leaves hardly any disinterested scientists who are devoid of conflicts of interest and can be trusted to evaluate and critique proposed scientific models or their practical implementation without raising suspicions of pursuing financial interests. As the biotechnology industry expands its reach, the health hazards and environmental pollution it produces are added to those chemistry and physics bequeathed us during the twentieth century.

In this chapter I have tried to hint at the complex dialectical relationships among the material, ideological, social, political, and economic dimensions and implications of the supposedly scientific gene concept. The gene, in fact, is a prime example of what Niels Bohr referred to as complementarity.

Bohr initially formulated this concept to denote the fact that electromagnetic radiation can be pictured as both waves and particles. Contradictory as those representations seem, it is not one or the other, but both at all times. Which representation constitutes the appropriate description simply depends on what instruments are used to detect the radiation. Similarly, genes are DNA molecules, but they also are symbols of health and disease, of hopes and fears for the future, of scientific fame and dishonor, of business fortunes and failures, and no doubt much more. To ignore any of these aspects leaves the gene concept incomplete. A central icon of our time, the gene is simultaneously a material object and an ideology, full of political, economic, spiritual, individual, and societal content.

CHAPTER TWO

Evolution Is Not Mainly a Matter of Genes

STUART A. NEWMAN

THE 200TH ANNIVERSARY of Charles Darwin's birth and the 150th anniversary of the publication of his *On the Origin of Species by Means of Natural Selection,* both falling in 2009, focused the attention of scientists, philosophers, historians, and substantial portions of the general public throughout the world on the phenomenon of organic evolution. Scholarly and popular books, museum and television shows, technical and public conferences, and organized pilgrimages to Darwin's country house and to the Galápagos Islands, a key venue in his scientific development, all attest to the iconic status of this thinker and particularly to the concept of natural selection, which the philosopher Daniel Dennett has called "the single best idea anyone has ever had."[1]

The theory itself can be summarized in various ways, but for the purposes of this chapter it is useful to parse it into eight independent propositions:

1. Organisms present themselves as "types," perpetuating themselves (in the words of the Bible) "each according to their kind."
2. Each organismal type, however, is represented by actual individuals that are all somewhat different from one another.
3. Part of this variability is also passed on from one generation to the next: offspring are not only recognizable members of their type but also carry on some of their parents' particularities.
4. As external circumstances change, by a rise in the ambient temperature or depletion of a certain foodstuff, for example, subpopulations of the group with particular quirks, or differences from the

norm, will survive or thrive to a better extent than average, contributing disproportionately higher numbers of descendants to succeeding populations.

5. Later generations within these subpopulations will thus have different average properties from earlier ones.
6. After enough generations have passed, the original type may no longer be recognizable in the selected subpopulations; a new type of organism will have emerged.
7. If no individuals of the new type can productively interbreed with any individuals of the originating population, "speciation," the smallest step of evolutionary significance, will have taken place.
8. The conditions and processes described in propositions 1–7 constitute the mechanism by which new biological forms arise over time; all large-scale differences, for example, between plants and animals or between insects and mammals, were generated by a series of many small species-level diversification events.

The observations contained in propositions 1–5 were uncontroversial in the nineteenth-century European context in which Darwin (and his contemporary coformulator of the hypothesis, Alfred Russel Wallace) presented the idea, and they are accepted even among present-day creationists. Only with items 6 and 7 does the mechanism of natural selection emerge, and even then there are few who would disagree with its implications.[2] Proposition 8, without which natural selection cannot explain large-scale evolution, or macroevolution, is the one that has engendered the most controversy. The most prominent opponents are denialists who reject the idea that macroevolution has even occurred. But there is also a growing number of evolutionary biologists who believe that macroevolution was the result of mechanisms other than natural selection. This disagreement with the standard model revolves to a great extent around the purported role of genes in evolutionary change.

It has frequently been noted that Darwin and his contemporaries had no understanding of the mechanisms of inheritance. To address this, mid-twentieth-century evolutionary biologists combined Darwin's mechanism of natural selection with the then-emerging understanding of the variation and transmission of genes. Since genes were conceived as the medium by which the variability referred to in proposition 3 could be conveyed from one generation to the next, the resulting modern evolutionary synthesis ultimately became a theory of the dynamics of gene frequency in populations, encapsulated in the doctrine that evolution is mainly a matter of changes in genes.

Even more relevant to assessing the role of natural selection in a comprehensive theory of evolution, however, is the fact that Darwin's theory in both its original and modern forms does not specify any mechanisms of development. Before modern biology became consumed with the doctrine of genes as both the cause and the subject of evolution, it was generally considered that the most interesting aspects of life's history were changes over time in the phenotypes (i.e., forms and functions) of organisms. But the phenotypes of multicellular organisms (Darwin's main concern) are generated by developmental mechanisms, formative processes that use the products of genes but are not coextensive with genes, their products, or their interactions.[3] Moreover, since developmental mechanisms shape and pattern tissues, which are parcels of pliable, chemically and mechanically responsive matter, they involve the physics of materials,[4] and these are in no fashion reducible to genes.

The unsuitability of tracking gene frequencies as a surrogate for evolutionary change can further be inferred from considering the actual function of genes. Genes specify the sequences of protein and RNA molecules, which have only an indirect impact on phenotypes. Sometimes a genetic mutation leads to dramatic change in the phenotype, but often mutation or even deletion of a gene leads to little or no detectable phenotypic alteration. Gregor Mendel's discovery of what we now refer to as genes was based on alleles (gene variants) of "large effect"—different versions of an underlying factor associated with smooth or wrinkled plant seeds, for example. When the modern synthesis was formulated, however, it was realized that (1) alleles of large effect will be rare in natural populations, and (2) if evolution actually proceeded by using such genes, proposition 8, the distinguishing premise of Darwin's theory, would be irrelevant. Seeking to preserve, and indeed enshrine, Darwin's mechanism, the architects of the synthesis took the route of constructing a self-consistent mathematical theory of gene-frequency dynamics that confirmed the efficacy of natural selection by assuming that genetic change was the driving force of evolution, and (contrary to what Mendel had found and what is well established in modern molecular studies) that alleles of large effect could be ignored.

Focusing on the transmission of genes rather than on the transmission of phenotypes via developmental mechanisms led to an impoverished evolutionary theory. Developmental mechanisms, in fact, are frequently of large effect. This is a consequence of their "nonlinearity"; that is, small continuous changes in the variables they comprise may lead to discontinuous outcomes (an analogy is the phase transition from liquid to solid that occurs when water is cooled to 0°C). Their actions, moreover, may be "plastic," that is, of variable outcome depending on environmental effects (another water-based analogy is that water in its liquid state can form waves or vortices under different conditions). Developmental mechanisms also often lead to stereotypical or otherwise predictable morphologies, a characteristic (termed "orthogenesis") that is considered anathema to a notion of evolution based on selection of unconstrained variation. In particular, if selection merely nudges organismal form along paths that are already laid out in advance by inherent material properties, it would hardly be (as asserted by a popular writer on the subject) "the only creative force in evolution."[5] The recognition that phenotypes are inherited via nonlinear, plastic, orthogenic developmental mechanisms, and not simply by collections of genes, implies that some of the major conclusions of the modern synthesis must be incorrect.

The history of the sciences of the nineteenth and early twentieth centuries can provide some insight into how the most widely accepted version of evolutionary theory came to stand on such dubious ground. Although Darwin's theory of the generation of organismal diversity was a decisive break with previous biological ideas, it was very much in keeping with the scientific understanding of its time regarding the physical world. Materialist science was on the rise when Darwin and Wallace were developing their ideas in the 1850s. The theoretical successes of Isaac Newton, Robert Hooke, Robert Boyle, Christiaan Huygens, and others in the areas of mechanics and optics and of Antoine Lavoisier and Friedrich Wöhler in chemistry were particularly impressive, as were practical advances in engineering of devices such as clocks, pumps, and mechanical reapers, which showed the versatility and capability of complicated machines. Many people were now willing to accept the proposition that living things were exclusively material objects. This tenet, however, was by no means inconsistent with their having been designed by a supernatural creator. As William Paley (1743–1805) famously pointed out in his *Natural Theology*,[6] anyone stumbling on a piece of metal-gear machinery as complicated as a pocket watch would immediately assume that it had a designer. Why should it be any different for an organism?

The problem, then, for nineteenth-century evolutionists with materialist aspirations was finding a naturalistic mechanism for the origination of complex forms and, once they existed, for their transformation. Newtonian mechanics, though universally acclaimed, did not supply an apt model for biology. For Newton, matter was inert, changing its state in a continuous fashion and deviating from its track only when acted on by external forces. Living matter, in contrast, is "self-realizing" and protean in its transformations. Indeed, although the philosopher Immanuel Kant was disinclined to invoke supernatural explanations, he saw no reasonable possibility of mechanical accounts of life and doubted whether there would ever be a "Newton of the blade of grass."[7]

Naturalists before Darwin had struggled with this problem. Jean-Baptiste Lamarck (1744–1829) studied physical as well as biological questions, and when he eventually concluded against his earlier beliefs that biological forms can undergo transmutation, he proposed models based on purported novel physical principles governing biological matter. He postulated a "power of life," an inherent tendency of living systems to become more complex over time, what scientists now refer to as "self-organization," and the "influence of circumstances," what we now encompass under the concepts of physiological adaptation and phenotypic plasticity.[8] As we will see later, a comprehensive theory of evolutionary change needs to incorporate both of these ideas, but Lamarck's inability, given the physics of his day, to explain the basis of self-organization, or how plasticity-mediated changes in phenotype could be inherited, led to his contributions being marginalized and even ridiculed as Darwinism consolidated its hold on evolutionary theory.

Other Continental thinkers, such as Johann Wolfgang von Goethe (1749–1832), Étienne Geoffroy Saint-Hilaire (1772–1844), and Lorenz Oken (1779–1851), advocated the idea that living matter is organized by "laws of form."[9] But although this perspective—"rational morphology"—strove to provide a scientific counterpart to the laws of physics and chemistry gaining currency during the same period, it inevitably had an air of mystery and impenetrability about it. The scientific revolutions that would bring the physical speculations of Lamarck and the rational morphologists into the mainstream of physical science were still to come.

The great selling point of the Darwin-Wallace theory was that it contained no hypotheses about how matter arrived at the point of being subject to natural selection. (The origin of life is still a mystery as of 2012)[10] but the scandalous obscurity of the foundational processes of biology hardly rattles the dismissive self-assurance of gene-centric modern synthesizers).[11] Nor did the theory of natural selection depend on any knowledge of the source of variation, either inherited or noninherited, in organismal form and function. The fact that a material object (e.g., an organism) would have properties slightly different from those of other copies of the same thing was simply the default science and manufacturing wisdom of the time.[12] That small differences can in some circumstances be advantageous is just common sense.

The leap of mind required to turn these ordinary observations into a (materialist) theory of evolution was imagining continuous trajectories of change between present-day organisms and their ancestral forms based on the small differences that appear in each generation. Although it might take a very long time, little by little a wormlike ancestor could change into a lobster or a lion by gradual transformation.

Such continuous pathways of organismal change are not supported by the fossil record, however, nor, as we now appreciate, are they required for physical plausibility. Although Darwin's theory is ostensibly concerned with changes in form over many generations, the only way a new form can arise is if an actual developmental mechanism is capable of generating it. To return once again to the water analogy, just as the physics of this substance allows it to form waves and vortices, but not honeycomb-type structures, and (if the external conditions change) to become vapor or solid, the materials that generate the bodies of animals—parcels of matter consisting of clusters of cells—also have preferred structural motifs. These include hollow and multilayered forms, forms that are elongated and segmented, and forms that contain branched tubes and appendages.[13] All the forms that characterize animal body plans and organs are in fact the physical manifestations of a type of substance described by physicists as soft, chemically and mechanically excitable matter.[14] The recognition of this fact, a result of theoretical and experimental advances of later nineteenth-century thermodynamics and twentieth-century nonlinear dynamical systems and condensed-matter physics,[15] can in some ways be seen as the realization of the program of the rational morphologists. Furthermore, the prediction by this "physicalist" model that large-scale evolution could have occurred relatively rapidly, with only minimal change in the molecular components of the underlying developmental mechanisms, has been borne out.[16]

Many recent studies in the field of developmental biology focus on the association between physical and genetic mechanisms in generating tissue structures. Gastrulation, for example, an early event in animal development, is the means by which the embryo becomes organized into distinct layers of tissue. This process has been shown to result from a combination of adhesive and mechanical forces, leading to a "sorting-out" effect among populations of cells that is very much like the phase separation that occurs in a suspension of oil and water.[17] Subtle physical differences between cells, at or just beneath their surfaces, will determine whether they act like molecules of "oil" or "water" in a particular mixture.[18]

Another such example is the formation of segments (termed "somites") in the embryos of vertebrate animals.[19] These are blocks of tissue that emerge in sequential left-right pairs to either side of the embryo's central head-to-tail axis, with the pair closest to the head forming first, followed in order by the ones increasingly closer to the tail. The somites eventually differentiate into the cylindrical vertebrae and associated muscles. Consideration of this case in some detail will illuminate a number of the points mentioned earlier.

The formation of somites depends on several gene products and at least two distinct physical processes. First, a gene product (FGF8) produced at the tip of the embryo's tail diffuses away from its source (physical process 1) so that it forms a concentration gradient that fades to low levels close to the head end. At the same time, other gene products (Hes1 and several additional ones) undergo synchronized oscillations in the presegmented tissue (physical process 2), constituting a "segmentation clock."

When the clock, acting within a domain of tissue, "strikes" a critical hour, it can cause the tissue to pull away partially from its surroundings, but the pulling away happens only in tissue not exposed to high levels of FGF8. That is, it happens only at the low end of the gradient, far from the tail tip. By the description above, this will first occur near the head. But as the embryo grows and lengthens, the low end of the FGF8 gradient will be found at positions that, although still far from the tail, are also progressively farther from the head. By this complex system of physical-genetic processes, involving interactions among a gradient, a clock, and tissue growth, the vertebrate body becomes segmented.

Several evolutionary implications emerge from this developmental picture. First, gene products can oscillate in concentration only when their production is part of a dynamical system with appropriate positive or negative feedback relationships among their respective genes and other molecules.[20] The mere presence of the genes does not create the clock. In fact, all the gene products thought to be involved in the segmentation clock are also found in nonsegmented animals.[21] This means that small alterations (due, for example, to mutation or environmental change) in a rate or feedback parameter can immediately turn a genetic "network" with no oscillatory properties into an oscillatory one. If all other ingredients were in place, a nonsegmented animal could spawn segmented ones.

Additionally, since the balance among several rates—clock tempo, the pace of embryo elongation, FGF8 diffusion speed—corresponds to a given number of somites, the number can change quite precipitously if one of these rates changes.[22] Snakes, for example, can have several hundred somites, but mice or humans have fewer than seventy, and the difference can be attributed to the different rates of embryo elongation in reptiles versus mammals.[23] Since rates of biochemical processes depend on temperature, we might expect to find that the number of segments in cold-blooded vertebrate species would vary with the temperature at which their eggs are incubated, and many such examples have indeed been described.[24]

From the gastrulation and segmentation examples and others, one can see that developmental mechanisms and not genes are the means by which biological characters are transmitted across generation lines.[25] Genes, in the usually understood form of DNA sequences,[26] are part of all developmental mechanisms, so the heritability associated with allelic variation will enter into all traits, but often in highly complex, indirect ways. Because developmental mechanisms exhibit, variously, nonlinear, plastic, and self-organizational properties, evolutionary transitions can be "saltational" (phenotypically abrupt), rapid, and influenced by environmental change in a direct (i.e., Lamarckian) fashion and not just as a consequence of selection of marginally favorable variants.

The fact that evolutionary theory became hooked on genes in the early stages of its formulation is an understandable consequence of the uneven development of the biological and physical sciences. But now that we have the means to analyze how living forms are actually produced and inherited, there is no excuse not to move on.

CHAPTER THREE

Genes as Difference Makers

EVELYN FOX KELLER

GENETICS IS A FIELD OF STUDY, a branch of biology, but what, in fact, is it about? How one answers this question, of course, depends on when and where one looks. But for classical genetics, and especially for the paradigmatic school of T. H. Morgan, genetics was about tracking the transmission patterns of units called "genes." What was a gene? No one knew, but notwithstanding this ignorance, a gene was assumed to be a unit that could be identified by the appearance of mutants in wild-type populations. That is, a phenotypic difference in some trait (a mutant) was taken to reflect a difference (a mutation) in some underlying gene associated with that trait. But to argue for such an identification between phenotypic difference and underlying gene in fact requires a two-step move. First, change (a "mutation") in some underlying entity (the hypothetical gene) is inferred from the appearance of differences in particular phenotypic traits (e.g., white eye, bent wing, narrow leaf), and second, the existence and identity of the gene itself is inferred from the inference of a mutation. In other words, the classical gene was identified, on the one hand, *by* the appearance of phenotypic differences (mutants), and on the other hand, it was simultaneously identified *with* the changes (mutations) that were assumed to be responsible for the mutants. Thus the first map of the mutations thought to be responsible for the observed phenotypic differences in *Drosophila* was called not a map of mutations but a "genetic map," a map of genes.

This is the sense in which the classical gene is often said to be a "difference maker."[1] But a gene was not only taken to be a difference maker; it was also assumed to be a trait maker. It was both the entity responsible for the difference observed and (at least implicitly) the entity responsible for the trait that had undergone a change—that is, the trait in which a difference had been observed.

One might say, then, that a certain confounding of traits and trait differences was built into the science of genetics from the very beginning; moreover, one might argue, necessarily so. The occurrence and frequency of trait differences were what geneticists had observational access to; by examining phenotypes, they could detect phenotypic differences that, in turn, were taken as indicative of changes in some underlying, internal entity. The locus of such changes could be mapped through breeding. As Horace Freeland Judson has observed, "In 1913, Alfred Sturtevant, a member of Thomas Hunt Morgan's fly group at Columbia University, drew the first genetic map—'The linear arrangement of six sex-linked factors in *Drosophila,* as shown by their mode of association.' Ever since, the map of the genes has been, in fact, the map of gene defects."[2]

Indeed, the same slippage persists in contemporary discussions of what Lenny Moss (and now numerous others) call "Gene-P," or "phenotypic gene." Moss writes that "Gene-P" is a "phenotype predictor" and cannot be defined by its nucleic acid sequence. But in fact, "Gene-P" is neither a gene nor a predictor of phenotype; it is a phenotype-difference predictor. Indeed, as Moss himself acknowledges, the reason that "Gene-P" cannot be defined by a specific sequence is that "invariably there are many ways to lack or deviate from a norm."[3]

Wilhelm Johannsen, the man to whom we owe the word "gene," was clearly worried about this problem when he asked: "Is the whole of Mendelism perhaps nothing but an establishment of very many chromosomal irregularities, disturbances or diseases of enormously practical and theoretical importance but without deeper value for an understanding of the 'normal' constitution of natural biotypes?"[4]

But it is hard to imagine that the slippage was entirely accidental. To think of genes simply as difference makers would have been to detract from the very power of the gene concept. Mapping "difference makers" and tracking their assortment through reproduction may have been all that the techniques of classical genetics allowed for, but the aims of these scientists were larger. What made genes interesting in the first place was their presumed power to mold and to form—in a word, their presumed power to act. "Gene action" was the term invoked to refer to the process by which genes exerted their power in the development of the characters or traits themselves. But for illuminating the nature of this process (the developmental process), studies of trait differences would not suffice by themselves. In fact, neither successful mapping of the locus of the factors (difference makers) presumed to be responsible for such differences nor analysis of their emergence and their intergenerational patterns of transmission taught us anything about the causal dynamics of the developmental process by which the traits themselves came to be. As John Dupré puts it, "Classical genetics was about invisible features that could trigger different developmental outcomes, but not about the causal explanation of these outcomes."[5] Furthermore, classical geneticists were, for the most part, well aware of this distinction. Nevertheless, the easy slide between genes as difference makers and genes as trait makers perpetuated the illusion (as widespread among geneticists as it was among their readers) that an increased understanding of the effects of gene differences would enhance our understanding of what it is that the entities called "genes" actually do.

What Is a Disease?

A similar confounding of the etiology of traits with that of trait differences pervades virtually all the current literature of medical genetics. Indeed, the very notion of a disease as an individual trait—in the sense, for example, that brown eyes are a trait—already incorporates this confusion. We may commonly speak of an individual as "having a disease," much as he or she might have brown eyes, but in fact, and as has long been understood by many writers, disease is a state that exists only in relation to another state already established as normal.[6] In his inquiry into the scientific rules for distinguishing the normal from the pathological, written more than a hundred years ago, Émile Durkheim stressed that "a trait can only be characterized as pathological in relation to a given species"—in other words, in relation to a standard of normality or state of health that is itself inextricably confounded with the norm of a species. "One cannot even conceive, without contradiction, a species that could, by itself and in virtue of its fundamental constitution, be irremediably sick. [The species] is the norm par excellence and, accordingly, can harbor within itself nothing of the abnormal."[7]

It is true that the French philosopher of science Georges Canguilhem, following Kurt Goldstein, made valiant efforts to internalize the diagnostic criteria of pathology, locating them within the individual,[8] but for all his efforts, by far the most common understanding of disease has continued to rely on comparison (or contrast) with a preestablished conception of "normal." English-language dictionaries routinely define disease as a relational state: it is a dis-ease, "an abnormality of the body or mind" (Wikipedia); "a departure from the state of health" *(Oxford English*

Dictionary); "a deviation from or interruption of the normal structure or function of a part, organ, or system of the body" (*Dorland's Illustrated Medical Dictionary*); "an interruption, cessation, or disorder of a body, system, or organ structure or function" (Medilexicon). Indeed, French dictionaries do the same. In virtually every dictionary I have consulted, a *maladie* is an "altération de l'état de santé." In French, as in English, an animal can be said to have brown eyes whether or not a comparison with other animals is at hand, but it cannot, without such a comparison, be said to have a disease.

Like medicine, genetics too might be said to be a comparative science. Comparing organisms with differing phenotypes and attempting to correlate these phenotypic differences with corresponding genetic differences (mutations) have been the bread and butter of geneticists from the earliest days of that science. But genetics aims beyond comparative judgments; it seeks an understanding of the developmental dynamics, and as I have tried to show, its language invites us to lose sight of the complex moves—first, in attributing the cause of a phenotypic difference to a genetic mutation; second, in the assumption that the presence of a mutation automatically signals the presence of a gene; and third, in attributing responsibility for the trait in question to the gene in which the mutation is assumed to have occurred—that are routinely made in effecting this shift from comparative to individual. It seems therefore no accident that the adoption of a lexicon of illness that refers to disease as an individual attribute comes with the emergence of a medical science grounded in genetics.[9] A disease, I am arguing, is not a trait but a trait difference. The attempt to associate disease states with genetic mutations responsible for "inborn errors of metabolism" can be seen simply as part of the more general effort to associate mutations with particular phenotypic differences, and it is subject to exactly the same sorts of conflation.

Today, the genetic basis of a disease is more likely to be associated less with a genetic difference—with a mutation, a departure from a presumed normal genome—than with a putative gene or genes; in short, medical geneticists seek the cause of a disease in a defective gene. Classical geneticists relied on "gene maps" to identify the gene presumed to be involved; today, medical geneticists tend to rely more on analysis of nucleotide sequences. Although they continue to talk about genes, in actual practice the identification of one or more such differences may or may not point to a defect in a particular gene (however that term is defined); these days, the direct object of interest is the change (or defect) located somewhere in the nucleotide sequence that is correlated with the expression of the disease. Despite widespread talk of "disease genes" or "disease-causing genes," it is the DNA itself that has become the focus of investigation. I claim that the notion of a gene "causing" a disease (or even of a particular sequence "causing" a disease) has exactly the same status as the notion of a gene "causing" a mutation. But perhaps more important is the fact that for diagnostic purposes (at least for diagnoses based on genetic tests), the attempt to correlate a disease state with an underlying gene is in many if not most cases largely irrelevant. Contemporary genetic medical diagnostics rely on the identification of aberrant or anomalous sequences, not on the causal pathways such anomalies may disrupt. Such anomalies may be anywhere in the genome; indeed, only rarely are they found in protein-coding sequences (that is, in the segments of DNA usually associated with genes).

For most of us, the crucial question is, can the identification of such sequences be useful in the treatment or prevention of disease, and if so, how? Most immediately (and perhaps most obviously even if it is generally left unstated), such information can be used to promote selective abortion. But if we are interested in therapeutic medicine, we need more than a simple correlation between aberrant sequence and aberrant phenotype. It is true that the early days of the Human Genome Project brought the promise that in time we would be able simply to replace defective sequences with normal ones (gene therapy), but that hope has failed to materialize, and at least one of the reasons for this is that the relation between DNA sequence and phenotype has turned out to be far more complicated than originally expected. As for the possibility of other kinds of treatment (and sometimes of prevention) in a particular individual carrying the aberrant sequence, this depends on understanding something about the biological function that has been disrupted by the change in sequence that has been identified. Such a quest takes us beyond the analysis of phenotypic differences induced by mutant forms. Indeed, it requires an altogether different and, almost always, far more difficult kind of analysis.

There are, however, examples that have proved to be relatively simple. Phenylketonuria (PKU) is one, and it is probably the most celebrated case of therapeutic intervention in the history of medical genetics. Indeed, it is everyone's canonical example, mine as well. Phenylketonuria is a disorder (now recognized as genetic) associated with a range of disabling symptoms, including mental retardation, and caused by the inability of the body to properly metabolize the essential amino acid phenylalanine. A major breakthrough in the treatment of this disease came with the recognition that its symptoms can be significantly alleviated if the affected individual adheres to a (carefully monitored) low-phenylalanine diet for his or her entire life.[10] However, the development of a strategy to treat PKU had nothing to do with either the identification or mapping of the gene(s) or genetic sequence(s) involved. Today we know that this disorder is caused by one or more mutations in the gene encoding the enzyme that breaks down phenylalanine (to date, as many as 400 such mutations have been identified). In point of historical fact, however, neither our understanding of the (at least proximal) cause of this disease nor the development of a therapeutic intervention depended on any sort of genetic analysis.[11] It is now believed that the phenotypic expression of PKU (the disabling symptoms) is the direct consequence of the accumulation both of high levels of phenylalanine and of toxic intermediates resulting from its faulty metabolism, but we learned this from direct biochemical analysis; ordinary medical observation showed how dreadful the symptoms can be. Furthermore, we have now acquired the ability to characterize precisely many of the mutations responsible for the absence of the necessary enzyme, and doing so has certainly been instructive. The bottom line, however, is that thus far, that ability has not significantly added to the possibilities of therapeutic intervention.

Of course, it need not have happened that way. Identification, location, and characterization of the guilty mutation(s) could have provided the starting point for a research program that over the course of time led to an understanding of the disease and possibilities for treatment. But whatever the sequence of historical developments, there is no way in which the genetics of difference could have achieved that understanding by itself. What was required for a causal analysis of the disease—and hence for the possibility of therapeutic intervention—was a biochemical analysis of the metabolic pathway that had gone astray. That analysis might have begun with the identification of a particular gene (even though it in fact did not), but if it had, what would have been required would have been an understanding of the gene's downstream effects, of the particular role that the gene in which the mutation had occurred actually played in development.

What Genes Do

The challenge of attributing causal function to genes (rather than to gene differences)—of understanding what genes do—has in fact plagued genetics from the outset, and it was not until the advent of molecular biology that it seemed possible to address this question. What does a gene do? Defined as a discrete stretch of nucleotides, a gene was said to make, or "code for," a protein. But even with that insight, it was one thing to be able to track the causal effect of a genetic difference (mutation, or change in nucleotide sequence) on a particular trait, and quite another to track the causal influence of the gene itself (or, indeed, of the protein it was said to make) on the development of that trait.

To be sure, enormous progress has been made since the early days of molecular biology, and we now know a great deal more about the ways in which the process of development makes use of an organism's DNA. But what we have learned has not so much answered earlier questions as it has transformed them. We have learned, for example, that the causal interactions among DNA, proteins, and trait development are so entangled, so dynamic, and so context dependent that the very question of what genes do no longer makes very much sense. Indeed, biologists are no longer confident that it is possible to provide an unambiguous answer to the question of what a gene is. The particulate gene is a concept that has lent itself to increasing ambiguity and instability over the years, and some have begun to argue that the concept has outlived its productive prime.

DNA, by contrast, is a concrete molecule—an entity that can be isolated and analyzed, identified by its distinctive physical and chemical properties, and shown to consist of particular sequences of nucleotides. We know what DNA is, and with every passing day we learn more about the exceedingly complex and multifaceted role it plays in the cellular economy. It is true that many authors continue to refer to "genes," but I suspect that this may largely be due to the lack of a better terminology. In any case, continuing reference to "genes" does not obscure the fact that the early notion of clearly identifiable particulate units of inheritance that not only can be associated with particular traits but also can be taken as agents that "act" to produce those traits has become hopelessly confounded by what has been learned about the intricacies of genetic regulation. Furthermore, recent experimental focus has shifted away from the structural composition of DNA to the variety of sequences on the DNA that can be made available for (or blocked from) transcription (i.e., to gene expression). Finally, and relatedly, it has become evident that nucleotide sequences are used to provide transcripts not only for protein synthesis but also for multilevel systems of regulation at the levels of transcription, translation, and posttranslational dynamics. None of this need impede our ability to correlate differences in sequence with phenotypic differences, but it does give us a picture of a causal dynamic among DNA, RNA, and protein molecules so immensely complex that it definitely puts to rest all hopes of a simple parsing of causal factors. Because of this, today's biologists are far less likely than their predecessors were to attribute causal agency either to genes or to the DNA itself. They recognize that however crucial the role of DNA in development and evolution is, it does not do anything by itself. It does not make a trait; it does not even encode a "program" for development. Rather, it is more accurate to think of a cell's DNA as a standing resource on which it can draw for survival and reproduction, a resource it can deploy in many different ways, a resource so rich as to enable it to respond to its changing environment with immense subtlety and variety. As a resource, DNA is certainly indispensable—arguably it can even be said to be a primary resource—but it is always and necessarily embedded in an immensely complex and entangled system of interacting resources that collectively are what give rise to the development of traits. Not surprisingly, the causal dynamics of the process by which development unfolds are correspondingly complex and entangled, involving causal influences that extend upward, downward, and sideways.

Does It Matter, and If So, Why?

Those familiar with my past work on the history of genetics will recognize this chapter (adapted from *The Mirage of a Space between Nature and Nurture*) as belonging to a long series of efforts to clarify some of the conceptual confusions that plague contemporary biological discourse, and some readers may wonder why I persist. Certainly, biologists often complain that such efforts are irrelevant to their undertaking, and that they know what they mean, at least sufficiently to proceed with their research. Perhaps they do, but the problem that to my mind remains the most critical concerns the focus of their research, the questions that are posed, and the ways adopted to try to answer these questions. It is here that conceptual confusion does its dirtiest work, most conspicuously when the justification of research agendas depends, however unwittingly, on a hidden multiplicity of meanings. I have argued that this is especially the case for research focused on the nature-nurture debate. In this case my recommendation is clear: we need to reformulate the questions to conform to what the realities of biological development permit scientists actually to answer. Let us ask not how much of any given difference between groups is due to genetics and how much to environment, but rather how malleable individual human development is, and at what developmental age. There is no reason to privilege birth as a cutoff point. Development is lifelong, and so too is its plasticity. We may not share the interests of breeders in artificial selection, but both as scientists and as citizens, we are surely committed to trying to maximize the development of individual human potential. For this, we need a better understanding both of what resources can contribute to such development and of how they can best be deployed. What kinds of research can provide us with such information? I would put my money on the new studies of phenotypic plasticity we are beginning to see not only in developmental biology but also in neuroscience, physiology, and ecology. There is no shortage of scientific work that can productively inform us about the things we want to know, but we need to pose our questions in ways that researchers can meaningfully address.

CHAPTER FOUR

Big B, Little b

Myth #1 Is That Mendelian Genes Actually Exist

DAVID S. MOORE

WHILE I WAS WAITING to catch a flight out of Columbus, Ohio, I heard an evening news story about the discovery of a genetic mutation that supposedly allows affected individuals to get away with fewer than eight hours of sleep each night. Beyond the specifics of the story, there was nothing particularly special about it; these days, it is difficult to pass through a 24-hour news cycle without some reference being made to a new discovery in the realm of genetics. But the story drew my attention because there was a central assumption buried in it, one that most people now share, namely, that there are genes that determine aspects of our behaviors, appearance, and health. However, in general, most scientists who actually study the genetic material, DNA, no longer believe that genes single-handedly determine any of these sorts of characteristics.[1] Amazingly, there is also a growing consensus among these scientists that we need to rethink one of the assumptions at the center of that assumption: namely, that there are such things as genes in the first place.[2]

Our modern notion of genetics has a long and interesting history, but for most people, it is rooted in the work of Gregor Mendel, a monk who lived in the nineteenth century in a monastery in what is now the Czech Republic. Mendel's experiments with pea plants are famous because they are often described for us in school science classes when we are relatively young; in fact, for many people, this is the only work in genetics to which they are ever exposed.

Mendel's basic research question can be construed as follows: why is it that if parents are different from each other in one of their traits—say, their coloration—their child sometimes looks like only one of the parents, and not like a blend of the parents? Although it is certainly sometimes the case that the offspring of a darker-skinned father and a lighter-skinned mother will have a skin tone midway between those of her parents, it is also the case that the child of a brown-eyed father and a blue-eyed mother does not typically have eyes that are colored midway between brown and blue, but instead has eyes that are as blue or brown as those of one of her parents. In Mendel's day almost everyone believed that parental traits—for example, height, body shape, and the color of things like skin, hair, and eyes—were blended in offspring, but this view was not consistent with Mendel's observations.[3] A particularly troubling sort of question for Mendel was why plants with purely red or purely white flowers can sometimes be the offspring of parent plants that grow only pink flowers; if inheritance works by a process of blending, two pink-flowered plants should never produce offspring with red or white flowers.

To study this question, Mendel bred generations of pea plants and examined such characteristics as the colors of the peas and how wrinkled they were.[4] In order to explain his observations—that inheritance was not working by a process of blending—he felt the need to posit the existence of material entities that (1) could be inherited, (2) dictated the characteristics of the pea plants, and (3) were effectively indivisible. We have come to know these entities as "genes," and the idea that they are indivisible means that if you have blue eyes, your children's children's children could have eyes every bit as blue as yours, even if your children and your children's children all have brown eyes and find themselves reproducing with brown-eyed mates.

The standard story we encounter in school tells of "genes for brown eyes" and "genes for blue eyes" that are not equally strong; using the same terminology that Mendel (1866) chose to use (but translated into English), we say that the genes for brown eyes are "dominant." The idea that genes for brown eyes are dominant comes from the observation that when individuals in a group of randomly chosen brown-eyed people mate with the individuals in a group of randomly chosen blue-eyed people, most of the children of those unions wind up with brown eyes. Consequently, dominant genes are typically represented by capital letters—in the case of a dominant gene for brown eyes, "B." Nondominant genes, called recessive genes, are typically represented by lowercase letters—in the case of a recessive gene for blue eyes, "b." And because you get one set of genes from your father and one from your mother, a given person's genes for eye color can be represented as either BB (a brown-eyed person, because both their maternally contributed and their paternally contributed eye-color genes are for brown eyes), bb (a blue-eyed person, because both their maternally contributed and their paternally contributed eye-color genes are for blue eyes), or Bb (a brown-eyed person, because a gene for brown eyes was received from one parent, and a gene for blue eyes was received from the other parent, a combination that yields brown-eyed offspring because the gene for brown eyes is dominant over the gene for blue eyes). Interestingly, the standard story is that a brown-eyed person characterized by Bb genes can have eyes every bit as brown as a person characterized by BB genes. And it is because a fully brown-eyed person of the Bb type can have a blue-eyed child (provided that person's mate is either bb or Bb) that Mendel's conceptualization has been, over the past 100 years, able to completely sweep the older notion of blending inheritance from biologists' theories.

Mendel's conceptualization has been so successful at explaining observed phenomena that it is generally regarded as correct. It is at the core of many genetic analyses, and it is taught to schoolchildren no differently than we teach them that one plus one equals two. The only trouble with this situation is that on a concrete level, Mendel's conceptualization turned out to be simply wrong; in fact, there really are no such things as single genes that determine human eye colors.

As it happens, there have been geneticists who understood this from the start. As early as 1915, Alfred H. Sturtevant, discussing the red and white eye colors that are characteristic of fruit flies, wrote:

> Although there is little that we can say as to the nature of Mendelian genes, we do know that they are not "determinants." . . . Red is a very complex color, requiring the interaction of at least five (and probably of very many more) different genes for its production. . . . We can then, in no sense identify a given gene with the red color of the eye. . . . All that we mean when we speak of a gene for pink eyes is, a gene which differentiates a pink eyed fly from a normal one—not a gene which produces pink eyes per se, for the character pink eyes is dependent upon the action of many other genes.[5]

Likewise, modern genetic research has confirmed a similar understanding for human eye colors. As Sturm and Frudakis note,

> What is still commonly taught in schools today as a beginners guide to genetics [is] that brown eye colour is always dominant to blue, with two blue-eyed parents always producing a blue-eyed child, never one with brown eyes. Unfortunately, as with many physical traits, this simplistic model does not convey the complexities of real life and the fact is that eye colour is inherited as a polygenic not as a monogenic trait. Although not common, two blue-eyed parents can produce children with brown eyes.[6]

These researchers concluded that "the use of eye colour as a paradigm for 'complete' recessive and dominant gene action should be avoided in the teaching of genetics to the layperson, which is often their first encounter with the science of human heredity."[7] But it turns out that the Mendelian conception of genes not only fails to capture accurately what determines our eye colors; it actually fails to represent accurately how genes contribute to the development of *any* of our traits. In fact, it is no longer even clear that there really are such things as Mendelian genes contained in our DNA that determine the final forms of our biological or psychological traits.

It is well known that nearly a century after Mendel published his findings, Watson and Crick correctly deduced the structure of DNA, ushering in the modern age of genetics in the 1950s. At last it became possible to begin studying how molecules that could be passed from parents to their children could influence the children's characteristics. What is less well known is how exactly the "genes" that have since been identified in DNA are related to the "genes" Mendel effectively identified in the middle of the nineteenth century. Because both entities share the same name, it is generally assumed that they refer to the same things. But there is good reason to think otherwise.

First, segments of DNA—which are the kinds of genes that we typically hear about these days on the evening news—most definitely contribute to the observed characteristics of all living things.[8] However, unlike Mendelian genes—which to this day remain strictly theoretical—they do not *determine* those characteristics. Instead, biologists have learned that our characteristics always emerge following the process of development, which always entails interactions between DNA and environmental factors.[9] These factors include both the environment outside our bodies and nongenetic factors (such as hormones, for example) that are inside our bodies (and many of these nongenetic factors in our bodies can be influenced by the environment outside our bodies). Thus, although our traits are always influenced by genetic factors, they are always influenced by nongenetic factors, too; genes do not determine our characteristics, as Mendelian theory implies.[10]

Second, several recent discoveries have cast serious doubt on the idea that there are coherent entities in our DNA that can unambiguously be called "genes."[11] Perhaps the most important of these discoveries is related to a phenomenon known as RNA splicing. It turns out that genetic information is scattered among segments of DNA that do not have any currently understood purpose.[12] To illustrate, imagine for a moment that information in DNA represents an instruction for the development of a characteristic (but please note that this is an imaginary scenario; in reality, the situation is quite a bit more complex than this, and many theorists would now argue that DNA is best not thought of as containing instructions).[13] If the instruction we are imagining is "begin to grow an arm here," it would ordinarily appear in the DNA scattered among purposeless information, like this: "do baryell note beginner to red dog rowing ckjswnrt bell tag an arm legitimate shopping ampere." (In case the instruction in question appears to be completely absent in that stream of information, let me use italics to help bring it out: "do baryell note *begin*ner *to* red dog *row*ing ckjswnrt bell tag *an arm* legitimate s*h*opping amp*ere*.") Obviously, to serve any useful function, the meaningless information—for instance, the "opping amp" segment separating the "h" from the "ere"—needs to be cut out of the sequence, and the meaningful portions must be spliced together to produce the functional instruction "here." Crazy, right? But we now understand that this is how the system works.[14]

Perhaps even more incredible is the phenomenon known as "alternative splicing," in which a single segment of DNA can be spliced in many different ways, producing many different kinds of "instructions."[15] For example, the seemingly gibberish-filled sentence above could be spliced to produce the instructions "begin growing a leg here," "grow a leg there," or even "do not grow arms here" (if you look back at the sentence, you should be able to make out each of these sentences buried deep in the gibberish). Clearly, a single segment of DNA—a gene—can have a variety of different effects depending on how it is interpreted, and remarkably, the interpretation favored in any given situation is typically a matter of context. Given this reality, the "genes" that molecular biologists are discovering every day are very different sorts of things from those Mendel's followers were imagining. We now know that DNA cannot be thought of as containing a code that specifies particular predetermined (or context-independent) outcomes.[16] In fact, what this means is that the same segment of DNA can do two entirely different things in different bodies (because different bodies can provide different contexts for their genes). So as unlikely as it sounds, it is possible that a particular gene in John Lennon might have done something different than that same exact gene would have done in J. Edgar Hoover. Indeed, a large team of biologists recently concluded that the various protein products coded for by "individual mammalian genes . . . may have related, distinct, or even opposing functions."[17]

Of course, if alternative splicing were a relatively rare event, one could still maintain that Mendelian genes are the rule and alternative splicing the exception. But it has become clear that it is the other way around. In the late 1990s scientists were estimating that approximately 33 percent of our genes were subjected to alternative splicing, but by 2003 that number was up to 74 percent.[18] Now we know that alternative splicing is virtually universal, influencing the transcription of between 92 and 95 percent of our genes.[19] And alternative splicing is not the only phenomenon that has cast a large shadow over the Mendelian concept of the gene. Among other recently discovered phenomena that call this conceptualization into question is the finding that some gene products can function both as molecules used in protein production and as molecules that perform entirely different cellular functions.[20]

One consequence of this strange state of affairs is that there is currently debate among theorists about whether Mendel's gene concept has any applicability to DNA segments at all. Evelyn Fox Keller wrote in her book *The Century of the Gene* that "the concept of the gene [is on] the verge of collapse," and to date, there is still no agreed-on definition of the word "gene" in biological writings.[21] The fact of the matter is that in spite of the frequency with which the word is used these days, it does not actually refer to any one thing—or class of things—in particular.

What is clear is that the genes most of us envision inside us, calling the shots and determining our characteristics, are myths. There are no coherent entities in our cells that deterministically dictate how our bodies or our minds will develop. Instead, unprocessed, ambiguous lengths of DNA—which are not themselves single genes for specific traits—are cut up and combined in a variety of ways (depending on the context) to produce other molecules that then merely *contribute* to the construction of our traits.[22] Of course, DNA sequences can be altered as a result of exposure to, for example, radiation, and such mutations can contribute to the development of various disease states. But these mutated segments of DNA do not themselves produce diseases single-handedly, any more than a gene that is necessary for the development of blue eyes can single-handedly cause a person's irises to appear blue.[23] Even the symptoms of diseases like phenylketonuria, cystic fibrosis, and sickle-cell anemia—all of which are conditions that were once thought of as being directly caused by the actions of single genes—are now recognized as phenotypes caused by a variety of factors that interact in complex ways during development.[24]

The question then remains: why is Mendel's conceptualization still regularly taught in schools? The answer is that his 140-year-old approach still works to a reasonable extent when we are trying to predict the characteristics of offspring produced when particular plants or animals mate. This is very useful, of course: breeders looking to maximize a characteristic in an animal—say, the amount of milk produced by a cow—can use Mendel's conceptualization to help them do that. But just because a particular methodology can help generate relatively accurate predictions does not mean that the conceptualizations of the people using the methodology accurately reflect reality. Five thousand years ago in Neolithic Ireland, a temple was constructed at Newgrange in such a way as to allow people to predict the coming of the longer days of springtime, but this was centuries before Stonehenge or Egypt's great pyramids were built, and long before anyone had an accurate conceptualization of how the earth's revolution around the sun—coupled with its tilted axis of rotation—produces our seasons. The predictions were accurate even though no one at the time had any real understanding of how or why the system was working as it was. Similarly, although what we learn about genetics in school has some heuristic value—that is, it can serve as an occasionally useful mental shortcut that leads to accurate predictions in some cases—we must not make the mistake of thinking that our genes work as Mendel hypothesized, namely, in a deterministic if-you-have-the-gene-you-are-doomed-to-have-the-trait kind of way. Because genetic factors always interact with nongenetic factors in the construction of our traits, the experiences and environments we encounter as we develop always matter, even if developmental science is still so much in its infancy that we currently do not understand much about *how* these nongenetic factors contribute to the construction of our traits.

In 2003 Lenny Moss wrote a book titled *What Genes Can't Do*. As a man with doctorates in both biochemistry and philosophy, Moss is exceptionally well situated to analyze critically the concepts used by biologists. One of the conclusions he reaches in his book is that we must begin to distinguish between two very different types of "genes," specifically, the segments of DNA that actually influence the development of our traits, and the hypothetical entities posited by Mendel that geneticists find useful in spite of the fact that they appear not to really exist. Moss calls the former "Genes-D" because they can be thought of as resources that organisms use during development, when our eyes, personalities, and bodies actually take on their characteristics. He calls the latter "Genes-P" because they are imagined to determine our traits preformationistically, that is, before development. Thus, although geneticists might find it useful (for the purposes of prediction) to imagine the existence of a Gene-P for blue eyes—the recessive little "b"—it is now quite clear that there is no such thing as a DNA sequence (a Gene-D) that causes the development of blue eyes. As Moss puts it:

> The condition for having a gene for blue eyes or a gene for cystic fibrosis does not entail having a specific nucleic acid (DNA) sequence but rather an ability to predict, within certain contextual limits, the likelihood of some phenotypic trait. . . . Blue eyes are not made according to the directions of the Gene-P for blue eyes [because no such physical entity actually exists]. . . . Reference to the gene for blue eyes serves as a kind of instrumental short hand with some predictive utility.[25]

Griffiths and Stotz are among the other theorists who have joined Moss in his efforts to distinguish different possible meanings of the word "gene."[26]

Thus, although Genes-P are not actually physical things in our bodies at all, Mendel's notion can still facilitate prediction in certain well-controlled contexts (e.g., in greenhouses, scientific laboratories, and livestock-breeding facilities). In contrast, Genes-D actually are the real, material genes we inherit from our parents, but they do not determine our characteristics independently of the contexts in which we develop. In the real world of Genes-D, it is simply not the case that the presence of particular genes allows us to make unerring predictions about the characteristics an individual will ultimately develop. In this sense, then, there are no such things as genes for blue eyes, breast cancer, obesity, alcoholism, or anything else, including the ability to get away with fewer than eight hours of sleep each night. Although the DNA we inherit from our parents contributes to the development of all our characteristics, it determines none of them.[27] The environments in which we develop always matter too. So the next time you hear a news story about the discovery of a new gene for a particular disease, talent, or vice, be excited and curious, but be skeptical as well; the new discovery will most likely contribute to our understanding of the disease, talent, or vice in the long run, but when the complete story is finally told, there will be more to it than the gene alone.

CHAPTER FIVE

The Myth of the Machine-Organism

From Genetic Mechanisms to Living Beings

STEPHEN L. TALBOTT

THE GENE MYTH is not just a myth about genes. It is a story about the nature of the organism and the character of biological explanation. Inspired by our experience with machines, the story (in one of its versions) is narrated in a language of causal analysis, where some things make other things happen, and our investigation of a collection of parts, one by one, enables us to piece together a knowledge of the integrated whole. The continual elucidation of explanatory "mechanisms" has seemed to vindicate the story, supported further by promises of a better life for humans and a steady stream of stunning technical achievements in data gathering and manipulation of organisms. It is no wonder that the Human Genome Project aroused such high expectations.

But this story has now come to the end of its useful life. The loss of the gene at the head of a chain of causal mechanisms explaining the organism represents more than the loss of the master link in the chain. It exemplifies the failure of *every* link considered as machinelike. The seeming chaos of causal arrows now being documented under the heading of "gene regulation" repeats itself in every aspect of the cell. Researchers dutifully trying to follow arrows of causation end up chasing hares running in all directions. Is there any subdiscipline of molecular biology today where research has been reducing cellular processes to a more clearly defined set of causal relations instead of rendering them more ambiguous, more plastic and context dependent, and less mechanical? Consider a few examples.

Signaling Pathways

Signaling pathways are vital means of communication within and between cells. In the machine model of the organism, such pathways were straightforward, with a clear-cut input at the start of the pathway leading to an equally clear-cut output at the end. Not so today, as a team of molecular biologists at the Free University of Brussels found out when they looked at how these pathways interact or "crosstalk" with one another. Tabulating the cross-signalings among just four such pathways yielded what they called a "horror graph," and quickly it began to look as though "everything does everything to everything." In reality, we see a "collaborative" process that can be "pictured as a table around which decision-makers debate a question and respond collectively to information put to them."[1]

Even if you consider a single membrane receptor bound by a hormonal or other signal, you can find yourself looking, conservatively, at some 2 billion possible states, depending on how that receptor is modified by its interactions with other molecules. Obviously there is no simple binary rule distinguishing activated from deactivated receptors. "The activated receptor looks less like a machine and more like a pleiomorphic ensemble or probability cloud of an almost infinite number of possible states, each of which may differ in its biological activity."[2]

Demise of Lock-and-Key Proteins

According to the old story of the machine-organism, not only does a protein-coding DNA sequence, or gene, specify an exact messenger RNA sequence, but also the RNA in turn specifies an exact amino-acid sequence in the resulting protein, which finally folds into a fixed and predestined shape. "There is a sense, therefore," writes Richard Dawkins, "in which the three-dimensional coiled shape of a protein is determined by the one-dimensional sequence of code symbols in the DNA." Further, "The whole translation, from strictly sequential DNA ROM [read-only memory] to precisely invariant three-dimensional protein shape, is a remarkable feat of digital information technology."[3]

We now know how great a misconception this was. Through alternative splicing, one gene can produce up to thousands of protein variants, while unlimited additional possibilities arise from RNA editing, translational regulation, and posttranslational modifications. As for the finally achieved protein, it need not be anything like the rigid, inflexible mechanism with a single, well-defined structure imagined by Dawkins. Proteins are the true shape changers of the cell, responding and adapting to an ever-varying context—so much so that the "same" proteins with the same amino-acid sequences can, in different environments, "be viewed as totally different molecules" with distinct physical and chemical properties.[4]

Nor is it the case that proteins must choose in a neatly digital fashion among discrete conformations. In contrast to the old "rigid-body" view, researchers now refer to the "'fluid'-like nature of protein structures." Even more radical has been the discovery that many proteins never do fold into a particular shape but rather remain unstructured or "disordered." In mammals about 75 percent of signaling proteins and half of all proteins are thought to contain long, disordered regions, while about 25 percent of all proteins are predicted to be "fully disordered." Many of these intrinsically unstructured proteins are involved in regulatory processes and are often at the center of large protein-interaction networks.[5]

Fluid, "living" molecules do not lend themselves to the analogy with mechanisms, which may explain why the mistaken idea of precisely articulated, folded parts was so persistent, and why the recognition of unstructured proteins has been so late in coming. Indeed, this recognition has hardly yet dawned on the biological community as a whole, a situation that has led to this lament by some researchers:

> Experimentalists have been providing evidence over many decades that some proteins lack fixed structure or are disordered (or unfolded) under physiological conditions. In addition, experimentalists are also showing that, for many proteins, their functions depend on the unstructured rather than structured state; such results are in marked contrast to the greater than hundred year old views such as the lock and key hypothesis. Despite extensive data on many important examples, including disease-associated proteins, the importance of disorder for protein function has been largely ignored. Indeed, to our knowledge, current biochemistry books don't present even one acknowledged example of a disorder-dependent function, even though some reports of disorder-dependent functions are more than fifty years old.[6]

However, the situation is rapidly changing, with high-profile articles on disordered proteins appearing in major journals. But a continuing mechanistic bias is evident even in the negative terms "disordered" and "unstructured." The loose, shifting structure of a protein need be no more disordered than the graceful, swirling currents of a river or the nonmechanical movements of a flock of birds.

The Living Chromosome

Nowhere has the story of the machine-organism proved more misleading than in its narrative about chromosomes, DNA, and genes. The DNA in this story contains linear sequences of nucleotide bases that are seen as little more than passive coding symbols for proteins, along with other sequences encoding regulatory functions. Such a chromosome might as well be so much computer memory, with its ones and zeros representing machine instructions—Dawkins's "read-only memory." There is nothing here to suggest that unlike computer memory, "DNA is a living molecule, writhing, twisting and bending in response to the physical forces applied to it by genetic processes."[7]

The old story has been changing with startling rapidity. Gene-packaging materials and a diverse bestiary of regulatory factors are now routinely described as participants in a "delicate 'dance' in time and space," a "regulatory ballet," "an intricate dance of associations," and a "chromatin choreography." There has been a peculiar increase in the use of terms such as "balance," "tension," "context," and "plasticity," together with an exponential explosion during the first decade of the twenty-first century in the appearance of the word "dynamic" in reference to the chromosome, chromatin-remodeling proteins, nucleosomes, and the spatial organization of the nucleus—all pointing (in the words of two biologists) toward the "highly choreographed" events regulating gene expression, a "three-dimensional pavane . . . that controls our genome."[8]

Chromosomes writhe like a nest of snakes, coiling and uncoiling, with the two strands of the double helix twisting more tightly around each other at one location and, with an opposite twist, loosening at another. Sections of a chromosome stream out from the main mass, forming loops and contacting other sections in order to consummate a needed transaction, and some even traverse the nuclear space to "kiss" (as some researchers have expressed it) critical loci on another chromosome. The electrostatic characteristics of a given stretch of a chromosome, or its distinctive bending, or the insertion of small molecules or protein residues into one of the grooves of the double helix so as to change the compression of the groove, or the relocation of sections of a chromosome either to the nuclear periphery or else to the interior—these and many other factors are part of the expressive language through which our genes speak and are spoken. And the language becomes an entirely different dialect when one moves from one tissue type to another, from one stage of an organism's development to another, from one part of the cell cycle to another, or from disease to health.

So here is what we see: the nucleus is not a passive, abstract space filled with mechanisms but rather a dynamic, expressive space. Its performance is part of the choreography that many researchers speak of today, and the performance cannot be reduced to any sort of computer-like genetic code. The cell nucleus, in its plastic spatial gesturing, is more like an organism than a machine.

And so, despite the fact that "DNA is often mistakenly viewed as an inert lattice" onto which proteins bind in a sequence-specific way, the fact of the matter is altogether different. The chromosome, remarks Christophe Lavelle of France's Curie Institute, "is a plastic polymorphic dynamic elastic resilient flexible nucleoprotein complex." Proteins and DNA are caught up in a continual conversation of mutual influence and qualitative transformation.[9]

All this is why scientists using computers to scan the several billion nucleotide bases of the human genome in the search for significant features have more and more been viewing these nucleotide sequences not merely as bearers of a linear code but as indicators of sculptural and dynamic form at different scales. Scans based on "a one-dimensional string of letters,"[10] abstracted from physical form, have failed to find many of the regulatory elements that now appear crucial to our understanding of genomic functioning. A search based on various aspects of form—sculptural, electrical, or otherwise—is leading to rapid discovery of new functional aspects of the previously one-dimensional genome.

The chromosome, no less than the organism as a whole, is a living, continually metamorphosing sculpture. That is, it lives by and expresses itself in gestural activity. The truth here could hardly be further from the countless images transmitted through the popular media to a public that has no means to correct them.[11] Nor does it sit well with the ubiquitous references to "mechanisms" and "mechanistic explanations" by the very biologists making all these recent discoveries.

The Unmechanistic Organism

There is no way in a brief space to convey the biologist's seeming obsession with mechanisms of every sort—"genetic mechanisms," "signaling mechanisms," "regulatory mechanisms," and even "molecular mechanisms of plasticity." The single phrase "genetic mechanism" yields about 22,600 hits in Google Scholar as I write, and the number seems to be rising by hundreds per month. In analyzing some of the technical articles I have collected for my epigenetics research, I found an average of 7.5 uses of "mechanism" per article, with the number in a single article varying from 1 to 32. The figure goes even higher when cognate forms such as "mechanistic" and "machine" (as in "molecular machine") are included.

Preoccupied rather crudely as they are with machine imagery, biologists do not even rise to the level of the outmoded classical mechanism of the physicists. That mechanism was more concerned with a particular sort of lawfulness than with mechanical devices. In any case there seems to be no great interest among biologists in clarifying their usage. Given the contrast between the ubiquity of appeals to "mechanisms" in the technical literature, on the one hand, and the actual qualities of the organism being revealed by molecular biological research, on the other, the lack of discussion about what is meant by "mechanism" is remarkable. After all, there is no obvious similarity between a sewing machine or a clock or any other machine and, say, a twisting, gesturing chromosome—or, for that matter, a cat stalking a mouse.

The parts of a clock are put together in a certain way; the parts of an organism grow within an integral unity from the very start. They do not add themselves together to form a whole, but rather progressively differentiate themselves out of the prior wholeness of seed or germ. They are growing even as they begin functioning, and their functioning is a contribution toward their growing. The parts never were and never are completely separate, never are assembled. A specific bit of food taken in from outside never becomes some new, recognizable part, added to the rest; rather, it is metabolically transformed and assimilated by the ruling unity that is already there. The structures performing this work, such as they are, are themselves being formed out of the work. Does any of this sound remotely like a machine?

When, on the other hand, we build machines, we impose our designs on them from without, articulating the parts together so that by means of their external relations they can perform the functions or achieve the purposes we intended for them. Those same relations give us our explanation of the machine. If the behavior of one of the parts depends on internal workings, and if we cannot yet analyze those workings in terms of subparts and their relations, then we regard the part as a temporarily unexplained "black box."

With organs that have developed as the progressive elaboration of a preexistent whole, there is no such outwardly imposed design or purpose. The coherence and function do not arise through the intention of an external player who connects part to part. The intention, such as it is, and the wholeness are there from the beginning—for example, in the fertilized egg that proceeds to subdivide and differentiate itself. We see no point at which something is added from outside in a machinelike fashion.

Another reason that we cannot explain the organism through the relations among parts is that those parts tend not to remain the same parts from moment to moment. For example, as virtually all molecular biologists now acknowledge, there is no fixed, easily definable thing we can call a "gene." Whatever we do designate a gene is so thoroughly bound up with cellular processes as a whole that its identity and function depend on whatever else is happening. The larger context determines what constitutes a significant part, and in what sense, at any particular moment. Where, then, is any sort of definable mechanism?

Certainly there are reasonable analogies between, say, our bones and joints, on the one hand, and mechanisms such as levers and ball joints, on the other. Such analogies can be multiplied many times over throughout the human body. But to avoid falsehood, it is necessary to add that these are only analogies.

Bones and joints are not, in fact, mechanisms. Bones, for example, are continually undergoing an exchange of substances with their environment, and even after the main period of our development is past, they are still being shaped and reshaped by their use or disuse (think of the bones of astronauts) and by the boundless range of other bodily processes with which they are interwoven. It is certainly true that mechanisms such as ball joints, levers, and cogwheels also suffer change—for example, through wear and tear. But unlike bones, mechanisms are not continually reshaped through the seamless integration of their internal processes with those acting from without. Gears and levers are not maintaining themselves and being maintained in anything like the way an internal organ is.

If we really mean "mechanism" when we use this word, then we have no choice but to acknowledge the externally designed character of mechanisms. If we only mean "lawful," then we should use that word and quit pretending that we have reduced organisms to mechanistic terms. (I will have more to say about the lawfulness of organic processes later.) And if we mean something else altogether, someone should spell it out.

When two scientists write that "clock genes are components of the circadian clock comparable to the cogwheels of a mechanical watch," it ought to be scandalous.[12] As a strange aberration of thought, the biologist's intransigence in likening organisms or their parts to machines and mechanisms is extraordinary in the history of science and perhaps notable even by the standards of a prescientific age. The usage is universal and is heavily relied on by otherwise rigorous scientists in their attempts to explain the organism, but it has no obvious meaning. Although the limits of my literature survey may be severe, I have never come across a contemporary essay on molecular biology whose authors thought it worthwhile to explain what they mean by "mechanism" or "machine."

Fortunately, however, those same essays shout out the living truth in nearly every sentence for anyone willing to listen.

The Living and the Dead

The moment your pet dog dies, everything changes. First, a living organism, hungrily devouring a dish of food, playing boisterously, contending with rivals, investigating an endless world of smells, barking at strangers; then, a decomposing corpse. At the moment of transition, all the living processes normally studied by the biologist rapidly disintegrate. The corpse remains as fully subject to the laws of physics and chemistry as the live dog, but now, in the absence of the living being, we see those laws strictly in their own terms, without the distinctive organization, the integration, the unity, and the coordination of the living processes.

Of course, every biologist knows the difference between the two states, even if (oddly enough) the difference between life and death does not often figure explicitly in the technical literature that presumes to characterize living things. But whether we acknowledge it or not, the difference is always there. No biologists who have been speaking of the behavior of the living animal will now speak in the same way of the corpse's "behavior." Nor will they refer to certain physical changes in the corpse as "reflexes," just as they will never mention the corpse's "responses" to stimuli, or the "functions" of its organs, or the processes of "development" being undergone by the decomposing tissues.

Essentially the same collection of molecules exists in the canine cells during the moments immediately before and after death. But after the fateful transition, no one will any longer think of genes as being "regulated," nor will anyone refer to "normal" or "proper" chromosome functioning. No molecules will be said to "guide" other molecules to specific "targets," and no molecules will be carrying "signals," just as there will be no structures "recognizing" signals. "Code," "information," and "communication," in their biological sense, will have disappeared from the scientist's vocabulary.

The corpse will not produce "errors" in chromosome replication or in any other processes, nor will it "attempt" error "correction" or the "repair" of damaged parts. More generally, the ideas of "injury" and "healing" will be absent. Molecules will not "recruit" other molecules in order to "achieve" particular "tasks." No structures will "inherit" features from parent structures in the way daughter cells inherit traits or tendencies from their parents, and no one will cite the "plasticity" or "context dependence" of a corpse's response to its environment.

Sometimes the biologist's language reaches a remarkable crescendo, as when two researchers say that with current tools "we can begin to develop and test systems-level models of cellular *signaling* and *regulatory* processes, therefore gaining insights into the *'thought'* processes of a cell." The same two researchers speak of signaling networks as the "*perceptual* components of a cell," responsible for "*observing* current conditions and making *decisions* about the *appropriate* use of resources—ultimately by *regulating* cellular *behaviour.*" Or you can go back to Barbara McClintock's Nobel Prize address, when she surmised that "some *sensing* mechanism must be present . . . to *alert* the cell to imminent danger." In the future we should try to "determine the extent of *knowledge* the cell has of itself, and how it utilizes this knowledge in a *'thoughtful'* manner when challenged."[13] But even without references to thought and perception, it is clear that biologists cannot open their mouths without employing a language alien to physics and chemistry and drawn from our inner life—a language of recognition and response, of intention and directed activity, of meaningful information and timely communication, of errant actions and corrective reactions, of healthy development leading to self-realization, or ill-health leading to death.

At the outset I mentioned one version of the gene-centric story of the organism. It was a story of mechanisms narrated in a language of causal analysis. But there is a second version of the story, one that speaks of information, instructions, signals, and, indeed, all the purposive and communicative notions I have just now reviewed. The gene in this version easily becomes a gene *for* a particular trait.[14]

How were these two very different stories stitched together? By appeal to the machine-organism. Because we habitually project our own conscious purposes and, in fact, much of our inner life into the machines we design, it is easy for biologists to invoke causal "mechanisms" while at the same time freely employing the cognitive, intentional language—the language of sensing and striving and communicating—that is essential to all biological description. Somehow, the constant, ritual invocation of "mechanisms" is felt to naturalize and legitimate even the most radical departures from physicalistic language. But if, as we have seen, the organism is not a mechanism, then we are left with the problem of reconciling a properly causal understanding of the organism with the language the biologist cannot seem to avoid—a language drawn from our own psychic and voluntary life.

Going back more than two centuries, we find biologists routinely aware of the challenge presented by their distinctive language and subject matter.[15] But during the era of molecular biology, beginning in the 1950s, this awareness was largely lost—presumably because molecular biology was sooner or later going to reduce the language of directed intention to the language of mechanism, relieving the working scientist of any need to puzzle over the meaning of her own words.

The reduction never happened. Certainly a great deal has happened, but the problem of the living organism has, if anything, become more acute. If biologists are having trouble cashing in their terminological chips for a language minted solely from mechanisms and physical law, perhaps they need to accept some other coinage. One way or another, it is time to face the problem and begin to think about what they find themselves actually saying when they try to describe an organism scientifically.

The Whole Is Coordinated Movement

At least since Kant's day, biologists have recognized distinctive features of their attempts to explain life. Their language has been variously thought to presuppose the following, for example:

- *A peculiar sort of unity whereby the whole is in some sense the cause of, while expressing itself through, the part.* Everything going on in the organism as a whole comes to expression in the form and functioning of the parts, and therefore every part is, in its own way, a revelation of the whole.
- *Means-end ("purposive" or "final") relations: organic activities are carried out as if "with a view toward" or "for the sake of" some end.* The developing eye of the embryo cannot see, but it is preparing for later seeing; the robin builds a nest for hatching and raising its offspring; and the elaborate meiotic activity of the mammalian cell is directed toward formation of a gamete and subsequent union with another gamete. These purposes, however, must for the most part be distinguished from conscious human purpose such as we employ in the construction of a machine. Biological purpose is immanent within the unconscious activities manifesting it. Furthermore, it is misleading to speak of the purpose of a living activity. Is the acorn's purpose to become an oak tree, or to decay and fertilize the soil, or to feed a squirrel? Yet each of these possibilities, bound up with the larger unity of life, hinges on the acorn's having a specific character and realizable potential.

- *A causal ambiguity, which can appear as a mutual and reciprocal play of cause and effect.* In other words, there are no causes or effects in any strict sense of those terms. The shape of the leg bone affects the carriage and behavior of the mammal, but at the same time the carriage and behavior are shaping the leg bone. Genes have a role in the production of RNA, but RNA is at the same time modulating the activity of the genes.

All three of these features are at least suggested by the rather simpler statement that we find in every organism a meaningful coordination of its activities. Coordination implies a view from the whole toward the parts, with each part caught up in, and receiving its essential character from, its place in the overall pattern. At the same time, the patterned coherence achieved and sustained by the coordination can reasonably be described (with the qualifications mentioned above) as the purpose, function, or end of the organic activity. And it is by virtue of such coordination—the continual "bending" of local activities toward the requirements of the whole—that physical causes as we usually think of them become fluid and diffuse, losing all fixity. They are continually subordinated to, or lifted into the service of, the coordinating agency of the organism as a whole.

Is all this to flirt with an unscientific or even mystical view of the organism? Actually, it is the refusal of holism that leads to mysticism. The distinguished twentieth-century cell biologist Paul Weiss once wrote that the biologist's

> common habit of personifying compounds by calling them "regulators," "integrators," "organizers," etc., and crediting them verbally with the "regulatory, integrative and organizing" effects which one observes but cannot explain analytically, either intends to endow chemicals with spiritual powers up and above their ordinary properties, or else is wholly meaningless. To state it bluntly, it would be rather a reversion to the prescientific age if on observing, for instance, the spinning of a whirl of fluid, one were to invoke a special compound as "spinner."[16]

Weiss did not eschew such terms as "regulate," "organize," and "integrate." His objection was to the personification of mere things ("compounds" or "organizers") as the source of organization or as the essential givers of life. He saw in this "an obvious reversion in modern guise to animistic biology, which let animated particles under whatever name impart the property of organization to inanimate matter."[17]

The problem had already been recognized by German botanist Fritz Noll in 1903. In marine biologist E. S. Russell's summary, Noll pointed out how "the chief theorists have tried to solve the problem of development by assuming a material and particulate basis [today's 'gene'], without however attempting to explain how the mere presence of material elements could exert a controlling influence on development. They have been forced to ascribe to such abstract material units properties and powers with which they would hesitate to credit the cell as a whole."[18]

Russell himself objected to gene-centrism; he refused to attribute the cause of visible change in the developing organism to self-sufficient factors that themselves remained inert and unaltered. "Aristotle would have recognized in this almost mystical conception something strangely like his 'soul'!" Much later Nobel laureate Max Delbrück actually suggested that DNA could be conceived in the manner of Aristotle's first cause and unmoved mover, since it "acts, creates form and development, and is not changed in the process."[19]

The crucial and self-evident principle ignored by such errant thought was put decisively by Weiss: "Life is a dynamic *process*. Logically, the elements of a process can be only elementary *processes,* and not elementary *particles* or any other static units."[20] It never made any sense whatever to take genes or other molecular structures as the explanation for what happens in an organism. Processes cannot be explained as configurations of things. Much less can highly integral, unified processes be explained as configurations of just a few things among the many caught up in this unity.

Weiss offered an intriguing formula to illustrate the nature of the explanatory challenge when we do try to reckon with cellular processes in their own terms:

$$V_{\mathrm{T}} < \Sigma(v_a + v_b + v_c + \ldots v_n),$$

where V_{T} "denotes the total variance within a population of cells of a given type (or between successive stages of the same cell), and $v_a + v_b + v_c + \ldots$ are the variances of component cell activities. The formula represents an 'operational' description of what it is that makes the cell as a unity 'more than the sum of its parts.' "[21]

His point was that if you look at all the various processes going on in a cell and consider in each case the normally expected "random serial excursions from an average course," then these processes ought to become progressively disordered in their relations to one another. But in fact, the variance of the whole collection of processes is less than the sum of their separate variances. Despite the countless processes going on in the cell, and despite the fact that each process might be expected to "go its own way" according to the myriad factors impinging on it from all directions, the actual result is quite different:

> Small molecules go in and out, macromolecules break down and are replaced, particles lose and gain macromolecular constituents, divide and merge, and all parts move at one time or another, unpredictably, so that it is safe to state that at no time in the history of a given cell, much less in comparable stages of different cells, will precisely the same constellation of parts ever recur. . . . Although the individual members of the molecular and particulate population have a large number of degrees of freedom of behavior in random directions, the population as a whole is a system which restrains those degrees of freedom in such a manner that their joint behavior converges upon a nonrandom resultant, keeping the state of the population as a whole relatively invariant.[22]

Amid all its varying, never precisely repeated subprocesses, and amid all the novel, unpredictable influences from the larger environment, a given type of cell insists on maintaining its own recognizable identity with "unreasonable" tenacity. And so, with a touch of irony, "less" change is what shows that the whole cell is "more" than the sum of its parts. There is, in other words, an active, coordinating agency subsuming all the part processes and disciplining their separate variabilities so that they are caught up in a greater unity.

And crucially, as we saw Weiss pointing out earlier, this coordinating agency cannot be equated with material units of any sort, nor even with particular interactions. We are talking about an intricate coordination of interactions, and "logically, this 'coordinating' principle cannot be of the same categorical order as individual reactions themselves—just one more of them."[23]

Here is where the modern scientist's uneasiness arises. What can be meant by an immaterial coordinating principle? An entelechy? A soul? Are we not back in the swamp of vitalism?

I would like to suggest ever so briefly that the centuries-old conflict between mechanism and vitalism, having now resurfaced unexpectedly in the domain of molecular biology, has a surprisingly simple resolution—one our mechanistic habits of thought have never allowed us to see properly.

Causes Are Not Laws

Physicists have learned to put nature's lawfulness on display through carefully controlled closed systems. Billiard balls interacting on a smooth surface can rather directly illustrate certain laws of force and motion if one assumes that there is no unnoticed rupture of the surface, no sudden breeze through the windows, no earthquake, or the like. The idea of a closed experimental system is to rule out as far as possible (which is never absolutely) any such interference from the larger context.

The fundamental thing the physicist is after is a lawfulness implicit in the observed events—not to make the events themselves into laws or causes. That is why there is no violation of the natural order if a breeze does pass through the room. It can be found (with additional effort) that by taking this new contextual element into account, the laws still hold true, even if the initially predicted "causal" events do not.

Nevertheless, it is convenient to see particular laws put on outward display as clearly as possible, and this can be done by contriving systems that are more or less "closed." Then one can say, in an approximate or contingent sense, that one billiard ball "makes" the other move in such and such a way—as if the moving ball itself were a reliable cause—even though there is no absolute causal necessity in any given thing or process or experimental configuration. The "making" here really just amounts to our cleverness in arranging things so that physical laws will manifest themselves in a particular sequence of events, barring unforeseen interference.

The ability to achieve relative closure of a system is the reason we can make workable machines: the parts keep working and maintaining their predicted relationships until normal wear and tear, loss of power, a frustrated user's fist, or some other circumstance intervenes. A proper law may never be violated, but any given set of "causal" relations among things can always be rendered invalid through contextual change.

We can assume that physical laws remain unviolated in the organism, just as in the machine, regardless of what happens. But when, as with the billiard balls, we try to create a closed system and to reduce organic processes to something like a mechanical picture of those laws, so that one thing reliably and consistently becomes the "cause" of another thing, we fail. Why? Because in biology a changing context does not interfere with some causal truth we are trying to see; contextual transformation, rather than causation, is itself the central truth we are after. Or, you could say, in the organism as a maker of meaning, interfering is the whole point. The organism is a unified context interacting with a still-larger context and capable of adapting itself to that larger context through constant self-adjustment. The ongoing construction and evolution of such an organismal context, with its continually modulated causal relationships, are what the biologist is trying to recognize and do justice to. Every creature lives by virtue of the dynamic, pattern-shifting play of a governing context, which extends into an open-ended environment. This play is how it gestures and expresses its own character. The organism can change its proximal goal from moment to moment and thereby also change the contextual significance of the details of its life.

Any effort to tie down causes and effects—with one thing "making" another happen—proves grossly inadequate to deal with this dynamic expression of meaning. Whether we consider a person going to the store for a loaf of bread, a bird building a nest, or a cell dividing, there is no constancy of context that allows us to fixate on individual things as causes. In each of these three cases particular circumstances are encountered that were never encountered previously in the entire evolutionary history of the organism. They therefore demand an unprecedented pattern of response. For example, an intersection on the way to the store is blocked off by an accident, the nest suffers an unusual derangement in a windstorm, and the almost infinitely complex signaling "state" of the cell is unique. But the intention, the aim, of the activity is maintained through an agency that adjusts to the means at hand. "Causes"—really, we just mean "lawful relationships in their momentary configuration"—are reconstellated in ever-new ways consistent with the governing unity. To focus on isolated things—a gene and a transcription factor—and expect their encounter to have the same causal result every time is to forget the ongoing reconfiguration of the surroundings.

Biological explanation fails as genuine understanding whenever we attempt to freeze (or ignore) the mobile, metamorphosing context of the living being as a whole and pretend that local things or events can be seen as machinelike causes. Not even the physicist treats billiard balls and their movements as "mechanistic explanations" in this sense, because such explanations would be invalidated as soon as the conveniently closed system was breached. The physicist is way ahead of the biologist in this regard—which is strange, because the biologist has the organism as teacher, in which the breaching, the directed engagement with its environment, and the coherent responding to ever-new and unpredictable stimuli are life itself.

The Organism Insists on Its Own Way of Being

For the physicist, then, the point is to get at the physical lawfulness within events, which is far different from treating events themselves as laws or causes. We may not find, in the organism, any new laws such as the physicist looks for. But nothing tells us that that sort of law represents the only kind of order the world could present us with. That is a question to be resolved through observation, not ideology. If we find in the organism a coordination of physically lawful behavior so as to achieve organic purposes and functions, is this mystical? Is it unscientific?

The questions reflect a misunderstanding. There is no need to extrapolate beyond what we observe. All the routine biological language of regulation, integration, signaling, healing, and so on is the scientist's attempt to describe as clearly as she can the way organic processes hold together and make sense. The description, so far as it is accurate and provides a coherent picture, is the understanding we seek, and we need not go beyond it. But neither should we deny what those same descriptions are saying about the organism's distinctive unity, integration, coordination, and wholeness.

Certainly there are more questions we may ask, just as we may ask where the law of gravity comes from or what energy is. (Richard Feynman famously remarked that "in physics today, we have no knowledge of what energy *is*.")[24] And we should not forget that even the language of physical law is drawn from our inner life—in this case, from our highest abstract conceptualizing powers. Just as we cannot say much about why the world respects the mathematical concepts we arrive at through our own conscious activity, so, too, we may not now be able to say much about why the organism shows such remarkable patterns of coordination, reflecting in an unconscious or less conscious way something like our own conscious intentions. But our ignorance about the source of things has not prevented us from duly observing the actual lawfulness of physical processes involving gravity and energy. Similarly, if in the organism we observe an agency by which processes are continually coordinated contextually—if the organism manifests itself most essentially as a kind of intelligent, directed movement at the level of an integral, self-shaping context rather than as a mechanism with a defined structure—well, then, that is a type of order or lawfulness the world presents us with. Only observation can tell us what sorts of order may be found in nature. The laws of the organism are no more inexplicable or mystical than the laws of inanimate matter.

I suspect that the main problem many biologists have with all this is that they can conceive of explanation only as an elucidation of physical mechanisms, and so they want to explain the principle of coordination in the organism as if it existed at the level of that which is being coordinated. I doubt that true biological understanding can be very much advanced until the obsession with mechanistic explanation is finally abandoned.

Think, for example, of the effort, common nowadays, to explain an organism's form by referring to genetic switching networks. Developmental biologist Sean Carroll presents beautiful pictures of molecular patterns in the fly embryo that prefigure and map directly onto the later arrangement of larval segments, and similarly he shows how the spots on butterfly wings are prefigured by distinctive molecular patterns. He then suggests that complex arrangements of genetic switches explain these patterns and therefore also explain the eventual form of the organisms.[25]

But we now know from the vast literature on gene regulation (oddly, Carroll does not even mention epigenetics in his book) that those supposed switching networks are in fact penetrated and influenced by virtually everything going on in the cell. By the time we get very far in tracing the relevant interactions through the organism, we realize that we are witnessing, at the molecular level, the playing out of the very form, the patterns, that we had hoped to explain, but at another level of description.[26] If we really did need explanatory mechanisms, then we would still be left with the task we initially set for ourselves: to explain what governs, controls, or regulates the complex, interacting molecular patterns that we find as such vivid, directed, perfectly shaped presentiments of the developing morphology.

It is not that identifying a so-called gene switch—or calculating kinetic energies or measuring mechanical stresses on macromolecules—gives us no understanding. Of course such insights are important. But they become biological insights, as opposed to physical ones, only insofar as they find their place within the living, metamorphosing form of the organism. They do not explain the form. If anything, we should say that the form explains the physical interactions in the sense that it gives us an understanding of their pattern, their shape, and their direction and place within a functional whole, none of which can be deduced from the physical reactions as such. We can observe the patterns by tracing the physical interactions, but those patterns can never be arrived at merely by working out the implications of the physical laws and substances.[27]

Happily, all the excitement and ferment in molecular biology today are nudging biologists, even if despite themselves, toward a much more living picture than their predecessors could tolerate, and this picture does not present them with obscurity and mysticism. For all the difficulty in tying down biological cause and effect, and for all the new focus on dynamism, context dependence, plasticity, and interweaving processes, do we find the new descriptions less revelatory, less faithful to the living organism, than the old ones based on the mechanistic conception of genes?

The organism as a whole is not beyond our reach. It is just that the most natural way to understand a whole is to look at it, not to lose ourselves in isolated processes as if they were detachable from that whole. Have we ever learned more about organisms in a shorter time than during the era of the great naturalists—Humboldt, Darwin, Fabre—and did not their disciplined, receptive, and appreciative human encounter with living creatures produce remarkable scientific progress? But how many naturalists of this caliber are we training today?

It may be hard to hear, but much of our biology is technology rather than science in the broad sense. It is a search for powers of mechanical manipulation. Throw a wrench into a signaling network or insert a transgene into a chromosome and wait to find out whether the unforeseen effects overwhelm the predicted or desired ones. (If you think this description too harsh, see http://natureinstitute.org/nontarget/, maintained by The Nature Institute.) Certainly sophisticated trial-and-error methods bring a kind of understanding—a good deal of understanding when billions of dollars and the full panoply of modern technology are deployed, and even when organisms are being mistreated as machines.

But the main lesson of this understanding has been to steer us back to the primacy of the whole organism, not as a mechanism whose isolated, decontextualized parts can answer our questions, but rather as a self-directing agency—an agency, I should add (although this has not been my topic here) that lives as part of a still higher unity through mutual engagement with its environment.

If the organism manifests itself as a coordinated and directed activity, and if, from germ to maturity, we can recognize the unity of this activity[28]—its distinctive way of being—and if our descriptive biological language is a continual invocation of this unified being, then why should we coyly refuse to acknowledge what we have recognized, pretending absurdly that we are just looking at a machine?

Given the revelations now pouring forth from the world's molecular biological laboratories, we have no need to hold on to the old mechanistic images. Far from being reduced to something unrecognizable, the organism is being given back to us as we have always known it—whole, full of surprises, and singing a song in harmony with our own being.

PART TWO

Medical Genetics

CHAPTER SIX

Some Problems with Genetic Horoscopes

EVA JABLONKA

The Charms and Perils of Genetic Astrology

The view that dominated biological thought in general, and evolutionary biology in particular, from the 1940s until the beginning of the twenty-first century was focused on the gene. Development was seen as the product of genes' actions, and ecology as the context for the natural selection of genes. This view was well reflected in textbooks of evolutionary biology and in popular books like Richard Dawkins's *The Selfish Gene*.[1] The dominance of the gene and its material incarnation, DNA, led, as Nelkin and Lindee have noted, to the perception of DNA as the secular equivalent of the soul.[2] But because, unlike the soul, DNA can be sequenced, this view was soon translated into future-telling. In 1989, for example, James Watson declared: "We used to think our fate was in the stars. Now we know in large measure, our fate is in our genes."[3] And soon enough, the time of genetic horoscopes arrived.

GenePlanet is one company that provides a DNA analysis through which, it promises, you will "get to know everything you ever wanted to know about yourself: to which diseases you are susceptible, what effect do medications have on you, what are your talents and special abilities, and who are your ancestors." Again and again the following slogan appears: "Discover your genes, know yourself!" And then after a question (How hard is it for me to quit smoking? Which medications work for me? Will I gain weight if I eat fatty foods? Who are my ancestors?) a large announcement is seen: "It's all in your genes!"[4]

Knowing oneself does not require much meditation and introspection any more. We are modern people living in the twenty-first century, not Greeks in ancient Athens. Instead of discussing philosophy and studying geometry, you spit into a container and send it to the company. After a while, for the modest sum of 399 euros (special offer), you get your genetic horoscope. It is all there—what your health risks are, what your hidden and apparent talents are, and many other wonderful things. Wiser and more realistic, you return to your daily life, to live in a more responsible and fulfilling way. Future possibilities are unraveled: "It's all in your genes!"

But what exactly is in your genes? If we take two people with identical genes (i.e., identical twins) who live in more or less the same environment, will they have identical futures? If we clone a human being, will the cloned person develop an identical propensity to that of the originator of its DNA? And if not, why not? If it's not all in your genes, where else does "it" (you) come from?

I must start by stating very clearly that there are cases where genetic counseling on the basis of DNA testing is enormously useful. There are "monogenic" diseases—diseases for which the presence or absence of symptoms depends on which alleles of one particular gene are present. Knowing which allele one has can be very informative and can lead to important existential decisions (such as whether to abort a fetus that will develop into a terribly sick baby, doomed to die). Tay-Sachs, a disease that is relatively common among Ashkenazi Jews, is such a monogenic disease, and thankfully, simple blood tests help in giving important counseling to pregnant mothers. However, such simple monogenic diseases are not very common: they make up less than 2 percent of all the diseases that are known to have a genetic component. For the remaining 98 percent of "genetic" disorders, the presence or absence of the disease and its severity are influenced by many genes and by the conditions in which a person develops and lives. And it is for the complex metabolic diseases, as well as for the interesting cognitive traits that are influenced by many genes, that the genetic-testing companies promise to provide their horoscopes.

A few years ago a British journalist decided to see how compatible the answers were from three different genetic-testing companies (GeneticHealth, a British firm; deCODEme, based in Iceland; and the American group 23andMe) about the health prospects "written in his DNA." The results were a little baffling. This is part of what he discovered:

> DeCODEme said my risk of developing exfoliation glaucoma, which causes loss of vision, was 91% below average. Yet according to 23andMe, I was 3.6 times more likely to get it than average. For age-related macular degen-

> eration, deCODEme put my risk at 20% lower than average, while 23andMe said it was 62% higher.
>
> According to deCODEme, my risk of developing Alzheimer's was 74% above average, while GeneticHealth said my genes were associated with "a fourfold increased risk of developing Alzheimer's disease by your late 80s."
>
> According to deCODEme, my risk of a heart attack, angina or sudden cardiac death is 54.8%, which is 6% above average. By contrast, 23andMe said my risk of a heart attack between the ages of 45 and 84 was 17.5%, below average. Yet another assessment came from Paul Jenkins, the clinical director of GeneticHealth, who said my risk of cardiovascular disease was "low to moderate."[5]

Clearly the situation is not very good. There may be many reasons for that (for example, different databases used by the different companies). But it may be that the assumptions made about correlating genes to health outcomes are inherently problematic.

So let us make the assumptions of the genetic-horoscope companies clearer. The main assumption is that genetic differences among individuals, that is, differences among the base sequences of their DNA, are the most important reasons for the phenotypic, visible differences among them. If you have allele A1, you will be prone to disease X, and if you have allele B2, you are likely to develop skill Y. A difference in a given gene, it is assumed, makes a difference at the level of the phenotype. Of course, the gene's effects can be somewhat modified by the environment (e.g., diet, exercise). However, the gene represents a program that the environment only triggers. The gene is assumed to be switched on or off by different conditions (e.g., "on" in environment 1, "off" in environment 2). Adaptation to unanticipated novelty that is not already "embedded in the program" is not possible. Nor is developmental plasticity very extensive, for if it were, a lot more than the DNA sequence would be necessary to anticipate the future.

How extensive is plasticity? Does a difference in a single gene usually make a large phenotypic difference? Or do differences in many genes make a phenotypic difference? If so, how many? Do genetic differences make the same phenotypic difference under different conditions? What conditions? How many types of conditions? Whose conditions? Are the conditions your own developmental conditions? Your parents'? Your more remote ancestors'? These are crucial questions, so let us look at them. Let us first start with what we have learned about the complex relations between genes and traits.

Genetic Networks

We have learned quite a lot about gene interactions during the past two decades, and it is common for geneticists these days to talk not about single genes and their effect on traits, but about gene networks. The regulatory architecture of gene networks is a focus of study, and it is clear that the organization of the network is essential for understanding the relation between the genetic and phenotypic levels. The identity of a single network component may be unimportant in many cases. Even the absence of a network component may not make a difference at the phenotypic level. We know that many knockout mutations (mutations that delete or totally inactivate a particular gene) are selectively neutral, having no visible effect on the phenotype.

In *Evolution in Four Dimensions,* my collaborator Marion Lamb and I discussed a study of a genetically complex common metabolic disease, coronary artery disease, and the role played by some alleles at an important relevant locus, APOE.[6] The APOE gene codes for a protein (apoprotein E, or apoE) that helps carry fats around in the blood. This locus has three common alleles, allele 2, allele 3, and allele 4, that are associated with differences in the incidence of coronary artery disease. The three most common genotypes (two alleles per genotype) are 2/3, 3/3, and 4/3, and in this population, people with genotype 3/3 have a below-average chance of developing the disease, those with genotype 2/3 have an average chance, and people with genotype 3/4 are twice as likely as the average person to suffer from coronary heart disease.

Since the apoE molecule helps transport cholesterol around in the blood, and population surveys have shown that people with allele 4 do have, on average, high cholesterol, the conclusion that allele 4 leads to high cholesterol seemed logical. However, the combination of high cholesterol and allele 4 is not the worst possible combination. Having allele 2 and high cholesterol leads to the highest probability of coronary artery disease, while having high cholesterol and a 3/3 genotype leads to average disease risk (like that of people with a normal level of cholesterol). Moreover, APOE is just one locus out of more than 100 that affect the development of this disease, and environmental conditions (e.g., diet, exercise) also have an effect. The complex interrelationships among genes and environmental conditions in this case, as in others, mean that we cannot just add the average effects of genes together and from this predict what a person's strengths and weaknesses will be.

This is not the only problem, however. Let us look at a variable trait like height. Height is a straightforwardly quantifiable trait (unlike intelligence) and does not seem too complex. It has a significant genetic component: its heritability (the part of the variance in the trait that is attributable to genetic differences) is 80 percent, and twenty relevant single-nucleotide polymorphisms (SNPs) are known. On the basis of the twenty DNA sequences that we have found to be associated with height, can we infer how tall an individual is likely to be? An extensive discussion in the *New England Journal of Medicine* was centered on the predictive value of genetic association studies, which included the study of height.[7] In these studies alleles or SNPs are drawn from the population, and the frequency with which certain alleles or SNPs are present in different groups (e.g., short and tall people) is tested for association with the trait (shortness or tallness).

It turns out that in spite of the high heritability, the twenty SNPs explain only 3 percent of the trait's heritability. By extrapolation from this 3 percent, it was estimated that approximately 93,000 SNPs are required to explain 80 percent of the population variation in height. A similar result is found for type 2 diabetes and many other traits. David Goldstein, the director of the Center for Human Genome Variation at the Institute for Genome Sciences and Policy at Duke University, concludes: "If effect sizes were so small as to require a large chunk of the genome to explain the genetic component of a disorder, then no guidance would be provided: in pointing at everything, genetics would point at nothing."[8]

So does the sum total of the heritability reside in the specific detailed configuration of the whole genome of the individual? If so, maybe one should try to discover different types (not too many types) of network architectures rather than focus on the component units. Maybe when the number of identical genes is beyond a certain threshold, a particular network organization is generated, and this affects the trajectory of development. If so, the genetic architectures underlying distinct developmental trajectories are of interest, rather than the individual genes involved. And perhaps the formation of particular developmental trajectories and their heritability are not due just to DNA sequence variation but to another type of heritable variation as well: to heritable epigenetic variation, which may be the result of plastic, dynamic responses to critical environmental conditions in ancestors. If this is part of the answer, a few words about the possibility of inheriting environmentally induced (plastic) responses are necessary.

Plasticity and Epigenetic Inheritance

Plasticity is defined as the ability of one genotype to generate different phenotypes depending on environmental cues that act as inputs into the organism's development.[9] Genetically identical animals can have different appearances (morphs) depending on the environment to which they or their ancestors were exposed. Identical twins can be rather different and have different fates; for example, they can be discordant for various diseases, such as diabetes and schizophrenia. Why this is so is not always clear. Sometimes it is due to different childhood or adult life experiences, sometimes it is due to something that presumably happened in the womb, and often we cannot pinpoint the reason(s). But we are beginning to learn about the mechanisms involved in generating some of the persistent plastic responses.[10] They are called epigenetic mechanisms, and they can sometimes lead to epigenetic inheritance.

Epigenetic inheritance occurs when environmentally induced and developmentally regulated variations, or variations that are the result of developmental noise, are transmitted to subsequent generations of cells or organisms.[11] The term "epigenetic inheritance" is used in two ways: a broad way and a narrow way. Epigenetic inheritance in the broad sense is the inheritance of any developmental variations that do not stem from differences in DNA sequence or persistent inducing signals in the present environment. This includes cellular inheritance through the germ line and soma-to-soma information transfer that bypasses the germ line. For example, soma-to-soma transmission can occur through developmental interactions between mother and embryo or through observation-based social learning. Epigenetic inheritance in the narrow sense is cellular epigenetic inheritance, where the cell is the unit of transmission, and variations that are not the result of DNA differences are transmitted from mother cell to daughter cell. Cellular epigenetic inheritance occurs during cell division in prokaryotes, during mitotic cell division in the soma of eukaryotes, and sometimes during the meiotic divisions in the germ line that give rise to sperm or eggs. In this latter case offspring inherit epigenetic variations through the germ line.

Marion Lamb and I distinguish four types of cellular epigenetic mechanisms, or epigenetic inheritance systems (EISs):

1. *Self-sustaining feedback loops.* Gene products act as regulators that directly or indirectly maintain their own transcriptional activity. Such positive feedback can lead to the transmission of these products during cell division, resulting in the same states of

gene activity being reconstructed in daughter cells. Alternative and heritable cell phenotypes can be generated, and this is indeed seen in fungi, bacteria, and other microorganisms.[12]

2. *Structural inheritance.* Cellular structures act as templates for the production of similar structures in daughter cells, which then become components of daughter cells. This type of templating is seen in prion-based inheritance in fungi,[13] the inheritance of cortical structures in ciliates,[14] and the reconstruction of what Cavalier-Smith calls "genetic membranes."[15]
3. *Chromatin marking.* Chromatin marks are proteins and small chemical groups that are attached to DNA and influence gene activity. Relics of these marks segregate with the DNA strands after replication and nucleate the reconstruction of similar marks in daughter cells.[16] Chromatin marks include modifiable histone proteins that are noncovalently bound to DNA, methyl groups that are covalently bound to the DNA, and patterns of bound nonhistone proteins. Methylation patterns are different in different cell types and in imprinted genes. Identical twins are born having similar patterns of methylation, but as the twins age, they grow more and more different.[17] Methylation patterns and other chromatin marks can be also inherited between generations. The inheritance of a testis disease in male rats (for six generations for several traits),[18] following a single injection of their ancestral female ancestor with vinclozolin, is based on the inheritance of chromatin (methylation) marks. The heritable effects of parental under- or overnutrition also involve the transmission of chromatin (methylation) marks between generations.[19]
4. *RNA-mediated inheritance.* Transcriptional states are induced and are actively maintained through interactions between small, transmissible RNA molecules and the mRNAs or the DNA/chromatin regions with which they pair. Such interactions can be transmitted between cell and organism generations through an RNA-replication system or via the interaction of the small RNAs with chromatin, which leads to heritable modifications of chromatin marks.[20] An example is the inheritance of a heart disease induced by the injection of small RNAs, which are then transmitted for at least two generations.[21]

Between-generation epigenetic inheritance is not rare. In a survey Jablonka and Raz found over 100 cases of epigenetic inheritance in forty-two species:[22]

- Twelve cases of epigenetic inheritance in bacteria. Most were of self-sustaining loops, but examples of chromatin marking and structural inheritance were also found.
- Seven cases in protists (a diverse group of mostly unicellular eukaryote organisms). Most were in ciliates (protozoa with hairlike organelles), where structural inheritance (the transmission of cortical morphologies) is common, and all loci may be modified through the RNA-mediated EISs. The other two types of EISs were also found in protists.
- Nineteen cases in fungi, involving many phenotypes and loci. Examples of all four types of EISs were found.
- Thirty-eight cases in plants, involving many loci and many traits. Among the thirty-eight, cases, four were in plant hybrids, and in all of these many loci were heritably modified. Genomic stresses such as hybridization and polyploidization (chromosome doubling), especially allopolyploidization (hybridization followed by chromosome doubling), seem to induce genomewide epigenetic changes, some of which are transmitted between generations through the chromatin-marking and the RNA-mediated EISs. No evidence was found for gametic between-generation inheritance based on self-sustaining loops and structural templating.
- Twenty-seven cases in animals, some of which involved many loci. As with plants, stress seems to induce multiple epigenetic changes. Epigenetic variations were transmitted through the chromatin-marking and the RNA-mediated EISs, and there was no evidence for between-organism gametic inheritance based on self-sustaining loops and structural templating.

And this is just the beginning. Since we conducted this survey, many more additional cases have been reported, especially in plants. Studies of methylation inheritance in *Arabidopsis thaliana* show that a considerable part of the genome's methylation patterns can be inherited for many generations, while other regions show instability.[23]

In addition to epigenetic inheritance in cell lineages, there are also soma-to-soma routes of transmission that bypass the germ line. Soma-to-soma transmission includes transmitting substances that affect development through feces ingestion, through the placenta and milk of mammals, and through the soma-dependent deposition of specific chemicals in the eggs of oviparous animals—employing diverse developmental mechanisms that lead to transgenerational effects.[24] Physical constraints can also lead to transgenerational effects: a mother's morphological features, such as her overall size and the size of her uterus, can constrain offspring development and lead to self-perpetuating developmental effects, to offspring that match the size of the mother.[25] For vertebrates, socially learned behaviors that do not require the transfer of materials are an important source of transmissible variations. Famous examples are bird and whale dialects, which are transmitted in ways similar to the transmission of dialects in human populations and form "linguistic" traditions. Another example is the traditions found among different groups of common chimpanzees—thirty-nine different behavioral phenotypes in the chimpanzee population in East Africa are attributed to cultural evolution, since the behaviors are transmitted from one generation to the next via social interactions among members of a group.[26] Developmental interactions among organisms that form coherent and persistent symbiotic communities also contribute to soma-to-soma inheritance.[27]

The evidence suggests that organisms can respond to changing environments by changing their development and their phenotypes, and sometimes these plastic responses can be stabilized and inherited. Moreover, it seems that the responses are not "programmed" in any rigid way, and that the organism can respond to new conditions by launching exploration processes at the intraorganismic level (for example, at the intracellular level); when a response that relieves the stress of the changed environment is found, this response is stabilized and may even be inherited. This was shown by Braun and colleagues,[28] who confronted yeast cells with a rewired regulatory circuit with a severe and unforeseen challenge and studied how the population adapted to it. They placed the essential *HIS3* gene from the histidine biosynthesis pathway under the exclusive regulation of the galactose utilization system. In a glucose-containing medium the *GAL* genes are repressed, and so is (in these rewired cells) the essential *HIS3* gene. The rewired cells were then placed in a medium containing glucose but lacking histidine and were therefore severely challenged (they could not synthesize histidine). The cells never had the *HIS3* gene regulated by carbon sources availability, so a substantial adaptive response was required for them to survive in a medium that lacked histidine. However, the cells did manage. Over 50 percent of the population overcame the challenge within only a few generations, and the adapted state was propagated stably for hundreds of generations. The study suggests that a biochemical exploration process was initiated under these stressful conditions, and among the many regulatory networks that emerged, those that relieved the stress were stabilized. This exploratory response to a new stress, followed by stabilization when a solution is found, is similar to neural learning by trial and error. It is possible that this type of response is a general strategy that is applied at many levels of biological organization, within cells and within systems. This type of entirely novel plastic response is not "programmed" and is impossible to predict except in a very broad functional sense.

Back to the Future

If we return to the puzzle of the missing heritability in the case of height (and other similar cases), then maybe part of the answer to this puzzle is that it resides not in the genotype but rather in the epigenotype, in the cellular networks that underlie genetic and epigenetic memory and heredity. And if the epigenomic developmental networks are indeed the relevant level of analysis for studying the heritability of complex diseases and traits, then it is not surprising that the genetic-testing companies cannot produce very consistent results. For a reliable estimate, a very large number of tested individuals and loci is necessary, and once this testing is done, it is likely that even if a statistically reliable association is indeed found, the contribution of the SNP or even the actual gene to the trait's heritability is rather modest. This does not mean that we cannot study the developmental pathways involved in common complex diseases that will help us design new medications. We can study them either by looking, as Goldstein has suggested,[29] at very rare genetic variants that do have a large effect and change the architecture of the landscape, or by looking at the effects of severe environmental stressors that alter the developmental trajectory. According to this latter strategy, gene expression profiles in the stressful conditions would be the focus of study, and the epigenetic factors that contribute to the persistence of the developmental trajectories during ontogeny, or between generations, would be specifically targeted.

Very obviously, genetic astrology is in trouble—the complex relations between phenotype and genotype during development mean that the inheritance of phenotypic differences is often an attribute of differences between gene networks rather than differences in single genes. Moreover, there is more to heredity than the transmission of variation in the sequence of DNA. Hereditary variations are also the outcomes of our own experiences (beginning in the womb) and the epigenetic history of our ancestors, stemming from their past experiences and their past lifestyles.

CHAPTER SEVEN

Cancer Genes

The Vestigial Remains of a Fallen Theory

CARLOS SONNENSCHEIN AND ANA M. SOTO

> Most serious of all the results of the somatic mutation hypothesis has been its effects on research workers. It acts as a tranquilizer on those who believe in it and this at a time when every worker should feel goaded now and again by his ignorance of what cancer is.
>
> —Peyton Rous, Nobel Prize winner

> It is difficult to get a man to understand something, when his salary depends upon his not understanding it.
>
> —Upton Sinclair, author

RARE IS THE WEEK when the media make no reference to the genes that are supposed to control our physiological and even psychological functions. These interpretations misleadingly assign individual genes in our genomes to definitive phenotypic outcomes. In truth, the complexity of phenotypic development cannot be found in the DNA contained within each of an individual's many trillion cells. In this particular regard, according to the central dogma of molecular biology, it has been widely acknowledged that DNA "codes" for proteins; uncritically, however, many have reinterpreted this axiom as meaning that DNA "codes" for phenotypes. This inference has been reinforced by experiments in which "knocking out" a particular gene resulted in the expected phenotype. However, this outcome is by no means the rule because in many instances there is either no particular phenotype or a totally unexpected one. In spite of this, time and again a preposterous link between a gene and homosexuality, infidelity, lack of fingerprints, political leaning, or some other trait is made in leading scientific journals and is then disseminated by the media, reinforcing the idea that what we are is inexorably determined by our genes.

Explaining cancer has not been spared the abuses of genetic determinism. In the pages that follow, we offer a brief historical account of cancer biology followed by a discussion of the two competing theories that aim at explaining how cancer arises from the many causes assumed to generate its phenotype. The two theories we refer to are the somatic mutation theory (SMT) and the tissue organization field theory (TOFT). We will conclude by arguing that the TOFT provides a superior explanation of cancer genesis (carcinogenesis) because the SMT focuses on the gene(s) alone and is incompatible with evolutionary theory. From a theoretical point of view, therefore, the TOFT removes the gene from the driver's seat (genetic determinism) and introduces the organism and its ability to self-organize as the conceptual focus (organicism) of the biology of cancer.[1]

A Historical and Epistemological Primer

Genetic determinism and gene-centric reductionism have dominated biological thought for more than half a century, particularly since the publication in 1945 of Erwin Schrödinger's book *What Is Life?*[2] This view reached its uncontested zenith with the publication in 1971 of *Chance and Necessity* by Jacques Monod.[3] The paradigm articulated by Schrödinger, Monod, and many others placed the gene in the metaphoric "driver's seat" of an organism's development. Introducing the concept of DNA as a "developmental program" predisposed generations of students and researchers to believe that development was the mere unfolding of a "script" or a "program" that was "written" in our genes. This deterministic view is now challenged on numerous grounds.

The most widely appreciated anomaly of the determinist dogma is its inability to account for the dramatic divergence between an organism's genetic endowment and its phenotypic complexity. In other words, as a result of recent successes in sequencing the genomes of numerous organisms, it has been learned that the number of genes an organism has is too low to predetermine its course of development to adulthood. A second critique of the determinist doctrine has come by way of philosophy and therefore was taken into consideration by only a few experimental biologists. What philosophers revealed early on was that there is no one-to-one correspondence among a DNA "gene," the several strands of RNA produced from them, and the resulting proteins.[4] In other words, without a clear isomorphic relationship between gene and protein, neither reductionism nor genetic determinism could be made to square with empirical facts. This philosophical analysis recently entered the mainstream in articles where it was asked, "What is a gene?"[5] A third school of thought, less popular but with a long tradition, frequently referred to as organicism and characterized by its integrative and dynamic approach to biology, offered a robust analysis that exposed many of the flaws implicit in theories of genetic determinism.[6]

An obvious reason for the present revival of the organicist view can be found in the growing interest in and awareness of evolutionary and ecological developmental biology. Indeed, environmental conditions were consistently a focus of analysis in the work of embryologists in the nineteeth and early twentieth centuries because their work with wild species exposed them to the role of the environment and its impact on phenotype development. August Weismann showed that the spring and summer morphs of a butterfly species could be generated by manipulating the temperature to which they were exposed during larval development. The gradual move of embryology to the study of animal models that reproduce all year long and thrive in laboratories where the influence of the environment is minimized by keeping lights, temperature, and food constant also refocused attention on the genetic aspects of embryology to the exclusion of evolution and ecology.[7] This shift in research methodology facilitated the ascent and the preeminence of genetics and, later, of molecular genetics in biology textbooks and research funding priorities.

When recent epidemiological studies revealed that starvation during fetal development resets an organism for survival in an environment of nutrient deprivation, interest in environmental contributions to phenotypic outcomes once again became fashionable. Exposing these children to a plentiful supply of food increased their propensity to suffer from metabolic syndrome, obesity, and cardiovascular disease later in adulthood.[8]

How have "mainstream" molecular biologists responded to the resurgence of interest in the environmental dimensions of phenotypic development? For the most part, they shift their positions as little as possible away from genetic determinism. Thus it is still common to hear of a "genetic program," although it is now acknowledged that it is modulated by the environment. The molecular biologists' response to this challenge is to talk of "genes" and "environment" in terms that still privilege the genetic aspects of the process. By turning to phenomena such as DNA methylation to explain environmental effects on phenotypic outcomes, for example, mainstream cell biology preserves its genetic essentialist paradigm to explain an organism's development. This view minimizes the complexity achieved through the manifold causal interactions among levels of organization. As a result, little effort is made to explain one of the main events in developmental biology, that is, morphogenesis, the process by which shape is achieved. Instead, the influence of methylation on the expression of certain genes is being correlated with the appearance of a given phenotype—an approach conceptually equivalent to the deterministic view of genetics. Even the emerging field of epigenetics cannot claim that it is unencumbered by the genetic essentialism of its theoretical forbearers. The original meaning of epigenetics, as defined by Conrad Waddington,[9] is thus being reduced to little more than genetic modification.

Can Reductionism Explain Biological Complexity?

Toward the middle of the nineteenth century, the cell theory introduced the basic concept that the cell is the unit of life.[10] However, each of the cells in a multicellular organism does not have an existence independent of the whole. Organisms and their cells are ontogenetically linked—that is, they are linked through the processes by which they collectively developed into the organism. From the very start of embryonic life, the levels of biological organization are entangled; a zygote is both a cell and an organism. Under a reductionist perspective, only bottom-up causation is accepted, and thus the beginning of multicellularity in embryos is explained from the perspective that cells "make" the organism by means of cell proliferation. To the contrary, from a holist perspective, the organism makes cells by dividing, and thus causality is top-down. Still, the organicist view is not limited to one type of developmental causality because it regards the embryo as a dynamic open system in which there are bottom-up, top-down, reciprocal, and multiple causalities.[11]

The organism imposes global constraints, while at the local levels biophysical and biochemical interactions among neighboring cells, tissues, and cellular environment determine shape. Differential cell movement and differential cell adhesion are the products of physical forces within the developing organism. Morphogens (chemicals secreted in different areas of a developing organism) form a "concentration gradient" as they disperse; this "causes cells that receive different local concentrations . . . to enter different developmental pathways."[12] The system in question is not a thing but a process. This dynamic property of the organism results in level entanglements, exemplified by the dual nature of the zygote: a cell and an organism.

The conflicting biological perspectives between reductionism and organicism can be seen in stark relief in the field of carcinogenesis, that is, the study of cancer initiation. In order to clarify how these perspectives intersect when one is explaining cancer, we will first provide definitions of what cancer is, although we acknowledge that any definition of cancer is hindered by the fact that its initiation is still unobservable, as remarked by Theodor Boveri almost a century ago. Next, we will briefly discuss the history behind two different theories on the causes of cancer. Finally, we will critically evaluate how the predictions made by these theories compare with the empirical evidence thus far accumulated.

Definitions and Glossary of Cancer-Related Topics

Cancers principally retain the distinctive structure of the organs from which they originate, although their hallmarks are altered tissue organization and excessive local accumulation of cells. For these reasons, the microscopic (histological) images of tumors have been seen as "caricatures" of normal tissues. Ever since the nineteenth century, cancerous growths have been, are, and probably will be diagnosed by pathologists through examination of tissue samples with a light microscope. This remains so despite persistent attempts by molecular biologists who, considering that cancer is a "molecular" and genetic disease, have proposed a diagnostic genomic analysis instead of a histological one.

Cancers—also known as neoplasms—can be either solid tumors, like those in the breast, colon, prostate, and lung, or "fluid" tumors, for example, leukemia and those derived from related blood-forming tissues. From a clinical perspective, neoplasms can be either benign, when they are encapsulated, slow-growing, and usually non-life-threatening, or malignant, when they are invasive and usually deadly. Ninety percent of clinical cancers are carcinomas and adenocarcinomas, that is, cancers arising from epithelial tissues—tissues that cover the external and internal surfaces of the body. The remaining 10 percent are neoplasms of what is broadly considered the connective tissue, nervous tissue and muscle. Last, but not least, carcinogenesis—the initial stages of tumor formation—can result in either regression, dormancy, or progression to full neoplasia and eventual death of the patient. In sum, progression and death are not the inevitable outcome of clinical and pathological diagnosed cancers.

Classification of Cancers According to Their Proximate Causes

Decades of epidemiological data offer a fairly good idea of the types of cancers clinicians observe in their practices. According to these data, cancers can mainly be divided into two dissimilar groups: (1) those inherited by children from their parents and (2) the so-called sporadic cancers. Inherited cancers carry the same mutation in all cells of the organism. We have called them "inherited genetic errors of development," and they represent less than 2 percent of clinical cancers.[13] Examples of these cancers are retinoblastomas linked to *Rb* gene mutations, breast and ovarian cancers due to mutations in the *BRCA1* and *BRCA2* genes, and familial colon carcinomas. On the other hand, the so-called sporadic cancers are the overwhelming majority of all clinical cancers. They include two subtypes, namely, cancers owing to exposure to carcinogens either (1) before or (2) after birth. The former are generated in normal embryos and fetuses when mothers are exposed to a variety of environmental agents that deleteriously affect the highly plastic process of organ formation. These cancers were noted in recent epidemiological data and include breast cancer and clear-cell carcinoma of the vagina, the result of prenatal exposure to the synthetic estrogen diethylstilbestrol. Another kind of cancer found in this subtype is early childhood hematopoietic malignancies. At present it is unclear what percentage of sporadic clinical cancers diagnosed during adulthood are due to alteration of organogenesis during intrauterine life. The other subtype of sporadic cancers consists of cancers caused by exposure to carcinogens during postnatal life.

Carcinogens can be either physical (e.g., radiation, foreign bodies), chemical (e.g., tobacco smoke, environmental contaminants), or biological (e.g., *Helicobacter pilori,* schistosomas, bilharzias, herpes viruses, and human papillomaviruses [HPVs]). Some of the physical (radiation) and chemical (tobacco-smoke ingredients) carcinogens cause mutations in some cells, while others do not. Both types of carcinogens are considered as *causes* of cancer. One may surmise that most, if not all, causes of cancers are known. Therefore, other than the low percentage of cancers due to inherited mutations (again, less than 2 percent), the overwhelming majority of cancers can be prevented by reducing exposure to carcinogens. Preventive measures, however, are not the dramatic "cancer cures" of which patients, clinicians, and drug manufacturers dream.

We have concluded with developmental biologists that cancer is a disease due to an alteration of normal development and tissue repair that can occur throughout an individual's life. Nevertheless, in discussing cancer in the clinical or experimental context, most commentators claim that cancers are a genetic disease—a disease caused by mutations in so-called cancer genes. This belief implies a deterministic link, meaning that there are specific genes in the genomes of each of our trillions of cells that during our lifespan undergo some kind of change (mutation) that in turn is responsible for cancer.

Causes and Explanations of Cancers

Thus far we have addressed only the causes of cancers. We have not referred in detail to how those causes generate the symptoms and signs of cancer, prominent among them being the appearance of tumors. As early as 1775 the physician Percival Pott observed cancer on the scrota of English chimney sweepers—young men who were exposed to the soot of the chimneys they had cleaned since childhood. Pott theorized that "the disease . . . seems to derive its origin from the lodgement [*sic*] of soot in the rugæ [creased skin] of the scrotum."[14] Although contemporary physicians concurred with Pott's conclusion, "there was a relative lack of effective impact of his work on British public health practice during the succeeding century."[15] In the centuries since, little has changed. Today, however, abundant epidemiological and experimental evidence has signaled unequivocally that most, if not all, of the proximate causes of sporadic cancers are environmental (Table 7.1).

However, all the compelling data accumulated so far have yet to make a significant impact on public health officials, lawmakers, or the public at large. For these reasons, cancer should be viewed from a sociopolitical as well as an etiological perspective. When it has been reliably identified that the causes of cancer are not as mysterious as often portrayed, the

Table 7.1 Cancer causes and explanations

Causes	Explanations
Viruses (examples: HPV, hepatitis, herpes)	Somatic mutation theory (SMT)
Radiation (examples: α, γ, X-rays, non-ionic radiation [NIR])	
Environment (examples: tobacco, DDT, bisphenol-A [BPA], benz-pyrene [BP], asbestos, hormones)	Tissue organization field theory (TOFT)
Inflammations (examples: leishmania, schistosoma, *Helicobacter pilori,* Epstein-Barr virus)	

Source: Authors.

Note: The left column lists all the causes of cancer; the right column lists the two theories that have been proposed to explain how all those causes of cancer generate the neoplasms.

failure to deal with those causes is nothing short of a society-wide dereliction of duty.

We now examine the explanations of carcinogenesis, that is, how it is that those many heterogeneous agents increase the likelihood of tumor development. Researchers offer explanations of carcinogenesis by proposing plausible theories that are subject to exploration.

A Brief Description of Cancer Theories

For the sake of simplicity, we mention here only the two principal theories that attempt to explain how the disparate causes of cancer can result in comparable neoplasms (Table 7.2). We have classified these theories on the basis of the hierarchical levels of biological organization at which it is assumed that carcinogenesis occurs. By this criterion, the most popular theory places carcinogenesis at the cellular level. First formulated in 1914 by Theodor Boveri, it is called the somatic mutation theory (SMT).[16] The second, alternative theory of carcinogenesis is the tissue organization field theory (TOFT). Formulated in 1999, the TOFT places carcinogenesis instead at the tissue level of biological organization.[17]

The SMT holds that cancer is a disease of abnormal cell proliferation that is mediated by genetic mutations in a single cell. More specifically, those mutations supposedly affect the genes involved in the process of cell division. Because of glaring lacks of fit, proponents of the SMT have had to introduce significant ad hoc course corrections over the past century in order to conform to pragmatic clinical and experimental observations. Nevertheless, "The premises of the somatic mutation theory survive in the contemporary oncogene paradigm."[18] As the theory of record,

Table 7.2 Brief history of cancer theories

- 1850s to 1900s: Cancer considered as a *tissue*-based disease (Ribbert and others, precursor of the TOFT).
- 1914–present: Cancer considered as a *cell*-based disease (Boveri's theory, considered a precursor of the SMT).
- 1908–1925: Development of cell culture techniques.
- 1935–1950s: Cancer considered as a developmental disease (Waddington, Needham, precursor of the TOFT).
- 1953–present: Molecular biology revolution ("greedy" reductionism).
- 1960s–present: SMT became the dominant paradigm in cancer (oncogenes, suppressor genes).
- 1999: The society of cells (TOFT).

Source: Authors.

the SMT is taught at colleges and universities worldwide, and in turn, this popularity trickles down into common daily discourse. It is not uncommon to hear that "cancer runs in the family," implying a genetic cause, while ignoring that such a family might be exposed to a shared deleterious environment (e.g., tobacco smoke, endocrine disruptors, radiation, asbestos).

Conversely, the TOFT directly challenges the core premises of the SMT by positing that carcinogenesis, like histogenesis (the formation of tissue) and organogenesis (the formation of organs), is a supracellular phenomenon, meaning that it occurs at the tissue level of biological organization. From this perspective, regardless of its many causes, cancer is development gone awry.

A further qualitative difference between the SMT and the TOFT relates to the premises adopted regarding the control of cellular proliferation in multicellular organisms. The default state is defined as the proliferation state that cells assume in the presence of abundant building blocks needed to synthesize new cells. Those siding with the SMT presume that quiescence is the default state of metazoan cells.[19] In contrast, on both epistemological and experimental grounds, supporters of the TOFT have adopted the evolution-relevant principle that—as is widely acknowledged to happen in unicellular organisms and plants—the default state of all cells is proliferation. In fact, "Among microbiologists, it is axiomatic to accept that proliferation is the default state of prokaryotes and unicellular eukaryotes."[20] Moreover, motility—the capacity for independent locomotion—as proliferation is also a constitutive, dominant property of all cells. Altered tissue architecture facilitates the expression of these two constitutive states that directly relate to tumor growth and metastasis. We have extensively commented on this subject elsewhere.[21]

In recent decades supporters of the SMT have reported an increased number of mutations required by a normal cell for it to become a tumor cell. Since 1982, when a single mutated gene was claimed to be sufficient to generate a tumor,[22] numerous reports have increased the number of those crucial mutated genes that vary widely among individual tumors of the same organ. To accommodate this unexpected finding, rather than identifying a few specific "oncogenes," as the SMT had envisaged, researchers now postulate that massive sequencing must be performed in an effort to organize these mutated genes into "networks" of large sets of genes. How was this conclusion reached? The raw data were from DNA derived from tumors, which is to say a conglomerate of heterogeneous cells—not from a single "cancer" cell. This diverse DNA was then subjected to "massively parallel" sequencing, in which there was no way to distinguish a priori which mutations were causal (driver mutations) and which were irrelevant (passenger mutations) because the cancer had already developed.[23] The distinction between these two mutation types is thus based on unverifiable inferences.

Where do proponents of the SMT go from here? In screening 518 genes in 210 diverse human cancers, researchers recently found 120 so-called driver mutations. This finding led researchers to propose an even more massive sequencing effort because it "implicate[d] a larger repertoire of cancer genes than previously anticipated."[24] Thus, given the failure to find a single or a combined set of somatic mutations that would consistently qualify for the assignment of being causally linked to a neoplastic phenotype, ad hoc modifications to the core content of the SMT are being proposed. Currently, a generously funded effort is aimed at sequencing the genome of hundreds of individual tumors belonging to each common organ/tissue location (breast, lung, prostate, and so on) with the hope that as more tumors are sequenced, a consistent pattern will emerge. In this regard, while using the most advanced technology aimed at making personalized medicine more affordable, a recent report generates additional doubts on the relevance of the SMT in coherently explaining carcinogenesis. Through exome sequencing, chromosome aberration analysis, and ploidy profiling on multiple spatially separated samples obtained from primary renal carcinomas and their metastasis, the authors identified gene-expression signatures of good and poor prognosis in different regions of the same tumor. Equally remarkable, they also uncovered extensive intratumor heterogeneity "with 26 of 30 tumor samples from four tumors harboring divergent allelic-imbalance profiles and with ploidy heterogeneity in two of four tumors."[25]

Departing from this unsustainable position adopted by the SMT, the TOFT proposes instead that cancers are tissue-based diseases; that is, they are the result of alterations of the communication among cells and tissues that affect tissue architecture, for example, histogenesis and organogenesis. These alterations cause the abnormal tissue structure observed in neoplasms, which in turn causes the cells in the neoplasm to express their default states, namely, proliferation and motility. More explicitly, "The TOFT proposes that carcinogenic agents generate a disruption in the reciprocal interactions between cells that maintain tissue organization, tissue repair and local homeostasis."[26] What occurs inside any individual cell in a tumor is thus a consequence of that altered communication within and among tissues. Although the TOFT acknowledges that mutations might be generated throughout the carcinogenic process, they appear irrelevant to carcinogenesis and its progression.[27]

The Intersection of Cancer and Biology

Several decades ago the famed geneticist Theodosius Dobzhansky coined a very powerful maxim: "Nothing in biology makes sense except in the light of evolution."[28] Cancer, as a biological phenomenon, is not an exception to this rule. The histories of the SMT and the TOFT lend themselves to being considered within the context of biology at large. While the TOFT was formulated only a little over a decade ago, sufficient time has elapsed for adherents of the SMT to have demonstrated its veracity either on explanatory grounds or, more important, from its clinical impact. After all, the success of a theory is evaluated on the basis of its ability to resolve paradoxes, generate knowledge in unexplored fields, and provide practical payoffs—in this case, effective cancer therapies.

Following the organicist tradition of developmental biologists of the first half of the twentieth century, we have conducted experiments aimed at addressing whether the target of carcinogens resides in the epithelial tissue or in the stroma (connective tissue) of the mammary gland. Using a theory-neutral experimental strategy, we observed that the tissue recombination of stroma exposed to a carcinogen with normal unexposed epithelial cells resulted in neoplasms.[29] The reverse combination did not. This observation suggested that the stroma, rather than individual cells in the epithelium, was the target of the carcinogen. As a follow-up, and in a complementary experimental approach, we tested the possibility of "normalizing" (i.e., reversing) the tumor phenotype of rat mammary-gland cells by inoculating these tumor cells into rats of different ages and observed that in adult rats those tumor cells generated phenotypically normal mammary ducts.[30] These results fit the expectations of the TOFT and challenge those of the SMT. The normalization of the neoplastic phenotype is supported by a body of literature gathered during the past four decades, now confirmed and strengthened by the use of tools that permit us unequivocally to identify the normal cells that once were cancer cells.[31]

Science, Society, and the Cancer Puzzle

Historically, it has become obvious that the allocation of resources (grant monies) have had a great impact on what subjects have been researched and to what extent. The war on cancer is probably the best example of this trend in biomedicine. Decades after the political decision was made to invest on cell-based and subcellular research, it is now widely acknowledged that the returns on that investment have not met expectations.[32] Despite this failure, much has been learned about cells and subcellular phenomena. It is thus time to reassess what future investments ought to merit the support of political and public health authorities and the public at large. Expectations from a novel, evolution-relevant approach in cancer research now appear more realistic.

A Final Note

A question frequently put to us after we give a seminar or in informal scientific exchanges is worded more or less as follows: "Do you mean to say that the thousands of articles and reviews that for decades have been headed by the statement 'It is generally acknowledged that cancer is caused by somatic mutations' or a slight variation on this theme are baseless?" Our bibliographic search and the experimental work performed in our labs for the past four decades led us to conclude just that. Now comes the slow, messy process of paradigm switching in the minds of our colleagues, funding agencies, and the people (the innocent bystanders) who uncritically adopted premises to conduct and support research on cancer that were essentially incompatible with evolution-relevant principles. In this chapter we have contributed to challenging the rationale supporting the original deterministic misconceptions on which research policies and the experimental agenda have been based for the past half century.

On the positive side, it is now clear that the panorama is changing both in cancer research and in the biology of complex phenomena.[33] First, the perception by molecular biologists that those phenomena cannot productively be explored one gene at a time is resulting in the massive entry of biomathematicians into various fields of biomedical research. This interest has been motivated by the new "omics" technologies, aimed at producing an "instant picture" of the messenger RNA and protein expression profiles of cells, and has resulted in the rebirth of systems biology, a discipline with roots in the Vienna school of Bertalanffy and Weiss.[34] Contrary to the theoretical content of the original systems biology, most of the practitioners of this updated version are adopting pragmatic principles and bottom-up causality.[35] Others, however, are searching for a theoretical framework whereby organicism and both bottom-up and top-down causality are considered.[36] Second, the new disciplines of tissue engineering and biomaterials research, aimed at producing tissues for transplantation, have brought a revival of biophysics and biomechanics to the study of form (morphogenesis) in development and cancer.[37]

Third, philosophers and theoretical biologists have revealed the insurmountable problems posed by metaphors that attribute information content to genes, on the one hand, and the ability of programming the organism to the genome, on the other.[38] Thus it is now clear that the exploration of alternative views and the resurfacing of emergentism are announcing a most welcome change in research on biology at large and on cancer in particular.[39] We argue that information metaphors should be replaced by a theory solidly grounded in mathematics and physics that can produce a new observable for the quantification and the understanding of biological organization.[40] This means, in short, the articulation of a theory of organisms.

We dedicate this chapter to the revered memory of Dr. Murray R. Blair (1928–2010), former dean of the Sackler Graduate School at Tufts University School of Medicine (1980–1982), who was an admired intellect, an enlightened leader, and a generous and loyal friend.

CHAPTER EIGHT

The Fruitless Search for Genes in Psychiatry and Psychology

Time to Reexamine a Paradigm

JAY JOSEPH AND CARL RATNER

THE JUNE 2009 edition of the *Journal of the American Medical Association* reported the results of a meta-analysis by Neil Risch and colleagues.[1] These researchers showed that a 2003 study by Caspi and colleagues, where the investigators believed that they had found a genetic variant associated with depression when it was combined with stressful life events, did not stand up to replication attempts. Caspi and colleagues' original study had been widely reported in the media and elsewhere as constituting a major genetic discovery in psychiatry.[2]

However, to the critical observers of genetic research in psychiatry and psychology, including those who had pointed to several glaring problems in Caspi and colleagues' study, the failure to replicate these results came as no surprise at all.[3] This study merely suffered the same fate as other gene-finding claims in psychiatry over the past forty years, such as the much-publicized but subsequently nonreplicated claims of a generation ago for bipolar disorder and for schizophrenia.[4] Clearly, some type of systematic error is common to these subsequently unsubstantiated findings.

Previously, a group of leading psychiatric genetics researchers had recognized in 2008, "It is no secret that our field has published thousands of candidate gene association studies but few replicated findings."[5] In the same year behavioral geneticist Robert Plomin and colleagues could not cite any substantiated gene findings for personality or IQ (cognitive ability).[6] A pair of personality-trait molecular genetics researchers wrote in 2009 that their field "has evidently not escaped the conundrum of non-replication that continues to plague the genetics of complex human phenotypes."[7] The authors of a 2010 article on cognitive ability and genetics noted, "It is difficult to name even one genetic locus that is reliably associated with normal-range intelligence in young, healthy adults."[8] By 2012 the situation remained the same.[9] Risch and colleagues concluded that "few if any of the genes identified in candidate gene association studies of psychiatric disorders have withstood the test of replication." They further concluded:

> Despite progress in risk gene identification for several complex diseases, few disorders have proven as resistant to robust gene finding as psychiatric illnesses. The slow rate of progress in psychiatry and behavioral sciences partly reflects a still-evolving classification system, absence of valid pathognomonic diagnostic markers, and lack of well-defined etiologic pathways. Although these disorders have long been assumed to result from some combination of genetic vulnerability and environmental exposure, direct evidence from a specific example has not been forthcoming.[10]

Thus the fields of behavioral genetics and psychiatric genetics are rapidly approaching a period of crisis and reexamination. In the words of a leading group of psychiatric genetics investigators, writing in 2012 about the decades-long failure to uncover any genes that cause schizophrenia (the most studied psychiatric disorder), these negative results "suggest . . . that many traditional ideas about the genetic basis of SCZ [schizophrenia] may be incorrect."[11]

There are two broad explanations for the ongoing failure to discover genes in psychiatry and psychology. The first, which is favored by genetics researchers and their backers, is that genes for "complex disorders" exist (although each gene may be of small effect size) and will be discovered once researchers improve their methods and increase their sample sizes. The second explanation, rarely considered in mainstream works, is that genes for psychiatric disorders and for normal variation in psychological traits do not exist. The latter explanation is consistent with Latham and Wilson's position that apart from a few exceptions, "genetic predispositions as significant factors in the prevalence of [most] common diseases are refuted," and that the "dearth of disease-causing genes is without question a scientific discovery of tremendous significance."[12]

Over the past two decades both the popular and the scientific literature have been filled with discussions of how improved methods in molecular genetics research will lead to gene discoveries. Although we cannot rule out such possibilities, our purpose here is to suggest that the misreading of previous kinship studies of families, twins, and adoptees has led the scientific community to the premature conclusion that genes for psychiatric disorders and psychological trait variation must exist.

In the past few years molecular genetics researchers have adopted the position of "missing heritability" as an explanation for their failure to discover genes.[13] The missing heritability interpretation of negative results has been developed in the context of the ongoing failure to uncover most of the genes presumed to underlie common medical disorders and virtually all of the genes presumed to underlie psychiatric disorders and psychological trait variation. In 2008 Francis Collins, current director of the U.S. National Institutes of Health and former director of the National Center for Human Genome Research, stated that missing heritability "is the big topic in the genetics of common disease right now."[14] Subsequently, the topic has grown even bigger.[15]

Heritability is "missing," according to one group of prominent researchers, because genomewide association (GWA) studies "have explained relatively little of the heritability of most complex traits, and the [gene] variants identified through these studies have small effects."[16] In 2009 a prominent group of researchers (including Francis Collins) headed by Teri Manolio, director of the U.S. Office of Population Genomics, published an article in *Nature* titled "Finding the Missing Heritability."[17] This article has since served as a reference point for molecular genetics researchers, including those in psychiatry and psychology, who have attempted to come to terms with decades of negative results. Manolio and colleagues recognized that only a few gene variants had been discovered for nonpsychiatric medical conditions, and they pointed to "the lack of variants detected so far for some neuropsychiatric conditions." They had no doubt that the problem is missing heritability, as opposed to nonexistent heritability, because "a substantial proportion of individual differences in disease susceptibility is known to be due to genetic factors." Manolio and colleagues saw missing heritability as the "'dark matter' of genome-wide association in the sense that one is sure it exists, can detect its influence, but simply cannot 'see' it (yet)."[18]

The reason that scientists are certain that "missing" genes exist and await discovery is their belief that previous family, twin, and adoption studies have provided conclusive evidence that genetic factors play a major role. But even if researchers eventually discover specific genes that play a role in intelligence or personality, or that predispose some people to develop psychiatric disorders, society could still choose to focus attention on mitigating psychologically unhealthy family, social, and political arrangements that impede human growth and learning and contribute to emotional problems and psychiatric disorders. Genetic-determinist ideas divert society's attention from these environmental conditions and shift blame onto people's brains and bodies. Even in the case of medical disorders such as type 2 diabetes, where poverty and malnutrition are well-known causes, supporters of genetic determinism continue to press for research dollars to be directed toward genetic research, as opposed to improving social and health conditions.[19]

Kinship studies of families, twins, and adoptees are known collectively as "quantitative genetic research." Although family studies constitute a necessary first step, they are widely seen as being unable to disentangle the potential roles of genetic and environmental factors. Because family members share a common environment as well as common genes, a finding that a trait "runs in the family" can be explained on either genetic *or* environmental grounds. As Plomin and colleagues recognized, "Many behaviors 'run in families,' but family resemblance can be due to either nature or nurture."[20] They concluded, correctly in our view, that "family studies by themselves cannot disentangle genetic and environmental influences."[21]

Twin Studies

Twin studies and adoption studies, which have been carried out since the 1920s, constitute the main quantitative genetic results cited in support of genetics. We will touch on some problem areas in adoption research later, but for now we focus on twin studies, which provide the most frequently cited evidence in support of important genetic influences on psychiatric disorders and variations in "normally distributed" traits such as IQ and personality. There are two main types of research studies of twins: studies of twins reared together and studies of twins reared apart.

Reared-Together Twins. Studies of twins reared together, which use a technique called the "twin method," compare the trait resemblance of reared-together monozygotic (MZ) versus reared-together same-sex dizygotic (DZ) twin pairs. If MZ pairs resemble each other more than DZ pairs (on the basis of correlations or concordance rates), twin researchers conclude that the trait has a genetic component and then go on to calculate heritability estimates based on the magnitude of the difference. They reach this conclusion on the basis of several theoretical assumptions about twins, the most important and controversial of which is the assumption that MZ and same-sex DZ twin pairs experience roughly equal environments. This is known as the "equal-environment assumption" (EEA). The logic appears straightforward, since MZ pairs share a 100 percent genetic similarity, whereas DZ pairs share only 50 percent of their genes on average.

There is, however, a fatal flaw in this logic: The EEA of the twin method is obviously not correct, since most research in this area finds that MZ twin pairs experience much more similar environments than do DZ pairs.[22] In addition, because they are more similar genetically, MZ pairs anatomically resemble each other more than DZ pairs, and this clearly will elicit more similar treatment from the social environment.[23] Therefore, a plausible interpretation of twin method findings is that the greater psychological trait resemblance of MZ versus DZ twin pairs, a result found by most twin researchers, is completely explainable on the basis of nongenetic factors related to MZ pairs' greater environmental and treatment similarity. From the standpoint of environmental confounds, the twin method has precisely the same problem as family studies because in both, the comparison groups experience far different environments. Moreover, new research findings have called into question several long-standing assumptions in the science of genetics, which raise even more questions about the validity of twin research.[24]

Interestingly, most contemporary twin researchers recognize that the environments experienced by MZ pairs are more similar than those experienced by DZ pairs.[25] However, on the basis of two main arguments, they continue to hold that the EEA is valid and that the twin method reliably measures genetic influences.

The first argument is that although MZ and DZ environments are different, these environments must be shown to differ in aspects relevant to the trait in question.[26] Furthermore, twin researchers often implicitly or explicitly suggest that twin method critics bear the burden of proof that these admittedly unequal environments differ on trait-relevant dimensions.[27]

The second argument twin researchers put forward in defense of the EEA and the twin method is that MZ pairs tend to "create" or "elicit" more similar environments for themselves by virtue of their greater genetically caused similarity of behavior.[28] For example, according to one group of behavioral genetics researchers, although MZ twins "may well be treated more similarly" than DZs, "this is far more a consequence of their genetic similarity in behaviour (and of ensuing responses by parents and others) than a cause of such similarity."[29] And in 2009 behavioral geneticists Segal and Johnson wrote, "It is important to note that if MZ twins are treated more alike than DZ twins, it is most likely associated with their genetically based behavioral similarities."[30]

Regarding the first argument, the proponents of a scientific theory or technique, rather than their critics, bear the burden of proof that their theory or technique is correct.[31] Although twin researchers have carried out a series of tests of the EEA, these studies have done little to uphold the validity of the twin method.[32] Ironically, although EEA test researchers usually conclude that their findings support the EEA, most find that MZ twin pairs experience much more similar environments than do DZ pairs. What they fail to understand is that the differing environments that automatically and without qualification invalidate genetic interpretations of family studies also invalidate genetic interpretations of twin method data.

We have seen that the second argument modern twin researchers put forward in defense of the twin method is that the environments of MZ twin pairs are more similar than those of DZ pairs because MZs "create" more similar environments for themselves on the basis of their greater genetic similarity. However, researchers putting forward this "twins create their own environment" position use circular reasoning because they assume the very thing they need to demonstrate. According to *The Penguin Dictionary of Psychology,* circular reasoning is "empty reasoning in which the conclusion rests on an assumption whose validity is dependent on the conclusion."[33] Twin researchers have used empty reasoning of this type since the 1950s to validate the twin method; they circularly assume that twins' behavioral resemblance is caused by genetics in order to conclude that twins' behavioral resemblance is caused by genetics.[34] Thus the only relevant question in determining the validity of the EEA and the twin method is whether—not why—MZ pairs experience more similar environments than those experienced by DZ pairs.[35]

Buried within the twin research literature on schizophrenia, which is frequently cited in support of a genetic basis for the condition, is a finding that the pooled concordance rate for same-sex DZ twin pairs is two to three times greater than that of opposite-sex DZ pairs (11.3 percent versus 4.7 percent).[36] Because the genetic relationship of same-sex and opposite-sex DZ twin pairs is the same, and because schizophrenia rates among males and females are roughly equal, from the genetic standpoint we should find no significant difference between these pooled rates.[37] Moreover, the pooled schizophrenia concordance rate for DZ twins is almost double that of ordinary (nontwin) siblings, despite the fact that the genetic relationship between DZ twins and ordinary sibling pairs is the same.[38] These findings are consistent with nongenetic explanations of the causes of schizophrenia, since pairs who share the same degree of genetic relatedness, but who experience more similar environments and a closer emotional bond, are consistently more concordant for schizophrenia than are pairs who experience less similar environments and a weaker emotional bond. These results provide additional evidence that—as we have seen with family studies—the twin method is unable to disentangle potential genetic and environmental causes of schizophrenia and other psychiatric disorders.[39]

Thus there are two main conclusions one can reach on the basis of twin method data:[40]

1. *Contemporary twin researchers' conclusion:* The greater resemblance of MZ versus same-sex DZ twin pairs provides solid evidence that a sizable portion of the population variance for psychiatric disorders and psychological traits can be explained by genetic factors.
2. *Twin method critics' conclusion:* The twin method is a faulty instrument for assessing the role of genetics, given the likelihood that MZ versus same-sex DZ comparisons measure environmental rather than genetic influences. Therefore, all previous interpretations of the twin method's results in support of genetics are potentially wrong.

We argue here that the available evidence calls for the acceptance of conclusion 2, and we agree with three generations of critics who have written that the twin method is no more able than a family study to disentangle the potential roles of nature and nurture. As the Nobel Prize–winning chemist Wilhelm Ostwald wisely lectured his students in the early twentieth century, "Among scientific articles there are to be found not a few wherein the logic and mathematics are faultless but which are for all that worthless, because the assumptions and hypotheses upon which the faultless logic and mathematics rest do not correspond to actuality."[41]

Reared-Apart Twins. Because many scientists and commentators have had doubts about the validity of the twin method, some have pointed to studies of twins reared apart (TRA studies), such as the Minnesota TRA research published by Bouchard and colleagues.[42] These investigations look mainly at psychological traits such as IQ and personality. However, several reviewers have outlined problems with the methodology and underlying logic of these studies.[43] Problem areas include the following: (1) it is doubtful that most reared-apart MZ pairs (known as MZAs) deserve the status of having been "reared apart," since most pairs had significant contact with each other for many years; (2) in several studies there were biases favoring the recruitment of MZA pairs who resembled each other in behavioral traits more than MZA pairs as a population; (3) there is controversy about whether "intelligence" and "personality" are valid and quantifiable constructs; (4) the Minnesota researchers failed to publish life-history information for the twins under study and then denied independent reviewers access to raw data and other unpublished information; and (5) there was likely researcher bias in favor of genetic explanations of the data.[44]

Perhaps the most important problem is the original TRA researchers' failure to control for several critical environmental influences shared by MZA pairs, including even those extremely rare cases in which studied MZA pairs were reared apart from early life and grew up without knowing that they had a twin sibling.[45] In the study containing the highest percentage of MZA pairs of this type, the author found,

> In all 12 pairs there were marked intra-pair *differences* in that part of the personality governing immediate psychological interaction and ordinary human intercourse. . . . The twins behaved, on the whole, very differently, especially in their cooperation, and in their form of and need for contact. Corresponding with these observations, the twins gave, as a rule, expression to very different attitudes to life, and very divergent views on general culture, religion and social problems. Their fields of interest, too, were very different. . . . Those twins who had children treated, on the whole, their children differently, and their ideas on upbringing were, as often as not, diametrically opposed. Characterologically, the twins presented differences in their ambitions and in their employment of an aggressive behavior. Emotionally, there was a deep-going dissimilarity with regard to the appearance of spontaneous emotional reactions or to the control of affective outbursts. Various traits of personality found their expression in differences in taste, mode of dress, hair style, use of cosmetics, the wearing of a beard or of glasses.[46]

Original descriptions of this type have not prevented the authors of numerous books attempting to popularize genetic research, exemplified by Steven Pinker's *Blank Slate* and Judith Harris's *Nurture Assumption,* from claiming that TRA studies and individual stories reported in the media "suggest that genes can cause striking similarities in personality characteristics, even in the face of substantial differences in rearing environments."[47]

Environmental influences shared even by perfectly separated MZA pairs always include common age, common sex, common ethnicity, common physical appearance, and common prenatal environment, and usually include common socioeconomic class and common culture. Reared-apart twin pairs (as well as genetically unrelated people born at the same time) are subject to the social and historical influences of their birth cohort. As behavioral genetics researcher Richard Rose once observed, "Were one to capitalize on cohort effects by sampling [genetically] unrelated but age-matched pairs, born, say, over a half-century period, the observed similarities in interests, habits, and attitudes might, indeed, be 'astonishing.'"[48]

Thus, for reasons unrelated to heredity, we should expect to find a much higher correlation in video-game-playing behavior in the United States among pairs of randomly selected 11-year-old middle-class Caucasian boys than we would expect to find among randomly selected pairs drawn from the entire 11- to 100-year-old male and female population of the United States.[49] This hypothetical example illustrates one of the central fallacies of TRA studies. (Bouchard and colleagues were the first TRA researchers to address age and sex confounds, but their adjustments were inadequate to deal with this problem.)[50]

On purely environmental grounds, therefore, we would expect MZA pairs to correlate well above zero for psychological and behavioral traits.[51] This means that the appropriate control group with which to compare MZA correlations would be a group consisting of genetically unrelated pairs of strangers matched on the environmental influences experienced by MZA pairs.[52] Most previous MZA studies, however, mistakenly used reared-together MZs as controls.[53] Thus we see that like the twin method, studies of twins reared apart are subject to their own set of invalidating environmental confounds and other biases.

Adoption Studies

Although twin research has been called the "'Rosetta Stone' of behavior genetics," adoption studies are also used to assess the role of genetic influences on various traits and disorders.[54] Adoption studies investigate people who receive the genes of their birth parents but are reared in the family environment of people with whom they share no genetic relationship. Adoption research originally focused on IQ and was extended to include personality and psychiatric disorders such as schizophrenia, attention-deficit/hyperactivity disorder, and bipolar disorder. In particular, the Danish American adoption studies are widely cited as having established schizophrenia as a genetic disorder.[55] Several commentators, however, have pointed to a number of crucial errors and biases in these studies.[56] In Tienari and colleagues' Finnish schizophrenia adoption studies, the researchers concluded that both genes and family environment play important causative roles.[57]

Like family and twin studies, adoption studies are subject to their own set of environmental confounds and biases that cast doubt on their ability to separate the potential influences of nature and nurture. Included among these biases are late separation (and accompanying attachment disturbance), range restriction, whether adoptees and family members are representative of their respective populations, and the selective placement of adoptees.[58]

Tienari and colleagues investigated the adoptive families of Finnish adoptees whose biological mothers were diagnosed with schizophrenia (index adoptees) and the adoptive families of control adoptees whose biological mothers were not so diagnosed.[59] Although 7 percent of the index adoptees were diagnosed as psychotic, in contrast to 1 percent of the control adoptees (which can be accounted for by selective placement factors; see below), Tienari and colleagues' analysis of the families of index adoptees diagnosed as psychotic reveals that 6 of 43 adoptees (14 percent) who were reared in "seriously disturbed adoptive families" were diagnosed as psychotic. In striking contrast, none of the 48 index adoptees reared in "healthy or mildly disturbed adoptive families" were diagnosed as psychotic. Moreover, 19 of the 32 adoptees (59 percent; index and control combined) raised in "severe disturbance" Finnish adoptive families developed a major psychological dysfunction (which included "character disorders," "borderline syndrome," and "psychotic"), whereas none of the 15 adoptees reared in Finnish "healthy" adoptive families developed such a dysfunction.[60]

If we look more closely at the "no-selective-placement" assumption of adoption studies, psychiatric adoption researchers must assume that factors relating to the adoption process did not lead agencies to place certain groups of adoptees into environments contributing to a higher rate of the disorder in question. However, the evidence suggests that adoption studies of schizophrenia were confounded by environmental factors on the basis of the perceived genetic undesirability of adoptees with a biological family history of mental disorders placed in early to mid-twentieth-century Europe.[61]

For example, Finland (like Denmark) had a long history of eugenics-inspired legislation aimed at curbing the reproduction of "hereditarily tainted" people.[62] The Finnish government created a commission in 1926 to study the possibility of sterilizing people seen as "mentally retarded," "mentally ill," or epileptic. In 1935 the Finnish parliament passed the Sterilization Act, which allowed the compulsory eugenic sterilization of "idiots," "imbeciles," and the "insane," which included people diagnosed with schizophrenia and manic depression. Compulsory eugenic sterilization was not abolished in Finland until 1970. The Finnish adoptees Tienari and colleagues studied were born between 1927 and 1979 and were therefore placed in an era in which the biological offspring of people diagnosed with a psychotic disorder were seen as undesirable, "tainted" adoptees. Clearly, few prospective Finnish adoptive parents would have wanted to adopt such a child.

Selective placement has also been identified as a confounding factor in IQ adoption research, since adoption agencies frequently attempt to match adoptees and adoptive families for socioeconomic status, in addition to matching on the basis of the assumed intelligence potential of the adoptee.[63]

Thus, despite adoption studies' theoretical potential to disentangle genetic and environmental influences, most adoption studies published to date have been plagued by methodological problems and potential environmental confounds. However, it is possible that a well-designed adoption study could separate genetic and environmental factors and put the nature-nurture issue to the test. The researchers performing such a study must, at a minimum, (1) choose as participants only those adoptees who were placed into their adoptive homes at or shortly after birth; (2) determine in advance, and publish or submit to a research register before undertaking the study, the specific hypotheses, methods, definitions, and comparison groups that will be used; (3) make a serious attempt to come to grips with problems such as selective placement and range restriction, and be willing to refrain from concluding in favor of genetics if such problems are found; (4) publish or place with a research register raw case-history information and data relating to participants and make this information and data available to qualified reviewers for inspection; (5) ensure that all interviews, tests, diagnoses, and ratings are performed blindly; and (6) study only those traits and disorders whose reliability and validity have been demonstrated by previous research.[64]

Conclusions about the Genetic Paradigm and the Need for an Alternative

We have suggested that the body of quantitative genetic research in psychiatry and psychology is contaminated by environmental factors.[65] In addition, these studies contain many glaring methodological problems and other biases. Although the relatives in these studies frequently manifest traits and disorders in patterns predicted by genetic theories, these patterns usually match the predictions made by theories of nongenetic causation as well.[66] Thus it is likely that family, twin, and adoption studies have been unable to disentangle the potential roles of genetic and environmental influences on traits and disorders, and that the investigators who typically perform this research have greatly underestimated the potential role of environmental confounds. It has been left to critics to focus on these problems, but their voices have been lost in the vast literature produced in the past few decades by authors claiming major genetic influences on these traits.

We call on behavioral scientists, particularly researchers in psychiatry and psychology, to suspend the current "missing heritability" interpretation of negative molecular genetics findings and to embark on a serious reassessment of the validity of twin and adoption studies. Latham and Wilson concluded that a reasonable interpretation of the failure to identify genes is that "heritability studies of twins are inherently mistaken or misinterpreted," and that the "dark matter" of missing heritability "becomes simply an artifact arising from overinterpretation of twin studies."[67]

In 1994 behavioral geneticists Robert Plomin, Michael Owen, and Peter McGuffin wrote in *Science* about a genetic variant associated with Alzheimer's disease and continued, "We predict that QTL [quantitative trait loci, or genes of various effect sizes] associations will soon be found for other complex human behaviors."[68] However, this prediction turned out to be wrong.[69] Indeed, three genetically oriented Nobel Prize–winning researchers and their colleagues, in a 2010 *Science* "Policy Forum" article, recognized the "frustrating lack of progress" in understanding the genetics of mental disorders.[70]

A final issue to consider is the broader context of genetic research into psychiatric disorders. This context includes not only scientific and social issues that form the assumptions that guide this work but also the scientific and social consequences of this work. This inquiry into the context of research is a branch of philosophy of science known as social studies of science. It is pursued in the Society for Social Studies of Science and in journals such as *Social Epistemology* and *Social Studies of Science*. The social and intellectual context affects the quality of particular research (just as social context affects all behavior). It is also an important avenue for assessing the plausibility and validity of the research.

Research into possible genetic causes of psychiatric disorders partakes of a system of superordionate scientific issues. These include the nature of human psychology and its relation to biochemical mechanisms. If genes cause psychiatric disorders in some specific manner, then they must work through biochemical mechanisms. This raises the broader, superordinate question whether psychiatric disorders are caused by (reducible to) biochemical mechanisms—that is, how can biochemical mechanisms cause someone to experience particular symptoms of, for example, depression, eating disorders, risky behavior, or social phobia? This question about biochemical mechanisms and psychiatric disorders depends on a still-broader, superordinate question whether psychology in general is determined by biochemical mechanisms. Research into these two superordinate questions strengthens or weakens the genetic hypothesis of psychiatric disorders. If research demonstrates that psychology is not determined by biochemical mechanisms, then psychiatric disorders cannot be determined by genes.[71] Conversely, research into the subordinate issue of whether genes cause psychiatric disorders strengthens or weakens the superordinate issues. Researchers and the authors of authoritative textbooks who claim that genes play a major role in causing psychiatric disorders help strengthen superordinate theories that these disorders have biochemical causes, and that psychology has biochemical causes.

Although pursuing these corollary issues is outside the bounds of this chapter, we recommend that readers do so in order to better understand the question of genetic causes of psychiatric disorders. We believe that research into these issues will support the rejection of the genetic paradigm of psychiatric disorders and will give grounds for an alternative paradigm that emphasizes the role of familial, social, cultural, and political influences.[72]

In a 2000 article titled "Three Laws of Behavioral Genetics and What They Mean," behavioral geneticist Eric Turkheimer concluded, mainly on the basis of twin studies, that "all human behavioral traits are heritable."[73] At that time, behavioral genetics and psychiatric genetics researchers believed that the completion of the Human Genome Project would rapidly lead to gene discoveries.[74] So did Turkheimer, who wrote that "behavior geneticists anticipate vindication" by the discovery of genes causing behavioral variation. On the other hand, wrote Turkheimer, "Critics of behavior genetics expect the opposite, pointing to the repeated failures to replicate associations between genes and behavior as evidence of the shaky theoretical underpinnings of which they have so long complained."[75] A dozen years later the critics indeed appear to have been vindicated, and the real problem may well be, as Turkheimer described it, the "shaky theoretical underpinnings" provided by genetic theories based on family, twin, and adoption studies.

CHAPTER NINE

Assessing Genes as Causes of Human Disease in a Multicausal World

CARL F. CRANOR

When my home water system is working properly, a faucet is turned on, and the water pressure drops below 25 pounds, the pressure pump turns on. When the pump builds up pressure, it turns off at 50 pounds. Thus one might think that the cause of the pump turning on and off is the change in water pressure. Although we are likely to say this and to attribute causation to a drop in water pressure, this is an oversimplification. For the pump to switch on, many components and conditions must be present and functioning well. The electricity and the breaker switch must be on, the pressure switch must be working properly, the wiring must be sound, the bladder in the pressure tank must be intact and holding water with integrity, the check valve from the holding tank to the pump must hold water pressure in the pump, and the pump impeller must be working properly, among other things. A drop in pressure is just one causal component or causal factor contributing to the pump's operating; it is a necessary factor among a set of nonredundant other factors and conditions that together are sufficient to make the pump run. If the water pressure dropped below 25 pounds, but the electricity was off, the wiring was broken, the pump was defective, or there was a general electrical blackout, it would not run. In fact, if the pressure were to drop below this level and the pump did not work, a repairman would begin to diagnose its failure by looking at the necessary components that are part of the total set of conditions needed for its proper functioning.

Similarly, if the pump were turning on too frequently, one would suspect that at least one necessary component of the set sufficient for it to function properly was missing or not functioning as it should for a well-regulated water pressure system. Recently my pump began running almost continuously even though no faucets were on to cause the pressure to drop; it dropped for some other reason. This put the pump at risk of burning out. A repairman would examine the set of conditions sufficient for its functioning well in order to determine which elements were defective. For example, is the bladder in the pressure tank intact? Is the check valve partially open when it is supposed to be closed, causing the pressure to drop? Is the pressure switch functioning properly?

This is not a discourse on properly functioning pumps. However, biological systems are somewhat similar to pumps, although they are much more complicated. When the biological systems of humans or other species are functioning properly, what in the normal course of things are the components that make this happen? If a biological system is malfunctioning because of diseases, dysfunctions, or disorders, what contributed to its malfunction?

An Intuitive Causal Model

The generic causal model for pumps is also appropriate to human and other biological systems. On this model a causal factor for an outcome is a necessary but nonredundant element of a set of conditions sufficient to produce the outcome. I shall call it the "NESS" (Necessary Element of a Set of conditions Sufficient for an outcome) model. Focus on poorly functioning biological systems. If a biological system fails to function properly, what factor(s), condition(s), or event(s) that occurred (or failed to occur) produced the malfunction in the form of disease, dysfunction, or death? Which conditions or causal factors that are sufficient for healthy functioning have failed? A comparatively self-contained and well-understood (at least by pump experts) water-pressure system is considerably simpler than a biological system. A pressure-pump system has few component parts, and they are relatively easily located and comparatively easy to test to see whether they are functioning properly.

The causal model I have in mind is a very commonsense one that applies to mechanical pumps, light switches, and biology. As Michael Scriven puts it, a cause of an event is "a non-redundant [and functioning] member of some set of conditions which is jointly sufficient for the effect . . ." selected from a complex set of conditions that together are sufficient to produce the consequence. His more elaborate summary is the following:

> Perhaps one might sum up this account of causal explanation in the following way. We begin with a *context* which determines (a) the *type* of factor amongst which we are hunting a cause, e.g. physiological or ecological, and (b) the *kind* of factor within that type, e.g. childhood variation, proximate cause, whatever accounts for the contrast with what happens in a certain other family of cases (the "contrast class"), etc. We also begin with a book of basic causal connections, established or confirmed by direct observation . . . or by inference from theory: if the effect is *E,* we can turn to the page labeled "Possible causes of E, type . . . kind. . . ." Our task is a kind of maze-solving or path-finding one. We have to find a "permissible" route from some cause to the given effect. (The historian and detective best exemplify this enterprise.) The whole task is a pattern-recognition task or a complex of them, not unlike a chess-problem: we are trying to find a configuration of events which meets certain constraints represented by the circumstances and modus operandi and possible cause/possible effect data.[1]

In Hart and Honoré's words, causal factors of an event are "necessary elements in a set of conditions generally connected through intermediate stages" with the event.[2] Causal factors may fall into any of several categories: actions, events, states of affairs, failures to act, background conditions, the nonoccurrence of an event, and the nonoccurrence of a condition. This commonsense notion of causation has been followed by other philosophers of science, epidemiologists, legal scholars, philosophers of law, and numerous others.[3] In each case the search for causal understanding is something of a pathfinding or detective explanation of what occurred and, with the best explanations, why it did so. This intuitive, pragmatic model of causation has been the topic of some discussion and refinement, but reviewing these would take us too far afield and not assist this discussion.

Finding one or several causal factors that prevent a pump from turning off properly is a comparatively easy project, but even that can take some time and have complications. The pertinent experts on biology are much more in the explanatory dark about causes. The number of component parts of well- or poorly operating biological systems that function invisibly and beyond current understanding is very large. Major aspects of a healthy or, in contrast, a defective biological system may be known, but others are simply not understood. If a biological system is not working properly—muscles are inflamed when usually they are not, lungs have cancer when healthy lungs do not, breasts or prostate glands are cancerous when normally they are not—what conditions or factors that are ordinarily present for normal, healthy functioning are missing or not doing their jobs properly? For many adverse conditions, scientists simply do not understand them, but we have a term for them: they are called "idiopathic" causes.

In attributing causation to an event or outcome, there are a number of motivations that might drive the search. Sometimes, ordinary people or scientists seek to explain the normal course of events: the movement of planets around the sun, the precession of Mercury, or the development of a healthy immune system. Other causal inquiries may seek to understand departures from normal events, causal dysfunctions, as it were: especially heavy rains in dry Southern California or California winters with abnormally low rainfall, the crash of aircraft when the overwhelming number of commercial airplanes fly quite safely, or children born with extreme and visible birth defects when most children are not born with such features. Feinberg usefully calls these considerations "lantern" criteria for shining light on and understanding something.[4]

Sometimes we seek more than mere understanding and aim to identify which of the numerous causal factors are controllable features of complex events; call these "engineering" or "handle" criteria.[5] When rivers overflow their banks because of too much rain (which we cannot control, but which explains the flood), we might look for factors that can be engineered so that towns and people are not harmed or we can reduce their risks. Or consider another example: it turns out according to the most recent science that all or virtually all childhood leukemias begin in the womb with a translocation of two chromosomes. For these translocations to result in the appearance of disease, there typically has to be a second "hit," now thought for the earliest childhood cancers to consist in a serious infection for which the body is not prepared. Of the many factors that contribute to childhood leukemias, which should be picked out as the most easily controlled and manipulated in order to prevent the disease? Researchers believe that environmental contamination from toxicants or perhaps infections cause the initial translocation of chromosomes, the first step in the disease. However, it might be possible for parents to engineer avoidance of the second hit by exposing their children to various minor childhood diseases early in life by sending them to preschool. This way their immune systems are gradually prepared to respond, rather than facing a major disease at a somewhat later time, which might trigger an immune response that constitutes the second hit in leukemia. Understanding some major causal contributors to childhood leukemia permits recommendations about which factors might be manipulated to prevent the disease. Trying to correct a chromosomal translocation seems an unlikely candidate. Of course, avoiding the kinds of exposures that seem likely to cause the chromosomal translocation in the first place by good toxicity testing of chemical substances before they enter commerce would be the best "engineered" outcome if Congress changed current laws and public health agencies could control exposures.[6]

For legal or moral purposes (or perhaps for metaphorical variations on them), observers might pick out a particular causal component for assigning responsibility or blame to an event that they regard as especially untoward. The law, for instance, makes a place for one causal factor in complex causes by permitting homeowners whose houses were burned by the conjunction of a prairie fire of unknown origin and one set by a passing steam engine by emitting embers to bring a legal action against the railroad if the second fire made a "material" contribution to burning down the house.[7] The purpose served is to provide some legal recourse where otherwise there would not have been any because of the multiple contributors to burning down the house. Although both the train-set and naturally caused fires contributed to burning down the house, the train's fire, while material, might have been a minor contribution to the overall event.

The NESS model of causation in a multicausal world opens up the possibility of making causal attributions for different purposes. Indeed, causal ascriptions are often based on different interests. Thus we should use causal language with some care. We have just seen how causal attributions might be motivated by understanding, engineering, blaming, or assigning legal responsibility. Moreover, understanding divides into various subcategories—we can seek to understand phenomena for social, biological, or other purposes. Which factors we select from the set sufficient to produce the event depend on the context and our interests.

Consider a broken window. If slamming a door with eight glass panes in it breaks one of them, several different causal inquiries might interest us. If the simple inquiry is why the glass broke, the door's slamming may be a sufficient explanation. If we are interested in why one specific pane rather than the seven others broke, we might ask what distinguished the broken pane from those that did not break. What is different about this pane? Was it the fragility of this pane, its location in the door, or the framing around the glass? If we are interested in why the door slammed in the first place when it rarely does, we might look to the wind or Sam's anger, or, further back in causal history and for a more psychological explanation, we might review Sam's relationship with his father and how they interacted just before Sam slammed the door. How far back one goes in seeking contributing factors that break the window is a further question. It may not be particularly revealing or interesting in most cases that Sam's relationship to his father led to his anger and slamming of the door.

Causal ascriptions have some plasticity and are dependent on the components of a sufficient set of conditions for an event, outcome, disease, or dysfunction (to the extent these are known), on context, and to some extent on the interests of those making causal attributions. But we should scrutinize causal claims because there are some broad constraints on their use. One might think of the first aspect of this as the "causal factual base"—just what are the factors that are, or are likely to be, sufficient to produce a disease or dysfunction? Understanding this would be one of the main scientific tasks. Unfortunately, very often many of these components are simply unknown. Once there is some understanding of the causal factual base, what features of it seem critical for avoiding adverse outcomes? Even if several are known, individuals or various groups may assign causation for their own purposes that may or may not be related to a more impartial assessment of causal contributions. Thus, just as with a slamming door shattering a glass pane, to what predecessor factor do we call attention for what purposes? We should be especially sensitive to this in assessing genes as causes of diseases or dysfunctions. Causal contributions also need not be simultaneous. For example, animal studies have shown that mice exposed to one pesticide in utero do not develop Parkinson-like symptoms, but when they are exposed to a second pesticide as adults, they exhibit signs of Parkinson's disease.[8]

The Causal Role of Genes

Phenylketonuria is one disease or disorder that appears to have a major genetic component, but even here background conditions shape how significant it is biologically. "Phenylketonuria (PKU) is a rare condition in which a baby is born without the ability to properly break down an amino acid called phenylalanine," and this deficiency is traceable to a recessive trait in the genes. Because this amino acid does not break down, over time it builds up in the body and causes mental retardation, brain damage, and other problems. Children with this genetic deficiency typically must avoid foods that contain phenylalanine, including milk, eggs, and other common foods.[9] Although this diet sounds simple enough, apparently it is not an easy regimen for the affected person or his or her family, especially if it must continue for a lifetime.

Even though PKU seems like a comparatively clear instance of a disorder traceable to an unusual and defective gene, Kenneth Rothman notes that there is a substantial environmental factor, namely, diet. If we lived in a world in which most people ate a quite different diet, the defective gene might not receive mention, and the disorder might even be nonexistent. The disorder is a function of both a certain defective gene and a diet high in a particular amino acid. Without that amino acid in the diet, there would be no PKU even if all or most of the population had the defective gene. Thus in a statistical sense a gene has a "strong" causal role to play in a population because in a large proportion of cases of PKU (perhaps all), given a typical human diet, that gene (or perhaps a similar gene) is present. Similarly, smoking has a strong causal role in lung cancer because in a large proportion of cases smoking is a causal factor. If one smokes one pack of cigarettes per day, one is at a risk of lung cancer ten times higher than that of people who do not smoke. However, smoking one pack per day will not necessarily cause lung cancer; only about 10 percent of such smokers contract the disease (but they may contract other cancers or heart disease). In contrast, if the entire population stopped smoking, some other factor, such as radon, might be considered a strong contributor to lung cancer because statistically it would have a greater causal role to play in lung cancer cases than smoking would if there were no smokers.[10]

Thus Rothman argues that whether a factor has a strong or weak causal role to play in bringing about a disease is not necessarily a stable fact about the world but depends on the presence of other conditions or factors that function with the chosen causal factor, for example, smoking, as a sufficient set to produce the adverse outcome. As long as the other conditions needed for lung cancer remain constant, smoking will be fingered as a major causal factor or even as "the" cause of lung cancer. However, if the prevalence of smoking were to change—to no smokers—something else would be a stronger causal factor for lung cancer; he suggests that a plausible candidate might be radon.

Current knowledge about breast cancer presents a more complicated picture of the contributions of genes and the environment. This serious disease affects about one in eight women over a lifetime. About 5 percent of women with breast cancer can trace their disease to an inherited defective tumor- suppressor gene, although not all women with the gene necessarily contract the disease.[11] Tumor suppressors produce proteins that reduce or prevent the division of cancer cells that are dividing. If this cell division cannot be inhibited, the cancer cells are uncontrolled, proliferate, and lead to tumors.[12]

What contributes to the other 95 percent of breast cancer cases? Andreas Kortenkamp notes "avoidable contributions" that include "work place exposures, food contaminants, pharmaceuticals, chemicals in consumer products, air, water and soil, and physical factors such as radiation." About one-third of breast cancers in identical twins are ascribable to genetic factors, and two-thirds result from avoidable environmental contributors, not obviously from genes.[13]

Natural or synthetic estrogens are postulated as strong contributors. The more estrogen to which women are exposed, the greater their chances of breast cancer. Women who give birth later in life increase their risks of breast cancer, while those who have children earlier reduce their chances. Late menopause (continuing estrogen cycles longer in life) and earlier puberty (starting estrogen cycles earlier) both increase breast cancer rates, while the opposite of each lowers the risks (less estrogen in women's bodies over a lifetime).[14]

Female children who receive higher exposures to their mothers' estrogens in utero because they are either identical or fraternal twins have higher breast cancer rates later. And exposure to the synthetic estrogen diethylstilbestrol (DES), which caused unexpected instances of vaginal cancer when these children were about 20 years of age, also increases their breast cancer rates in middle age. Data from estrogen-based hormone replacement therapy further supports the theory that greater exposure to estrogen increases breast cancer rates.[15]

Overall, in only a small percentage of cases are genes notable contributors to breast cancer. When a woman has deleterious tumor-suppressor genes (named *BRCA1* and *BRCA2*), she is at a greatly increased risk of breast cancer. However, researchers otherwise largely point to nongenetic factors for elevated breast cancer rates. Of course, even when a defective tumor-suppressor gene is a substantial contributor, many other aspects of a woman's biology must be part of the causal path ending in cancer. That is, the gene is not the only biological event leading to cancer; it is just a particularly important one is a small number of cases.[16]

For both 5 percent of breast cancer cases and perhaps 100 percent of PKU cases, genetic contributions are strong contributors to disease, and calling attention to this serves numerous purposes. Testing and finding the problematic gene can identify those at future risk. They may be able to take some preventive or precautionary steps with regard to the ultimate disease. Genetic treatment of either is a much more radical step that is beyond current medicine and perhaps should remain there (more on this later).

A further complication in the genetic causation story merits comment but not extended discussion. Some substances, such as DES or bisphenol A, can alter the expression of the genetic code without altering the genetic sequence itself. The idea is that although the genetic sequence in a person's body does not change, environmental influences can cause "altered gene expression or altered protein regulation associated with altered cell production and cell differentiation that are involved in the interactions between cell types and the establishment of cell lineages."[17] As a result, the genes do not express themselves as they normally would or express themselves inopportunely, and this ultimately leads to dysfunction or disease. Thus diseases can arise because an exogenous agent modifies how a gene expresses itself. In experimental animal studies researchers expose the animals to chemicals that alter how one or more genes express themselves, and this results in disease. To what does one ascribe the causation? Because one or more external exposures trigger how the gene functions, causation is attributed to the toxic exposure.

Whether genes are causes or make notable contributions to disease are questions in need of understanding. Both advocates and critics of genes in a causal role often seem tempted to distort causation in order to make their points. Indeed, the topic of genes as causes is one part of the intellectual discussion over which there are disputes. This idea can take a central role in considerations about various research strategies, as well as therapeutic options. To an outsider it often appears that discussions focus on genetic causation rather than on the pros and cons of different research strategies or of various therapeutic strategies. I argue later that this focus should shift. Understanding causal notions against the background of more ordinary ideas of causation seems an important prerequisite for considering broader research and therapy issues.

The Temptation to Overemphasize the Causal Role of Genes

Now that the promise of the Human Genome Project has become reality, advocates of these various genetic therapies often seem to overemphasize the causal role of genes in diseases and in social behavior. P. A. Baird argues that physicians should recognize the "internal causes" of diseases and develop a new model for them. The "manifestations" of disease should be deemphasized, and the causes should receive greater attention. The results of this for diagnosing and perhaps treating diseases seem clear. As he puts it, "Genetics will increasingly enable us to interfere earlier in the cascade of events leading to overt disease and clinical manifestations." He further suggests that one level of strategy would be to identify genetic markers so that "all single gene disorders" can be detected and a family may take this into account in considering reproduction choices (presumably including abortion). Similarly, people who have the gene might take some appropriate action (only partially specified) "to avoid the complications" of the disease. Such action could include treatment by pharmaceuticals or, as with PKU, changes in diet.[18] Some courses of action would be one step up from merely making diseased people comfortable because the source of the disease would be understood. He further argues that attacking a disease at its source might well be much less expensive and more efficient than addressing manifestations later (here one might think of vaccinating children against measles rather than comforting and treating them once they contract the disease).

His quick assessment is not easy to accept, however. Moreover, a proper analysis would include not only the relative ease of addressing the disease but also the background superstructure, machines, and experts needed to identify genes easily and then treat the diseases they cause. Will it be possible to treat single-gene diseases in a manner that is as easy and inexpensive as vaccinations to prevent measles? This is an issue that would need detailed examination by scientists, economists, and philosophers to compare the pluses and minuses not only of treating the causes versus more traditional strategies, but also of the superstructure needed for each. Addressing this issue is beyond the topic of this chapter.

Some go further and urge a form of gene therapy, such as deleting, modifying, or augmenting the offending gene or substituting a healthy gene for it. Theodore Friedman claimed that research "has firmly established mutant genes as uniquely appropriate targets for therapy, for at least some genetic disorders." This would direct "treatment to the site of the defect itself—the mutant gene—rather than to secondary or pleiotropic effects of the mutant gene products." In presentations he has gone further to try to persuade audiences by his choice of terminology. Friedman has candidly acknowledged that he "calls something a genetic disease to [get us to] think of particular kinds of therapies. It's a kind of tool to convince people to focus on certain therapies. Nothing is simpler than treating a disease at its cause."[19] Although it may appear simpler, it could be quite expensive, greatly complicating research efforts, or lead to ignoring promising and inexpensive alternative therapies. Moreover, his choice to use language in this manner is misleading. Friedman asserts a conceptual claim, likely overemphasizing the causal contributions of genes, in the service of a research and treatment agenda.

At this point the analogy of the causal components of a water system seems helpful. To the extent that the biology of disease is more complex than the parts of a water system, with many components needed to produce the adverse effect, it may be less clear that trying to treat a gene or genes as some of the components is the best way to address the disease. Nonetheless, knowledge of predisposing genes might sometimes be helpful. The treatment of PKU has nothing to do with modifying or deleting the deficient gene or substituting a "healthy" gene for it. However, knowing that one has the PKU gene can lead to precautionary treatment: changing the foods whose proteins are harmful. For women with *BRCA1* or *BRCA2*, knowing that they have a gene that predisposes them to breast cancer gives them reasons for precautionary watchfulness, for example, more frequent breast examinations, perhaps beginning at an earlier time in their lives than for women without the gene. Knowledge of the gene's contribution provides one with warnings to be watchful or take precautionary actions, at least at this stage of knowledge. However, early warnings of some deleterious genes can pose difficult personal decisions even to know about such possible contributions.

Underestimating the Causal Role of Genes

Some commentators tend to underestimate the causal role of genes or even to deny that genes make causal contributions to disorders in order to make a normative point or perhaps in an exaggerated attempt to correct or refocus what they see as misguided discussions, such as those I have just considered. Thus some commentators deny that genes "control" or "program" phenotypic disease and challenge the biotechnological dream of manipulating genes in order to control the quality of infants born "because [such a dream] exaggerates the control genes exert over metabolism and development."[20] They are concerned that the "power of the particles [genes] is exaggerated, while the contributions of the systems in which they operate are undervalued or ignored."[21] Given the NESS model of causation, these points have much to recommend them. The view emphasizes the complexity of the biological system in which many components, conditions, or even nonevents contribute to a given disorder, and it hesitates to attribute causation to genes. Ruth Hubbard in another article describes many of the numerous features of the biological pathways from genes to diseases.[22]

It appears that genes rarely "control" or "program" disease, just as no single component controls or programs the operation of my water system. In addition, of course, genes are merely one part of a complex system with numerous components that must function together to produce health or disease, just as my water system has many components that must work together to successfully deliver water under pressure to the house. Critics of genes as causes have another modest point on their side. Even though a single gene is related to sickle-cell anemia, some people become seriously ill from the disease, while others reveal only "minor symptoms later in life."[23] Something is different in the causal path between two people, each of whom has the sickle-cell gene, but one is very sick and the other much less so. Genes cannot fully explain the disease spectrum and its different manifestations in two such people; something else that is part of the causal path must account for the difference. The biology of the gene or the supporting biological pathways is different. I concur that the "power of the particles [genes] is exaggerated, while the contributions of the systems in which they operate are undervalued or ignored."

There is considerable good sense, pragmatism, and truth in illuminating biological complexity, but this too can be overstated. Given the NESS model of causation, there are good reasons in some contexts to acknowledge that a gene is "the" or "a" cause of a disease. This can be done without exaggeration or strong metaphysical assertions about genes controlling or programming the diseases. Thus one can recognize that a human disease is the "outcome" of a complex set of causal conditions and at the same time draw attention to one of those contingencies for understanding or shedding light on the process. In this it does not differ from the failure of my pump system or smoking causing lung cancer or *BRCA1* contributing to breast cancer. A complex set of conditions, contingencies, events, or nonfunctioning components does not detract from one of them being considered a or the important contributor for a particular event compared with more common disease processes or with ordinary healthy biology. It is important that the biological complexity not be hidden. However, if an unusual antecedent in an otherwise normal biology and disease is a defective gene that most people do not have, this can be quite illuminating. What researchers do with such facts and how we treat them socially are quite different. Our understanding of the phenomena is assisted when we can correctly draw attention to factors that are normally absent, although they are present in a particular instance, or normally present but absent this time, and that "make a difference." It appears likely that single-gene diseases that are a comparatively direct product of a deleterious gene are rare. Thus, although they might be representative of "genetic" diseases, it appears that they are not frequent, as Hubbard and Wald assert.[24] Single genes may be seen as causes even in a multicausal process, even if they are rare.

What role this understanding should have for broader social and therapeutic purposes is a separate issue. Decisions about research and therapy should be carefully considered, since undue attention to a search for single genes for diseases may well ignore some of the complexity of the phenomena and "close off possibilities of other [areas of research]." Ian Hacking explains:

> A science can develop along many possible paths, bringing into being different phenomena. Phenomena that we create on one possible historical path might not be created on another historical path, say because we have invented neither the instruments nor the theory of the instruments through which we could recognize or control them. Moreover, any path that we do follow has its own momentum. Experimental techniques and instruments, when they produce what are taken to be stable results, themselves suggest further steps to take by analogy. Had we not started out on a path, we would not have created later on what we do in fact create.[25]

Hacking points to research risks from choices researchers might make and strategies they might adopt, and to how those could be seen as mistakes in retrospect. Similar points could be made about directions for therapies. Hubbard and others are quite concerned about this, but their point could be put more modestly. In a biological world that is clearly multicausal, as a field develops, researchers should not develop tunnel vision with a focus only on new technologies but should retain some sensitivity to wider issues. One should not blind oneself to the biological richness that a more complete causal picture offers because of a too-narrow focus on one feature of causation. Those who oversimplify by underemphasizing genes as causes importantly remind us of some of these points. The fact that there may be some single-gene-caused diseases should not blind us to the much broader complexity within which such genes are embedded, both for understanding and therapeutic purposes. Moreover, even if a single gene has a major role in a disease, choices of therapeutic strategies are separate issues. An appropriate and careful assessment of the pluses and minuses of replacing, manipulating, altering, repairing, or inserting genes to address single-gene-caused diseases should be undertaken before pursuing such strategies. Such an analysis should be conducted not only for developing individual treatments but also for broader institutional responses to diseases, especially when complex infrastructures and large amounts of money will be needed.

Overascribing causal influence to genes is an even greater risk for so-called polygenetic diseases or disorders—adverse effects that are traceable to two or more, perhaps a myriad of, genes. In such cases a focus on "a" or "the" genetic cause of adverse effects would be even more misleading than it might be for single-gene-caused diseases. This could even more significantly distort both understanding and attempts at therapies. Moreover, it seems exceedingly difficult to identify which particular genes out of a larger number play an important causal role in producing disease.

This last point suggests that skeptics about gene-caused diseases may have worried that even if there are some single-gene-caused diseases, other scientists or physicians may be too quick to jump from understanding the sources of diseases to engineering solutions to them or blaming genes for the disorders. Although "engineering" or "blaming" can be appropriate uses of causal language, they do not follow logically from understanding the source of a disease, and each requires a separate analysis for an appropriate application of these concepts.

As more of an aside, some researchers are searching for genetic causes of congeniality, violence, running speed, "undiscovered genes for [addiction to] pornography," and other conditions.[26] After one considers genetic contributions to disease, such claims seem far-fetched on the plausibility scale. Indeed, there is no doubt some genetic contribution, however tiny, in some of these areas, but is it any more than the contribution of genes, as Rothman puts it, to every disease? He notes that every disease has both an environmental and a genetic factor, and thus genetic contributions in most cases will have little more significance than the need for a functioning electrical system is for my pump to work properly. It would seem that in order to discover that a gene makes a notable contribution, epidemiologists would have to do a careful case-control study in which they examined everyone with the desirable or undesirable trait and then tried to find all biological contributions, including genetic ones, together with all personal, social, and other environmental contributions to the phenotypic trait, such as decisions individuals make, efforts they expend, and social pressures on them. That is what a good epidemiologist would do in trying to isolate the causal contributions to disease on the basis of a group of people who are sick with a particular disease. He or she would carefully examine the history of sick individuals, compare them with a comparable group of persons who are not ill, and then try to infer what conditions contributed to illness. The quick stories one reads about genes for violence, sociability, and running speed seem to lack this kind of detailed analysis. Moreover, even if such desirable or undesirable end points have some genetic components, they are likely multigenetic and numerous social and personal contributions. Even optimistic genetic researchers seem to doubt that they will have much success identifying major genes that contribute to a disease even when they know that it is a multigene disorder.

Genes are and always have been an integral part of biology. They are a component of the factual base of causal pathways leading to either health or disease. (How much they contribute to social behavior is a much more difficult issue.) New genetic sequencing technologies are now an important part of the research landscape. They may have led researchers to focus too much attention on genes in some cases or, at the other extreme, too little attention on genes. How substantial a contribution genes make to causing disease or keeping an organism healthy is a much more vexed issue. Although scientists now have elaborate sequencing equipment and statistical models for finding genes that may contribute to disease, too great a focus on them seems likely to distort understanding (shining light on diseases), therapy (engineering them), and even blaming them. It appears much better to take a more pragmatic approach toward either searching or not searching for genes in each of these areas. What research strategies seem most likely to yield understanding of a disease process? What are the costs in understanding, time, scientific resources, and equipment of one research approach versus another? Similarly, for therapies, what are major and minor costs and benefits of pursuing variations on gene therapies versus intervening in causal pathways at other points in order to bring about improvements in health? To what extent does evidence support or not support a prominent role for genes in diseases? Are they more like *BRCA1*, which is an important component of breast cancer, or like the myriad of genes that may influence the development of autism? As an intellectual matter these messier and more complex strategies should receive attention instead of overemphasizing or underemphasizing the causal role of genes.

In the end, this chapter is a resounding argument for pragmatism in understanding the causal role of genes in disease, treating the diseases, and blaming one part of the causal basis of disease rather than another. Scientists and the public alike should not be quick either to overemphasize or to underemphasize genetic contributions to diseases.

CHAPTER TEN

Autism

From Static Genetic Brain Defect to Dynamic Gene-Environment-Modulated Pathophysiology

MARTHA R. HERBERT

Autism Status Quo: Genes, Brain, Behavior, and Hopelessness

To put the myth of genetic reductionism bluntly: DNA makes the rules; everything else obeys. It follows, then, that a complete understanding of living systems will ultimately be achieved—in fact, can only be achieved—by completing our genetic understanding.

Autism would appear to be a good, although disheartening, example. Autism has been considered to be a genetically hardwired neurodevelopmental disorder and is defined in the American Psychiatric Association's *Diagnostic and Statistical Manual of Mental Disorders* by a set of behavioral dysfunctions: impaired social interaction, impaired communication, and restricted and repetitive behavior.[1] In plain terms, autistic people tend not to "get" how to interact with other people, are not good at figuring out what someone else is thinking or feeling, have all sorts of problems saying what they mean or understanding what another person means, and often do weird things like stacking up all the cans in the house over and over and over and having meltdowns when someone interrupts them. In fact, these behaviors occur in different combinations, proportions, and degrees, so that autism is now thought of as a "spectrum" of disorders.

This description, though standard, does not begin to convey how bad autism can be for some people and their families. Nor does it convey either the brilliance of many people with autism, even if they cannot talk, or the multisystem pathophysiology (prominently immune, gastrointestinal, and metabolic/mitochondrial dysfunctions) present in large numbers of persons with with autism.

Early on, autism was attributed to faulty parenting, but for some decades now the blame has been placed on genes and brain defects. The evidence to support this has been inferential but has seemed compelling. Twin studies indicated that identical twins are far more likely both to have autism than fraternal twins, which seems clearly to indicate that autism is the result of defects in a fetus's genes. Changes in the brain have been identified that have been considered "defects" and attributed to prenatal genetic causes, although the specific genes and mechanisms by which these changes could be created have not been delineated or, if they have been delineated, have applied to only a tiny minority of individuals with autism. Overall the belief has been that broken genes cause the autistic child to be born with a defective brain that, in turn, produces a lifetime of dysfunctional behavior. It makes sense: broken genes, broken brain, broken behavior—a life sentence, hopeless. Like a chromosome-abnormality-based condition such as Down syndrome, it is all set in stone before you are even born, and you are stuck with it.

That has been the theory, and that is certainly the way it can look when one is facing autism head-on. In a typical case, a 4-year-old boy named Jeff was diagnosed with autism. The doctor explained to his parents that there was no way to undo the damage that Jeff's brain had been born with, any more than one could turn the limbs of a Thalidomide baby into normally developed arms and legs. "At some point," the doctor warned them, "you'll need to put your boy in an institution. You simply won't be able to take care of him." He then handed them a brochure for an institution where he had already put Jeff on the waiting list.

Jeff's parents were shocked and, for a while, defiant. But eventually they came to the heartbreaking conclusion that the doctor had been right. Jeff was impossible. He constantly banged himself against the walls. He never slept through a night—and neither could his parents. He would spend hours fiddling with a plastic dish, just spinning it in noisy circles with an uncanny and irritating precision. For a while before his regression into autism he had at least been able to talk, but he lost that ability, and it never came back. Nothing ever got better. As Jeff's mother put it, "Jeff was screwed for life and so were we."

What is to be done? If it all begins with the genes, it seems clear that researchers need to find the genes that cause autism. It might then be possible to devise extremely precise molecular or genetic interventions that would undo the damage or at least provide a work-around. But after decades and many millions of dollars devoted to looking for the fundamental genetic explanation of autism, there has been no home run. Instead, things have gotten more complicated. Genetic testing has become much more sophisticated, and more (actually many more) candidate genes have been identified, but there has been little progress in therapeutics, and it remains to be seen how useful gene and molecular targeting of therapeutics will be in improving the level of function and quality of life of those with autism.

Anomalies Undermining the Genes-Brain-Behavior Model

Although genetics has increased in complexity, observations have been accumulating that autism involves an array of phenomena outside the genes-brain-behavior model.

Not a Static Prevalence. The incidence of autism by some accounts seems to have gone up tenfold or more since 1985.[2] Such things do happen when diseases become better known to the medical profession and to the public, but in this case the leap in incidence seems too high to be entirely due to a change in reporting. If autism is a genetic defect, why would it suddenly start occurring much more frequently and then keep doing so for twenty or more years? That is not how a genetic illness should behave. In spite of these numbers, a good, though finally diminishing, number of researchers do not believe that the increases are happening at all. Their refusal may have elements of circular reasoning: they attribute the statistical finds to the fact that autism has become much more widely known and is therefore much more commonly diagnosed than it ever had been, and they dismiss any possibility that there could be a real increase, on the apparent assumption that such an increase is impossible in a genetic disorder. But although many studies have shown that portions of the increase can be attributed to other things such as earlier age at diagnosis, changes in diagnostic criteria, or diagnostic substitution (i.e., labeling people autistic who previously would have been diagnosed with something else, such as mental retardation, attention-deficit/hyperactivity disorder, or epilepsy), no one has actually proved that the entire higher incidence is merely a reporting artifact and not at least in part a genuine increase.

In the past few years careful data reviews (including a review of every single reported case in California for an extended period) have parsed out cases that can be attributed to such other causes; they find that although a modest proportion can be attributed to diagnostic substitution, earlier diagnosis, and altered diagnostic criteria, from 40 to about 65 percent of the increase in numbers is an unexplained increase and could very well reflect environmental influences.[3] At this point one of the strongest arguments of those who claim that there is no increase appears to be that the science showing increases is not definitive.[4] The word "environment" is even starting to crop up more frequently in essays of geneticists. So the reported increase poses a public health challenge that, given the seriousness of the affliction, certainly deserves serious attention. Studying environmental influences is at long last being named as a fundable research priority.

Not Just Genes: Environmental Contributors. By now a growing list of environmental agents are implicated in increasing the risk of autism. There is evidence that proximity to pesticide application during pregnancy, proximity to freeways and other sources of air pollution, heavy-metal exposure, and maternal infections all contribute to increased risk.[5] A long list of further exposures is under consideration, including substances that can cause harm by their presence (e.g., flame retardants) or by their absence or deficiency (e.g., essential fatty acids [omega-3 fatty acids], which are pervasively low in the population, or vitamin D, commonly deficient, which is important to immune and endocrine function, as well as to DNA repair mechanisms, and is potentially pertinent to the de novo mutations being found in autism and now in other conditions as well).[6]

Not Just a Few High-Impact Genes: Hundreds of Mostly Lower-Impact Genes. Instead of finding a few genes, scientists have identified hundreds with possible influence. Even the small number of autistic children with identifiable genetic conditions include hundreds of different mutations, and the commonalities across these different genes that would produce autism are not yet clearly specified.[7] Meanwhile, the yield of academic investigations into autism genetics has been more thorough testing methods but very little therapeutics. Some drugs seem promising, but to date drugs with any demonstrated efficacy have only suppressed symptoms like aggression and have not improved core behavioral features of autism.

Not Just Inherited Genes: De Novo Mutations. Copy number variations (CNVs), which are alterations of the DNA of a genome that result in the cell having an abnormal number of copies of one or more sections of the DNA, have been found in autism, sometimes at a higher rate than in individuals without autism. A portion of these genetic changes are de novo variants not found in the parents of the affected individuals.[8]

Not Even Mainly Genes: Substantial Environmental Contribution. For years twin studies have been cited to prove the high heritability of autism. Concordance (shared diagnosis) figures of 90 percent of identical twins and 10 percent of fraternal twins were used to justify an intensive focus on genetics to the exclusion of environmental influence. In the largest twin study to date, lower monozygotic (identical twin) concordance and higher dizygotic (fraternal twin) concordance yielded a smaller gap between autism rates in identical as compared with fraternal twins, with 55 percent of the variance for strict autism and 58 percent for autism spectrum explained by shared environmental factors, with moderate genetic heritability of 37 to 38 percent.[9] The article's conclusion suggesting a greater role for environmental factors and a smaller role for genetics than heretofore presumed was hailed by some as a game changer, although its methodology infuriated many genetic researchers.

Not Just Brain Genes. For years geneticists looked only for "brain genes" in candidate regions of the chromosomes, but they may have been overly limiting themselves. Dan Campbell of Vanderbilt University, along with Pat Levitt, a neuroscientist now at the University of Southern California, identified the *MET* gene as significant in many people with autism and found that its impacts ranged from the brain to the immune system to the gut and beyond and that it was environmentally sensitive.[10]

Not Just Local, Modular Brain Disturbances: Whole-Brain Involvement. One of the most replicated findings in the brains of individuals with autism has no immediately obvious relationship to the core defining behaviors of autism. Autism brain research had hunted, not very fruitfully, for specific brain areas that might explain the "aberrant behaviors" of autism, but the field was surprised by an increasingly replicated finding of larger brains, identified by postmortem brain weight, head circumference, and MRI measures.[11] My MRI brain-volume analyses, as well as those of others, revealed that the extra brain volume was distributed throughout the brain rather than being centered in the regions of the brain known to be associated with the language and behavior capabilities affected by autism.[12]

Not Just Prenatal. Researchers have found that autistic children are not simply born with enlarged brains; they develop this enlargement postnatally.[13] This contradicts the previous orthodoxy that autism is caused by alterations to brain development in utero. There is a massive growth spurt for the first two years after birth, with some parts of the brain (like prefrontal and cerebellar white matter) being as much as a third bigger in some children with autism.[14] This is a huge size difference in an organ that is tightly spatially confined by the skull. Then the growth rate slows down, and by adolescence the brains of people with autism are, on average, slightly smaller than normal. In retrospect, the long and strongly held belief that autism was based on prenatal brain changes was supported only by a modest number of observations in postmortem tissue samples of older individuals. The interpretation that these cellular changes could have occurred only prenatally, and the grip it held on the field for so long, went far beyond what the data could truly support. Alternative explanations of observed cellular changes have been emerging that rest on postnatal (and often toxic) influences, as I will discuss later.

Not Necessarily Present at Birth. Autism was initially assumed to be present from birth. Parents' reports that their child "regressed" were dismissed as a function of the parents not knowing much about child development. However, regression is now an established phenomenon that occurs in many, though not all, children with autism, although it may occur differently—rapidly or gradually—in different children. Videotapes of children before their regression were rigorously analyzed, and the complete or near-complete absence of abnormalities before regression was verified in a number of cases.[15] Although it can be argued that regression is just the late expression of inborn genetic tendencies, it is also possible that postnatal influences may have contributed to the regression; this cannot be excluded a priori.

Not Just Behavior. In addition to the behaviors at the core of the psychiatric definition of autism, a plethora of other neurological problems are very common. These include seizures (present in 7 to 46 percent), brain epileptiform activity (present in possibly the large majority), significant sensory threshold and sensory integration problems, coordination and gait problems, oral motor dyspraxia impacting swallowing and speech, dyspraxia associated with other activities such as hand use, disturbed sleep, major alterations in speech prosody, and in some cases catatonia. These phenomena suggest underlying alterations in the brain's substrate that go substantially beyond the behavioral problems that are most obvious to our social intelligence.

Not Just the Brain. Many people with autism have chronic somatic medical problems that are not included in autism's definition.[16] It has been widely observed clinically that autistic people suffer more than their share of gastrointestinal disorders, including intractable diarrhea, severe constipation, gastroesophageal reflux and esophageal ulceration, and deficient pancreatic production of digestive enzymes.[17] There are also problems that interface between the gastrointestinal system and behavior, such as lack of toilet training and extreme "picky eater" behavior—narrow, nutritionally inadequate eating preferences (generally beige-colored foods like wheat and dairy) with food refusal. This has variously been assessed as being coincidental, not coincidental, and vanishingly unlikely to be coincidental.[18] Some of these disorders are quite painful or interfere in other ways with normal life, such as the painful gastroesophageal reflux and esophagitis that can disrupt sleep and lead to self-injurious behavior, particularly in nonverbal individuals who cannot use words to get help for their pain (although even highly verbal individuals with autism may have serious problems locating the specific source of pain, presumably because of sensory processing issues). An array of immune problems (allergies, recurrent infections, autoimmunity, and mothers with autoimmunity) are common as well. Yet until recently discussion of these nonbehavioral concerns was frowned on, and the corpus of research on these issues is only beginning to develop in earnest.

Not Just Deficit: Giftedness and High Intelligence. Although the assumption that autism is predominantly associated with mental retardation has been taken as a verity, it is now being called into question. First, the quality of the studies that claimed to ascertain this has been criticized.[19] Second, it appears that IQ tests relying on verbal instructions underestimate IQ—in one study the Wechsler scales of intelligence scored predominantly in the mental retardation range, whereas the Raven's Progressive Matrices (a strong measure of fluid intelligence) did not.[20] Many people with autism in fact demonstrate great creative skills and brilliance even when they cannot talk. Increasing use of augmentative communication devices, on the order of a social movement within the autism community, is yielding unexpectedly articulate communication from a growing number of nonverbal autistic individuals. The obstacle with speech may well be dyspraxia (impaired ability to initiate or coordinate verbal production) rather than a deficit of language capacity itself. All of this raises questions about the pertinence of the many genetic studies that hunt for genes associated with mental retardation in their study of autism. In addition, the behavioral meltdowns of nonverbal individuals can sometimes be due to sheer boredom and frustration with a primitive, repetitive curriculum far below their capabilities. In her memoir *Strange Son,* Portia Iversen relates that when her 9-year-old nonverbal son was given an assistive communication device, it was discovered that he could read and understand English and Hebrew and could do fourth-grade math, all without having received explicit instruction.[21] Others in similar positions can handle math or physics or write poetry.

Not a Life Sentence: Evidence of Remission and Recovery. The idea that autism is a static encephalopathy is being challenged by evidence that autism can remit. Some autistic children, for example, get a lot better when they have a fever.[22] They may make eye contact or even talk, which they had not been doing otherwise. This has been reported in significant numbers of autistic individuals, although the mechanisms underlying this phenomenon are still under investigation; hypotheses range from immune modulation to bioenergetic upregulation to altered neurotransmitter activity.[23] One mother told me that she was glad whenever her young son had a fever because he would become more communicative and, as she eloquently put it, she would "get to visit with her son." Even if such spontaneous remission is transient, it would be unsurprising behavior for a disease but is hard to explain for a condition arising from a brain whose incapacity is assumed to be permanent and unalterable. Yet this belief seems to inform medical advice given to parents; one parent told me, in tears, that her neurologist had found mild delayed myelination in her son's brain scan and had told her with authoritative definitiveness, "Your son will never learn," even though this grossly overstates the impact of this brain finding.

Beyond transient remission, there are also growing numbers of cases where substantial improvement in condition or even loss of autism diagnosis has been documented, usually after intensive treatment.[24] A small but growing body of literature is addressing what this means for how we think about autism. The National Institute of Mental Health is carrying out an autism remission study in which it is identifying such cases and doing exhaustive history and testing (neuropsychological studies, genetic studies, immune measures, brain studies, and others) to learn what may be different about these children as compared with those who do not remit or recover.[25] In summary, evidence is shifting the conception of autism from a genetically determined, static, lifelong brain encephalopathy to a multiply determined dynamic systems disturbance with chronic impacts on both brain and body.[26]

Dynamic Physiological Processes Implicated in Autism. The problems in the body beyond the brain have been a particularly challenging dimension to incorporate into the world of autism, in part because of the depth of the belief that autism is a brain or neurological disorder. The modular division of the body into distinct organ systems, each with its own textbook chapter, has difficulty linking body and brain. With the rise of systems biology, the pathways back and forth between brain and body are becoming elucidated in detail; they supersede the belief that the brain is protected from the body by a more or less impermeable blood-brain barrier and that the brain is immune privileged—that is, shielded from systemic immune problems.

A further barrier to considering the body's impact on the brain was the reaction to the work of Wakefield, who argued not only that there was a link between autism and vaccines but also that this link was mediated through the gastrointestinal system. For the better part of a decade any attempt to discuss gastrointestinal or immune issues with autism was construed as a support of Wakefield's vaccine hypothesis, and it was difficult to discuss, let alone get funding for, clinical or research observations about these problems.[27] One way around the essentially taboo character of somatic problems in autism was to treat them as coincidental symptoms. For example, one could talk about gut problems provided one made it clear that they did not cause the autism in the brain. Improvement after treatment of gut problems, which is often observed, would then be explained as a consequence of reduction of pain and discomfort, but not of any direct impact on core brain mechanisms generating autistic behaviors.

Systems biology is increasingly documenting phenomena that blur the boundaries across organ systems. It is becoming commonplace to discuss the impacts on the brain of immune cytokines, gut microflora, nutrition, and stress. This is helping dissolve the taboo in autism about considering body-brain relationships. Along with this more general shift in scientific framework, more specific documentation of disturbed systems biology in autism has accumulated. The following are some of the dimensions of this:

Immune Dysregulation. Hundreds of articles have documented immune abnormalities in autism, although not all have been of high quality.[28] Abnormalities include alterations in immune cytokines, presence of autoantibodies including antibrain antibodies, inflammation, and more. In 2005 Pardo and colleagues at Johns Hopkins University published evidence of innate immune activation in brain tissue.[29] In particular, they identified activated glial cells—astrocytes (or astroglia) and microglial cells, visible microscopically with special staining techniques but not macroscopically via brain imaging. Although the findings of this study electrified some segments of the autism world, they were shunned by others. More recently a number of other groups have replicated these neuropathological findings, and several studies of gene expression in brain tissue have supported the existence of this phenomenon as well.[30]

Mitochondrial Dysfunction. A role for mitochondria is being identified in a plethora of chronic diseases ranging from cancer and obesity to diabetes and neurodegenerative diseases. A higher rate of clear mitochondrial disorder in autism (about 5 percent) than in the general population and biochemical evidence of milder mitochondrial dysfunction in as many as one in three persons with autism have been identified in articles and meta-analyses.[31] The idea of mitochondrial dysfunction in autism has met with some consternation and avoidance, again to a significant degree because of the link to the vaccine controversy. In this case the link is particularly driven by the daughter of a neurologist in training at Johns Hopkins who developed autism after receiving nine vaccinations in one day; her family received a settlement from the U.S. Court of Federal Claims, which hears vaccine injury lawsuits. Officials argued that this was a rare occurrence, given that this child had a genetic vulnerability to mitochondrial disorder that was just triggered by the vaccines and would have happened soon enough anyway. More broadly troubling is the well-known exquisite vulnerability of mitochondria to a wide range of environmental insults.[32] The existence of debate about mitochondrial disorder versus dysfunction seems much more marked within the autism community than in other conditions where thousands of articles have been published about the occurrence of mitochondrial dysfunction even in the absence of established genetic causes of mitochondrial disorder.

Oxidative Stress. Oxidative stress is also being identified in many chronic illnesses. Free radicals are normally produced in small amounts by oxygen metabolism. These free radicals can damage DNA, molecules, cell membranes, and other structures, but under normal circumstances they are "quenched" by antioxidants. When the production of free radicals exceeds the body's antibiotic reserves, there is a buildup of oxidative stress. Various laboratory indicators of oxidative stress, as well as genetic vulnerabilities to this problem, have been identified in individuals with autism and also sometimes in their mothers.[33]

Methylation Disturbances. Disturbances of methylation have been identified in people with autism.[34] These disturbances potentially affect DNA methylation and hence epigenetics. They also affect other processes, such as neurotransmitter synthesis, cell membrane function, and silencing of viral genes. Associated with these disturbances are abnormalities in sulfur metabolism, which can affect the production of antioxidants and the elimination of toxicants.

Disturbed Gut Microbial Ecology. Data supporting disturbance of intestinal microbial ecology in autism have been accumulating. These include direct measurement (e.g., abnormal variants of *Clostridia,* presence of *Desulfovibrio* that can alter sulfur metabolism) and indirect evidence (such as animal models of autism-like behavior and brain changes with exposure to by-products of gut microbial metabolism).[35]

Hormonal Dysregulation. Indications of possible hormonal dysregulation include atypical growth patterns, abnormal patterns of autonomic arousal (both hypo- and hyperarousal), marked changes at puberty (e.g., sometimes seen are increases in aggression and severe premenstrual syndrome), sleep disturbances, hypothyroid, and seizures.

Active Pathophysiology, Genes, and Environment

All of these phenomena are not simply systems disturbances; they are active physiological processes, shifts in ongoing functional activities in the organism, that are by no means necessarily caused by genetics and are by no means necessarily related to hardwired brain changes. Immune problems are clearly not all genetic because the immune system mediates between organism and environment. Oxidative stress and mitochondrial dysfunction are known to be promoted by myriad environmental exposures. Genetic vulnerabilities to these problems are common in the general population, but when several occur in one individual, the likelihood of being autistic can increase greatly (in some cases manyfold). Even so, given the increased vulnerability to environmental impact conferred by these gene variants, the outcome of autism might be not so much caused as facilitated by such genes, with the degree of environmental exposure determining much of the risk and the severity of the outcome.

Active Pathophysiology and the Brain. Genetic research has focused on the synapse as the central locus of brain disturbance in autism spectrum disorders and has looked for synapse-impacting gene variants that might be at the root of autism. This fits the idea that genes, which are inborn, change the brain in a way that inevitably causes autism.

However, the active, dynamic pathophysiological processes listed above may expand that narrative: (1) they can affect the mechanisms and milieu of synaptic functioning, and (2) they have many triggers, not only genetic but also environmental, and so are not necessarily inborn and unalterable.

Impact on Synaptic Functioning. Immune disturbances are known to affect neuronal excitability, and pro-inflammatory cytokines can increase vulnerability to seizures, which are evidence of major synaptic dysfunction.[36] Mitochondrial dysfunction can be a consequence or a cause of seizures and may also alter the function of neurons and neuronal systems in more subtle ways.[37] Gut microbial by-products can have pharmacological and immunological impacts on the brain as well.[38]

Given these considerations, it can be argued that genes do not uniquely compromise synapses, but that environmental factors and their physiological consequences can also impact the molecular regulation of synapses and networks. But how important is that impact?

Could Active Pathophysiology Be Impairing Connectivity?

One of the core abnormalities in autism brain function appears to be altered, and usually decreased, connectivity and coordination among areas of the brain. By now there is a large literature that supports this.[39] The neurocognitive functions considered to be core features of autism involve complex information processing. Thus it would make sense that impairment in connectivity and coordination would lead to less nuanced function in these areas of complex processing.

From the point of view of the gene-brain-behavior model, showing the association of a gene variant with an alteration in connectivity would seem to imply that this gene variant causes the connectivity issue and therefore causes autism.[40] This would generally seem to imply that connectivity abnormalities, being genetic, are static, lifelong features of brain function in affected individuals.

However, a recent article by Narayanan and colleagues showed that brain connectivity could be increased in a matter of minutes after administering propranolol, a drug that reduces sympathetic nervous system activity (the “fight-or-flight” or stress part of the autonomic nervous system).[41] They theorized that sympathetic activity increases “noise” in the “signal-to-noise” ratio of the brain, and that reducing this noise allowed the brain to have greater signal—greater bandwidth for accessing more remote parts of its networks.[42] This finding opens the door to other types of nongenetic mediation of brain connectivity. In particular, brain excitotoxicity, a consequence or concomitant of many of the above-listed pathophysiological disturbances, may alter neuronal function and lead to altered network function.

Clinical phenomenology of ups and downs in some people with autism also suggests that the functional status of brain networks may be variable. In addition to transient improvements with fever, there may be marked improvements on a clear-fluids-only diet before medical procedures; moments of verbal lucidity in nonverbal individuals under conditions of emotional intensity; and striking though transient improvements in individuals receiving steroids for other reasons (e.g., asthma attacks) or other immune-modulating drugs or upon emerging from anesthesia. Children and adults under these circumstances may suddenly be socially interactive when they were previously remote, verbal when they were previously nonverbal, or spontaneously articulate when they were previously only repeating stock phrases. Conversely, some people with autism also show marked deteriorations with exposures to allergens, certain foods (e.g., gluten and dairy, or what some parents have called "pizza psychosis"), or various environmental agents. Thus the association of variability in function with environmental stimuli supports at least a degree of environmental modulation of brain function.

Does Active Pathophysiology Modulate Genetic Substrate, or Could It Be a Primary Cause of Brain Dysfunction?

If you believe the narrative that genes cause connectivity disturbances, then environmental modulation of synapses and networks can only be decoration on the cake, not the cake itself. On this view connectivity problems are inborn.

But how does a narrative implying inborn dysfunction explain autistic regression? How do toddlers who previously seemed normal become transformed into being autistic? Was the connectivity problem always there? If so, why did it turn into autism at a certain point? And if the connectivity problem was not always there, what happened to produce it? There are actually almost no pertinent brain data during the period before the diagnosis of autism when the brain events creating autism would presumably occur. On the basis largely of data from older individuals, there are an anatomical theory and a functional theory for how connectivity problems develop.

The anatomical theory is that there is early brain cellular overgrowth that alters the cellular substrate of networks, with hyperactive short-range and hypoactive long-range connections.[43] This overgrowth is assumed to be genetically driven. This model generally predicts increased cellular density. But brain-imaging data are contradictory on this point.[44] There are many reports of larger brain volumes, but measuring volumes alone does not tell anything about the cells or other substances that may be contributing to the volume. Other forms of brain imaging can look at tissue characteristics, and these imaging techniques have yielded data that do not necessarily support the assumption that "bigger" equates with "more cells." Brain imaging of metabolites (magnetic resonance spectroscopy) generally shows lower metabolite density—and, in particular, a reduction in the concentration of n-acetylaspartate (NAA), a marker of neuronal integrity or neuronal density. And brain imaging of white-matter integrity (diffusion tensor imaging) has sometimes shown that the white matter is less organized rather than more densely packed. These data suggest that the large brain and brain-region volumes need to be accounted for by something other than the presence of more cells.

The functional theory is that big brains come from ongoing cellular dysfunction. In particular, cellular dysfunction associated with inflammation and mitochondrial-bioenergetics-oxidative stress leads to cellular swelling as well as impaired cell fluid transport and increased extracellular fluid, and this set of cellular disturbances, rather than greater cell or fiber density, is what leads to bigger volumes.[45] In this model the larger size is not due to a greater density of optimally functioning neurons and networks but instead to dysfunctional tissue changes such as fluid buildup or even "swelling." These physiological problems may or may not require a substrate of genetic mutations to create vulnerability at the synapse. This theory is more consistent with the above-mentioned brain-imaging findings of lower density of metabolites and more poorly organized white matter.

The functional theory further posits that it is not so much the number of cells or fiber tracts that alters brain activity as the chemical milieu of the cells and synapses. The biochemical and immune environment at the cellular level can alter the communicative functions of neurons (and of glial cells, as I will discuss shortly). Particularly pertinent are excessive glutamate, an excitatory neurotransmitter, and the set of changes collectively known as "excitotoxicity."[46] These phenomena are known in other settings to contribute to brain excitation and seizures.[47]

A 2003 article by Rubenstein and Merzenich positing that an increase in the excitation/inhibition (E/I) ratio is at the core of autism was the first prominent statement of aspects of this functional theory.[48] Not coincidentally, this functionally oriented article was also the first forthright statement of the interactive roles of genes and environment and also of the multiple possible "combinatorial" pathways to this increased E/I ratio.

The idea here is that overly intense brain excitation could be a functional disturbance that has an environmental and active pathophysiological component, not just a genetic component. Documentation of brain inflammation and immune activation in autism changed the playing field because it became clear that we were not dealing with healthy tissue that was wired differently but rather with brains that were having health problems with their cells. These health problems, as mentioned, particularly related to glial cells. Although these cells were previously characterized merely as helper cells or "nurse cells," the critical nature of their functions is now being identified.[49] Glial cells outnumber neurons in the brain as much as ten to one. Unlike neurons, they readily generate new cells. Whereas neurons are shielded (by glial cells) from direct contact with the blood-brain barrier, microglia initially enter the brain through that route, and astrocytes wrap the blood-brain barrier. In fact, microglia and astroglia mediate environmental impacts in the brain.[50]

Astroglial cells not only support core metabolic features of neurons but also may even control the very formation of synapses.[51] They also perform a variety of other functions, such as modulating glutamate and sequestering toxins and heavy metals (until they get overloaded). Beyond these metabolic functions they also participate in networks—they are linked together by gap junctions to form vast syncytial networks communicating via calcium waves—and they have their own "gliotransmitters." Microglial cells are derived from monocytes, which are systemic immune cells, and they play an immune function in the brain.

Astroglia and microglia are activated by stressors such as infections and toxins. When they are activated, they perform immune and cleanup functions, but if the situation is not resolved, a chronic cascade of aberrant chemistry occurs, and they increase brain oxidative stress and excitotoxicity. Chronic activation of these cells is a major contributor to many neurodegenerative disorders.

From Genes and Neurons to Environment and Glial Cells

Just as fully understanding the causes of autism requires decentering from genes, fully understanding autism's brain mechanisms requires decentering from neurons. The identification of immune-activated astrocytes and microglial cells and the upregulation of glial genes in autism brain tissue raises not only the commonly asked question of what is causing this but also the more rarely asked question of how these immune changes might affect how these cells perform their basic functions. What happens to the effects of these glial cells on synapses and networks when they are immune impacted? This question is not at all specific to autism. But in neurobiology (which some have suggested should be renamed "neurogliobiology") it is not yet a well-developed research area.

Research is showing that glial cells can be critical contributors to severe brain derangements like seizures.[52] Less well investigated is the contribution of glial pathophysiology to milder but still vexing features of brain dysfunction. In autism these features include sensory hypersensitivity, sensory integration problems, attention deficits, sleep disturbances, dyspraxia, and motor coordination abnormalities.

Cause?

Including glial cells in brain models of autism would allow further exploration of the mechanisms of postnatal emergence of autism (particularly regression) and of the ups and downs that are hard to explain in a purely neural model. But even more, if active pathophysiological processes, mediated by glial cells, are sufficient to alter synapses and brain networks, is it still always necessary to have an underlying genetic "cause" for autism? Or could someone with no special genetic vulnerability develop autism just from an environmental exposure or from the cumulative impact of a series of possibly diverse environmental exposures? At present this is a rhetorical question, but I think that the grounds for asking it are strong enough that research should be pursued to answer it.

Conversely, if genes are in fact not absolutely necessary, are they still sufficient on their own to cause autism? This probably depends on the gene. But there are issues even with genes of strong effect. Here we can look at genetic syndromes where there is a high incidence of autism, such as fragile X and tuberous sclerosis. These diseases certainly have high rates of autism, but the rates are by no means 100 percent. Only about 50 percent of individuals with fragile X have autism.[53] What distinguishes them from the others with fragile X but without autism? At present this is not known. Could it be that these genes do not cause the autism but simply greatly increase the risk? Geneticists may assume that it is other genes that make some persons but not others cross the threshold into autism. Is it possible that those who do develop autism have it not only because of genes but also at least sometimes on account of an overlay of active pathophysiology? We do not know; we need to look. For this to happen, the genetic role in autism needs to be a question, not an assumption.

Smith-Lemli-Opitz syndrome (SLO) probably has the highest autism rates of any genetic syndrome.[54] Cholesterol synthesis blockage is a key element of this syndrome's impact, and when children with SLO are treated with cholesterol, their autism can go away. Furthermore, it is not unusual for children without SLO to have extremely low cholesterol (even below 100).[55] Such children have been reported to have their cholesterol go up as their autism difficulties reduce in response to other treatments, to improve in response to direct cholesterol administration, or both.[56] This suggests some final common pathway to autism that can be arrived at by many routes. It also suggests that capabilities are obstructed but become accessible when the obstruction is removed.

Modulating Severity by Treating Intermediary Metabolism

Inborn errors of metabolism are a category of disease caused by known or presumed genetic problems. In some cases, such as phenylketonuria (PKU), public health programs in place (for PKU, newborn testing followed by a low-phenylalanine diet) can largely prevent symptoms even though a gene is involved. Modulation of autism severity by treatment in cases of documented metabolic disorder was reviewed by Page in 2000,[57] and more examples have since been reported. Autistic symptoms associated with PKU are reduced by a low-phenylalanine diet,[58] in hyperuricosuric autism by a low-purine diet with or without allopurinol,[59] in patients with low cerebrospinal-fluid biopterin by biopterin supplementation,[60] in some hypocalcinuric autistic patients by calcium supplementation,[61] in some patients with lactic acidemia by thiamine and/or ketogenic diet,[62] in cerebral folate deficiency by folinic acid supplementation,[63] and (as mentioned) in Smith-Lemli-Opitz syndrome by cholesterol treatments.[64] Johnston offered a variety of mechanisms whereby metabolic disorders, sometimes with treatable aspects, might lead to neurological changes that could underpin autism.[65] Zimmerman framed his report of promising immunological treatments in terms of a need for a search for reasons for their apparent efficacy, at least intermittently.[66] James reported that her correction of oxidative stress and methylation abnormality profiles through intervention with methylcobalamin, folinic acid, and trimethylglycine was accompanied by qualitative clinical improvement,[67] and work is proceeding to quantitate the observed qualitative behavioral improvements. Symptom severity was reduced in a trial of high-dose vitamin C in autism.[68] In this regard but more generically, Ames and colleagues modeled high-dose vitamin therapy as a treatment approach for a range of genetic disorders characterized by decreased coenzyme binding affinity.[69]

These examples suggest that in some cases one might find a metabolic key to remove or compensate for obstructions from pathways that are currently functioning poorly or not at all because of problems such as a high need for some metabolic substrate, excessive production of a metabolite whose breakdown pathway is slowed or blocked by a genetic defect, or impairment of energy production through some genetic defect slowing mitochondrial metabolism. The cases of cholesterol and of the large numbers of children with autism and biochemical indications of mitochondrial dysfunction but no detectable mitochondrial gene mutations also suggest that errors of metabolism may be produced by environmental rather than genetic factors.

Obstructed Rather than Defective

Given the clinical observations of transient improvement, persistent remission or recovery, and response to metabolic intervention, it becomes necessary to ask whether the brain in autism is truly and intrinsically "defective" or is instead "obstructed," at least in many cases. These many clinical episodes indicate that the brain capacity is present, at least in many cases, but that there is a problem with organizing the means of expression, with organizing sensations into perceptions and constructs, or both. Autism from this point of view becomes more of an "encephalopathy"—an obstruction of brain function, possibly through an encephalopathy related to immune activation or metabolic dysfunction. If this is the case, research and care ought to be oriented much more to overcoming the encephalopathy so that people can express their full potential.

Note that I am not saying that people with autism should become "normal," just more fulfilled. Members of the neurodiversity community might object that autism is an identity that should not be eradicated or cured. The discrepancy between that position and what I am saying may be a matter of terminology rather than of fundamental disagreement. Overcoming the systems-biology-based difficulties associated with autism may or may not eradicate the creative brilliance of many with autism. I would expect not, but this would remain to be seen—that is, it is an empirical question that needs to be answered by evidence, not just debate. But eradicating those gifts is not the intent here. Achieving full potential, however that may look, is the expected outcome of success in overcoming active pathophysiological problems.[70]

A point of commonality is that both the position I am presenting and the neurodiversity community object to common educational and therapeutic approaches that are palliative—controlling behaviors, warehousing, restraining, drugging. As I write, the 16-year-old son of a dear friend is being transferred to a regular high school. A year ago he was nonverbal and having seizures and had never talked to his parents. Now he still cannot talk, but he can type, and he has a lot to say about things over many of the years of his life—it turns out that he has been observing and understanding a great deal. Had he not had the nonstandard opportunity to develop typing ability, he and his family and others would never have been able to add communication with language to their relationships with him and his with them. And who knows what creative contributions he will make to the world as he goes forward?

Environment: The Gift That Keeps On Giving

Epidemiology is the field of investigation that most often comes to mind as the route to finding causes of autism. It is important for finding associations between environmental factors and changes in autism rates, but it does not explain how these environmental factors cause autism. Pathophysiology is where the rubber hits the road. Pathophysiology is the locus of how genes and/or the environment can cause autism. It takes environmental impact beyond looking for association between environmental agents and autism incidence and looks at how the environment affects underlying physiological mechanisms.

The environment has inched into the discourse on autism, small steps at a time. To date, the environment-autism discourse has largely mirrored the genes-autism discourse: just as genes were said to create a different brain-wiring diagram, so environmental agents could contribute to altered brain development as well. In both cases the deed is done right from the start, and as Jeff's mother said, you are "screwed for life." But why should environmental effects be confined to the early developmental period? Why should it just be about the brain? There is no a priori reason that they should be. And why should they necessarily be a life sentence? Environmentally vulnerable physiology can have impacts far beyond those constraints.[71]

Hardware or Software? No simple metaphor can truly map onto the complexities of molecular biology, but roughly speaking, the metaphor shared in these "early brain development" models of autism is that hardware comes before software. There is presumed to be some kind of fundamental alteration in architecture, be it of a cell receptor, a brain region, or a network diagram, and be it caused by gene, by environment, or by both together.

But with reflection on the role of active pathophysiological processes in altering synapses and networks, it starts to become conceivable that software, or so-called fluid processes, could conceivably precede architectural alterations. That is, the chronic persistence of biochemical, metabolic, immune, and signal-processing perturbations could skew developmental trajectories in micro- and macroanatomy. This shifts the central locus from genes to molecules and pathophysiology.

From Developmental to Early-Onset Chronic Pathophysiology. Once one takes this step, it is easy to take the next step and consider whether environmental influences might continue to modulate the severity of these software or fluid alterations in an ongoing fashion. That is, why could they not affect function—particularly but not exclusively brain function—even after the period when brain development can be perturbed?[72] If this is so, then the distinction between "developmental disorder" and "chronic illness" starts to blur, and one starts to see an interacting dance between genes and environment that can begin from conception or even from preconception epigenetic influences.[73]

From a Fixed Unitary Phenomenon to Modifiable Manifestations of Complex Interacting-Systems Problems

At this point the category of "autism" starts to deconstruct into a complex web of influences on brain and systemic biology, starting anywhere from before conception to after birth. Moreover, if dynamic pathophysiological processes can contribute through multiple pathways to significantly compromising ongoing brain function, and if this compromise is not necessarily permanent, then the very diagnostic category of "autism" may turn out to be a reification of a set of systemically interacting dynamic processes. That is, it turns a concept into a thing that is then treated as a unitary entity. This produces scientific and conceptual chaos.

Autism as an Epiphenomenon or Emergent Property of a Challenged System. But from a systems-biology vantage point autism stops being a unitary category and instead becomes an epiphenomenon—the behavioral output of a system where many things are interacting to deplete the resilience and degrees of freedom of an increasingly vulnerable system. From a dynamic-systems point of view behavioral patterns are emergent properties of altered systems functioning—they are "patterns of behavior that 'fall out' of the structural and energetic status of the system without being represented in that system."[74] In the case of autism the structural and energetic status of the system is compromised. This leads the exquisite coordination of the brain to become, in some respects at least, less differentiated and less nuanced and to retreat to a less resource-needy mode of operation. It is as if the brain is conserving scarce resources. (This may also involve concentrating resources in certain areas where great creativity may emerge.)

If this is so, then giving the system more abundant resources (material, informational, or both) may allow the system to function more freely and flexibly. This underlying principle motivates a range of superficially different therapeutic strategies that seem to enable some with autism over time to lose their diagnoses. Resources may include supplying nutrient cofactors to overcome blocks in metabolic pathways, movement or communication support, biofeedback to learn stress reduction, communication support, and many other forms of medical, educational, behavioral, and neurophysiological intervention.

Specific Genetic Determinants or Final Common Pathways of Pathophysiology? Given the official DSM-IV definition of autism, many geneticists and others have believed that specific genetic determinants must exist to create such a replicable set of behaviors. More recently, though, with the identification of hundreds of genes, most of which are rare, that all contribute somehow to autism, it has become a challenge to find common themes across disparate influences. To address many disparate genetic mechanisms with a smaller number of evidence-based therapeutic approaches, one would need to identify a strategic node in some common network.

If one looks at the active pathophysiology contributors to autism, the picture is different. Many environmental contributors may cause or exacerbate inflammation, oxidative stress, or mitochondrial dysfunction, but it may not be necessary to identify precise molecular targets for treatments to have a major impact. The upstream environmental triggers of active pathophysiology—or its avoidance—can be influenced in fairly generic ways. Inflammation, oxidative stress, and mitochondrial dysfunction can all be significantly affected by diet, for example.

The idea that so-called autism is an epiphenomenon or an emergent property of a challenged system also extricates us from the constraints of needing to find specific determinants for specific behaviors, since from this vantage point the specificity of the behaviors does not arise from specific biological determinants but from the dynamics of the system.

Time to Get a Grip

If we look at biology through genetic lenses, then the terrain seems stable, and we have time to flesh out the mechanisms because they are not changing. But if we look at biology through dynamic lenses, where genes and DNA and RNA and molecules and physiology and environment are in a constant dance, then we do not assume stable mechanisms. In this setting we can conceive of environmental perturbation of molecules and pathophysiology, and we can begin to imagine that some perturbations could be dangerous.

Within the genetic narrative, autism is what it is (and has always been), and it will take science to significantly improve the quality of life of affected individuals. Therefore, our agenda should be to provide evidence-based therapies and proceed painstakingly over time to improve what is available through targeted scientific research.

Within the environment narrative, genetics may contribute risk, but the systems output we label "autism" plausibly is occurring more frequently on account of a growing number of environmental exposures, a compromised food supply that is depleting nutrients necessary for physiological resilience and contains additives that when processed by the body may drain its reserves or cause direct harm, major and even radical alterations in activities during infancy and childhood, and probably even epigenetic effects passed on to children from noxious exposures experienced by parents or grandparents. Within this explanatory framework it is to be expected that the increases are being driven by an epidemic of active pathophysiology that in infants and toddlers can lead to autism. This narrative yields an action plan to implement while we wait for targeted science because the processes it describes are amenable to modulation and reduction through measures such as eating a high-nutrient-density, high-antioxidant, and anti-inflammatory diet, encouraging more full-bodied motor activities during early development, and reducing exposures to toxins and infectious agents.

Within the genetic narrative, there is no plausibility either to the claim that autism numbers are rising or the claim that people with autism can lose their diagnoses. Any increase in numbers is dismissed as an artifact, and any child who loses the autism diagnosis is presumed not to have been truly autistic in the first place. When large studies do not yield decisive genetic information, the proposed solution is to increase statistical power by organizing even larger studies. As one of my geneticist friends said to me, "You environmental researchers should just wait for a few years. By then we'll have worked out what the key genetic features are of autism and then you will have a basis for pursuing environmental research to explain the few residual features not explained by genetics."

Within the environmental narrative, the increasing reports of reversals of diagnosis (including those in children with rigorous, high-quality clinical evidence of an initial autism diagnosis) are cause for examination of core assumptions and for reshaping research and clinical agendas. These reports not only give hope for individuals but also provide a strong impetus for an aggressive public health program right now.

Addressing an Apparent Epidemic through a Praxis of Environmental Pathophysiology

With autism rates of about 1 in 88 in children and 1 in 54 boys—and probably rising—anything we can do sooner rather than later to stem the tide ought to make eminent public health sense. Given that simple measures like diet, nutrition, avoiding toxins, and rethinking education and pedagogy may help reduce the active pathophysiology that may be aggravating if not causing autism, a strong argument can be made that implementing these measures ought to be primary in a public health approach to autism. Yet this approach is marginalized to a significant degree because it is implausible in a simple gene-brain-behavior narrative.

Many parents of children with autism have implemented these measures on their own and have organized widely to train other parents. They have grown frustrated with the professional care emanating from the genetic narrative's static encephalopathy-based model of autism. Many perceive that their child is "in there" and feel that physicians and other clinicians are not doing everything they can to rescue their child from what is trapping them in brain and often also body constrictions.[75] Standard medical and behavioral care does not help them with the daily, often excruciating difficulties with sleep, low threshold for intolerance of sensory stimulation, food refusal, and seizures. It offers behavioral therapies that sometimes help a good deal and too often only a little, and it offers pharmaceuticals that also help only sometimes and often lead to nasty side effects, like obesity and hormonal dysregulation.

Countless parents have been denied medical workups for conditions like constipation or esophagitis because "it's just part of the autism," with the implicit or explicit inference that since autism is, in the minds of the clinicians, untreatable, anything associated with it is untreatable as well. It should be noted that although doctors who make this claim are generally strong advocates for "evidence-based medicine," there is no evidence for the claims they are making, and parents often feel that they are actually saying these things to avoid dealing with difficult, unruly, time-consuming, and even perhaps viscerally disturbing patients.

Another issue is that even when professionals take on these complex patients, their treatment armamentarium does not usually include such measures as supporting challenged metabolic pathways except in certain cases (e.g., the use by some but not all specialists of a "mitochondrial cocktail" of high-dose vitamins and other cofactors in mitochondrial metabolism). The taboos around some of the alternative treatments used by parents have stopped many professionals cold from even familiarizing themselves with the methods and rationales of these approaches. Over time, as success stories have accumulated of children (and even some adults) greatly reducing the severity of their problems and sometimes even losing their diagnoses, some serious scientific attention has begun to be paid to these phenomena. As mentioned earlier, the basic principles of these therapies include tackling subcomponents of "the autism" as problems that can be solved and thereby reducing the stress on the whole system so that it has more of a chance to recalibrate.[76]

But even if medical professionals were to expand their repertoire and their ability to deal with these challenging patients, I think that the magnitude of the problem exemplified by autism is much bigger than what medical professionals can solve on their own. Today's typical modern diet and lifestyle are corrupted by poor food, lack of exercise, and chemical exposures through cosmetics, home-care products, household construction materials, and pesticides in the home, school, and workplace. All these and more may well be implicated in running down the body's systems and creating inflammation, oxidative stress, and mitochondrial strain and dysfunction until all of this finally affects the brain. Given this deeply unfortunate global state of affairs, once we take seriously the role of the environment, we need to come to grips with the magnitude of what we face. A much bigger response is needed than carefully targeted and specific treatments. Doctors do not have time to give transformative teaching on lifestyle changes or address public policy. This will need to involve not just highly trained professionals but also paraprofessionals as well as laypeople—even a social movement to transform unhealthy lifestyles into health-promoting (and sustainable) living habits. Such a broad-based transformational program will work best if we ever can move toward a reasoned dialogue about the predicament we are in.

Once we understand that autism is not genetically inevitable—that it is not a genetic tragedy but an environmental and physiological catastrophe—the point becomes not just to understand autism, but to change how we conduct our lives so that we support health rather than harm.

Beyond Autism

Clearly, gene myths are a problem in autism and are among the forces putting obstacles in the way of implementing a full-force public health campaign to reduce environmental risks. Clearly also, gene myths are not restricted to autism, although autism is an interesting case study. It is becoming painfully obvious that environmental risks are key players in epidemics of chronic illnesses from cancer and heart disease to diabetes and obesity, and that the active pathophysiological processes involved in autism also are present in these other conditions. The public health measures I am proposing to help reduce autism risk would also reduce risk for many other conditions. Since so many environmentally modulated illnesses are still often considered genetic in origin, people suffering with these problems remain confused and often do not become aware of the self-care measures that could potentially help them greatly.

Shifting from a static genetic model of autism as a lifelong brain defect to a dynamic model of active systems pathophysiology creates a platform for incorporating the contribution of genes into a broader, more inclusive, and more empowering framework. Successes in this shift will have much broader applications beyond autism.

I am grateful to John Elder for his assistance in framing and preparing this chapter.

CHAPTER ELEVEN

The Prospects of Personalized Medicine

DAVID JONES

THE ADVENT of the Human Genome Project created unprecedented enthusiasm for the prospects of genetic science and medicine. Francis Collins, who led the effort for the National Institutes of Health (NIH), showed little modesty about the project as it neared completion in 1999. Invoking the legacies of Lewis and Clark, Sir Edmund Hillary, and Neil Armstrong, he described the project's "audacious goal": "To wrest from nature the secrets which have perplexed philosophers in all ages, to track to their sources the causes of disease, to correlate the vast stores of knowledge, that they may be quickly available for the prevention and cure of disease—these are our ambitions."[1] For Collins, the "genetic revolution" was already under way. One aspect of the revolution had particular promise: pharmacogenomics.[2]

The basic idea of pharmacogenomics is simple and alluring. Doctors have long known that individual patients vary in their response to medications. Some of this variation, as with other forms of biological variation, must have a genetic basis. Pharmacogenomics seeks to characterize the genetic causes of interindividual variability in drug response, develop tests that will distinguish relevant genotypes, and then apply this knowledge to optimize the efficacy and safety of the prescribed regimen. This would allow doctors to personalize medicine. As Collins, who had since become director of the NIH, and Margaret Hamburg, director of the Food and Drug Administration (FDA), described in 2010, the goal is "the right drug at the right dose at the right time." The excitement over pharmacogenomics has eclipsed other ways of personalizing medicine. Hamburg and Collins described a "shared vision of personalized medicine" that was exclusively genetic and paid no attention to social and environmental determinants of treatment outcome. Pharmaceutical companies share some of this excitement. They hope that pharmacogenomics will identify novel drug targets and allow them to develop medications targeted for specific populations. Hamburg and Collins described how the federal government would help bring this to fruition. Inspired by another past triumph, the government was "building a national highway system for personalized medicine, with substantial investments in infrastructure and standards. We look forward to doctors' and patients' navigating these roads to better outcomes and better health."[3]

As with the other domains of genetic science and hope, it is worth taking a moment to step back and assess the appropriateness of pharmacogenomic enthusiasm. The ten-year anniversary of the completion of the Human Genome Project inspired a binge of assessments of the project's impact. Many of these were surprisingly guarded.[4] The situation for pharmacogenomics is no different. Even though the principles of its science remain sound, and even though scientists have identified an ever-increasing number of clinically relevant alleles and tests, pharmacogenomics remains on the periphery of clinical practice. Why has the revolution been so slow to come? Many factors have contributed, from the subtleties of the science itself to the uncertain benefit offered by existing pharmacogenomic tests. Meanwhile, the advent of pharmacogenomics has reinserted race into therapeutic practice and has drowned out other potentially promising ways to personalize medicine. Given this mixed legacy, it is essential that clinicians and researchers base their decisions on the actual utility of pharmacogenomics and not on its promised utility. This will facilitate wise decisions about where attention ought to be focused in pursuit of the best ways to personalize medicine.

A Primer on Pharmacogenomics

It is not hard to fathom why scientists and physicians have been so optimistic about the promise of pharmacogenomics. Physicians have known since the time of the Hippocratic corpus that different patients respond differently to the same treatment. Some of the art of medicine involves recognizing and managing this variability.[5] Sometimes the effects are subtle, such as finding the right dose to achieve a therapeutic outcome. Sometimes the effects are perplexing, as with the idiosyncratic side effects that patients experience. The same antidepressant, for instance, might cause diarrhea in one patient but constipation in another, or sedation in some and insomnia in others. Many factors contribute to the variability.

Sometimes the problem lies in the disease: different patients with the same diagnosis might have different pathophysiologies underlying their disease. Sometimes the variation arises from the patient. The optimal dose of a drug depends on a patient's age, weight, body fat, sex, kidney and liver function, and comorbid disease. Drug metabolism can also be influenced by environmental exposures, including medicines, foods, pesticides, and tobacco smoke. Most fundamentally, the impact of a medication depends on how reliably a patient takes it.[6]

Physicians have long suspected that heritable differences in drug response exist as well. Writing in 1902, Archibald Garrod suggested quite plausibly that the activity of bodily enzymes varied just as much among individuals as did the observable traits of human morphology: "The individuals of a species do not conform to an absolutely rigid standard of metabolism, but differ slightly in their chemistry as they do in their structure."[7] Following this lead, doctors in the first half of the twentieth century described several examples of seemingly heritable variation in drug response. Scientists working for the League of Nations grappled directly with this problem as they worked to standardize measures of the potency of hormones and realized that drug response varied from person to person. Interest in such variations surged after World War II. During the war American physicians recognized that primaquine, widely used for malaria prophylaxis, could trigger hemolytic anemia, especially in African American soldiers. As physicians deployed isoniazid in the first global campaigns against tuberculosis in the 1950s, they recognized that some patients suffered peripheral neuropathies, while others developed hepatitis. These side effects also had a racial distribution. Researchers traced both phenomena to specific drug-metabolizing enzymes.[8]

One line of work was especially revealing. After two patients died in Berlin after receiving routine doses of Procaine (novocaine), Werner Kalow began detailed studies of the enzymes that metabolize certain anesthetics. He found that one enzyme, now named butyrylcholinesterase, had different activity in different patients. He showed that the reaction rate depended on the affinity between the enzyme and the drug, something that depended on enzyme structure. Since genes determined the structure of enzymes, Kalow concluded that this variation in cholinesterase activity must have a genetic basis. Family studies and other evidence of Mendelian inheritance soon confirmed these suspicions.[9]

Researchers quickly elucidated the basic principles of the field. Arno Motulsky published a widely read review in the *Journal of the American Medical Association* in 1957 that alerted physicians to the potential importance of heritable variations in drug response. Friedrich Vogel coined the term "pharmacogenetics" in 1959. Kalow published a comprehensive review of the field in 1962.[10] Their work made it clear that humans—and other species—vary in their response to medications and that some of this variation is genetic.

But was pharmacogenetics relevant? Early discoveries implicated only a handful of drugs. When Kalow published *Pharmacogenetics* in 1962, physicians knew of six drugs influenced by specific drug-metabolizing polymorphisms. Over the next seventeen years they identified only two more. Despite this paucity of enzyme-drug links, researchers did accumulate further evidence of the relevance of the field in the 1960s. Kalow again provided a crucial example. His studies demonstrated the genetic basis of malignant hyperthermia, a rare condition in which routine anesthetics caused life-threatening complications. Twin studies, meanwhile, suggested that genetic traits influenced the metabolism of many drugs.[11] Concurrent work by pharmacologists characterized many of the specific enzymes involved in drug metabolism. The most common drug-metabolism pathway converts lipid-soluble molecules into water-soluble molecules so that they can be excreted more easily in urine and bile. This typically occurs in a two-step process: phase I reactions, including oxidation, reduction, and hydrolysis; and phase II conjugation reactions, including acetylation, glucuronidation, sulfation, and methylation. Between ingestion and excretion, drugs interact with many transporters, receptors, and metabolizing enzymes. The best-known example is a set of liver oxidases, the cytochrome P450 system. Scores of different P450 enzymes exist (e.g., P450-2D6 and P450-3A4), each of which contributes to the metabolism of different drugs, and each of which has alleles with variable activity.[12] By the 1980s, and with increasing steam in the 1990s, proponents of pharmacogenomics trumpeted the growing list of drugs for which specific genetic variants had been identified.[13]

Pharmacogenomic excitement appears in many prominent reviews. Collins, writing about genomic medicine in the *New England Journal of Medicine* in 1999, described the basic hopes: "There may be large differences in the effectiveness of medicines from one person to the next. Toxic reactions can also occur and in many instances are likely to be a consequence of genetically encoded host factors. That basic observation has spawned the burgeoning new field of pharmacogenomics, which attempts to use information about genetic variation to predict responses to drug therapies." He listed several promising examples, including genetic variants that influenced the efficacy or safety of tacrine for Alzheimer's disease, statins for coronary artery disease, and oral hormones for contraception.[14] William Evans and Mary Relling, writing that same year in

Science, acknowledged that diverse factors influenced drug effects but emphasized the genetic polymorphisms with "incontrovertible clinical consequences": "Despite the potential importance of these clinical variables in determining drug effects, it is now recognized that inherited differences in the metabolism and disposition of drugs, and genetic polymorphisms in the targets of drug therapy (such as receptors), can have an even greater influence on the efficacy and toxicity of medications." They hoped that it would soon be possible "to select many medications and their dosages on the basis of each patient's inherited ability to metabolize, eliminate, and respond to specific drugs." They provided an even longer list of examples than Collins had done, including chemotherapy agents, warfarin, codeine, and many others.[15]

The importance seemed to grow over time. In 2003 the *New England Journal of Medicine* published back-to-back reviews dedicated to pharmacogenomics. Although the basic ideas were similar, the science had more precision and depth, with ever-longer lists of alleles relevant for ever-longer lists of drugs. The rhetoric was again revealing. Evans, noting that it had been estimated that genetics "can account for 20 to 95 percent of variability in drug disposition and effects," emphasized the importance of genetics over the other factors. After all, "Unlike other factors influencing drug response, inherited determinants generally remain stable throughout a person's lifetime." New discoveries every year convinced proponents that "the potential is enormous for pharmacogenomics to yield a powerful set of molecular diagnostic methods that will become routine tools with which clinicians will select medications and drug doses for individual patients."[16] Richard Weinshilboum, from the Mayo Clinic, shared the excitement: "With the completion of the Human Genome Project and the ongoing annotation of its data, the time is rapidly approaching when the sequences of virtually all genes encoding enzymes that catalyze phase I and II drug metabolism will be known. The same will be true for genes that encode drug transporters, drug receptors, and other drug targets." Like Evans, he channeled interest away from the other factors and toward genetics: "Even though individual differences in drug response can result from the effects of age, sex, disease, or drug interactions, genetic factors also influence both the efficacy of a drug and the likelihood of an adverse reaction."[17]

By 2011 "major advances" necessitated another *New England Journal of Medicine* review. The list of relevant alleles and drug effects had grown once again, including prominent examples from cancer, cardiology, and infectious disease.[18] The science was, admittedly, more complicated than physicians would have liked. Although a few alleles exerted powerful monogenic effects, proponents acknowledged that most pharmacogenomics pathways involved complex polygenic interactions. This made it difficult to predict drug response from just a few tests. However, researchers remained confident that with better knowledge and more powerful sequencing technologies, success would soon be within reach. As Collins described in a celebration of the tenth anniversary of the completion of the Human Genome Project, such remarkable advances were "paving the way for an era of 'genomic medicine.'"[19]

Excitement extended, though with some ambivalence, into the pharmaceutical industry. Pharmaceutical executives worried about the potential market fragmentation that pharmacogenomics might cause if new drugs were tailored for small patient subsets. But Evans managed a positive spin. Traditional drug development sought "medications that are safe and effective for every member of the population, a strategy that aims to provide a marketing bonanza but one that is a pharmacological long shot because of highly potent medications, genetically diverse patients, and diseases that have heterogeneous subtypes."[20] Pharmacogenomics, in contrast, offered a surer route. Targeted drug development could "reduce attrition across the development pipeline and lessen the number of products withdrawn in the post-marketing period."[21] This would yield more drugs for smaller markets—not "blockbusters" but "minibusters."

For clinical and capitalist dreams to bear fruit, many pieces must fall into place. As Weinshilboum's group described in 2011, "A blend of scientific, regulatory, and psychological factors must be addressed if pharmacogenomic tests are to become a routine part of clinical practice."[22] The FDA had a crucial role here. By including pharmacogenomic information on drug labels, it could influence the standard of care and push the incorporation of pharmacogenomic tests into clinical practice. The FDA was eager to take on this role. As Hamburg and Collins described, in 2010 "about 10% of labels for FDA-approved drugs contain pharmacogenomic information—a substantial increase since the 1990s but hardly the limit of the possibilities for this aspect of personalized medicine." Increasing the number of labels with pharmacogenomic claims will take work. The FDA must define its standards of evidence for evaluating claims made by manufacturers that the specific alleles have relevant effects. It must ensure that pharmacogenomic tests perform reliably. The NIH, meanwhile, must assemble comprehensive information about the tests and their clinical relevance. Governance by standard setting had good precedents: "When the federal government created the national highway system, it did not tell people where to drive—it built the roads and set the standards for safety. Those investments supported a revolution in transportation, commerce, and personal mobility." Through a similar set of standards for pharmacogenomics, the NIH and the FDA planned to "make personalized medicine a reality."[23] The FDA, for instance, now hosts a website that lists all drugs that include pharmacogenomic biomarkers on their drug labels. As of March 2012 this included 105 drugs, from abacavir to warfarin. Why did clinicians need to know? The same old idea recurred like a mantra: "Pharmacogenomics can play an important role in identifying responders and non-responders to medications, avoiding adverse events, and optimizing drug dose."[24]

Pharmacogenomic Discontents

Despite the enthusiasm still evident at the tenth anniversary of the Human Genome Project, discontent had appeared. Observers expressed concern about the slow progress of the genome revolution in general and about pharmacogenomics in particular. For instance, the *New England Journal of Medicine* juxtaposed Collins's celebration of the Human Genome Project at its tenth anniversary with a much more tempered perspective from Harold Varmus. Genomics, Varmus noted, remained "more closely aligned with modern science than with modern medicine": "after the first decade of a postgenome world, only a handful of major changes—some gene-specific treatments for a few cancers, some novel therapies for a few Mendelian traits, and some strong genetic markers for assessing drug responsiveness, risk of disease, or risk of disease progression—have entered routine medical practice." As a result, "Only a few selected items of that new information are now widely used as guides to risk, diagnosis, or therapy."[25]

Pharmacogenomics had come up short for many reasons.[26] Hamburg and Collins described how researchers had "identified genetic variability in patients' responses to dozens of treatments."[27] Physicians, however, employ hundreds of drugs. Cost was once a problem for genomics, but with full genome sequencing soon to be available for under $1,000, this problem will dissipate. Understanding the data will prove far more difficult. As Varmus explained, "Physicians are still a long way from submitting their patients' full genomes for sequencing, not because the price is high, but because the data are difficult to interpret."[28] Many of the relevant therapeutic outcomes are under polygenic control and require detailed knowledge of complex genetic regulatory systems before predictions based on sequence data can be made. Doctors and patients, unfortunately, often have only a limited knowledge of genetics. Proponents hope to black-box the process with point-of-care tools embedded in electronic medical records so that physicians will have access to pharmacogenomics recommendations at the bedside and in the clinic as they make their prescribing decisions with patients.

But even in the best-case scenario, will pharmacogenomic knowledge have clinical utility? Thus far many studies have had disappointing results. One study analyzed 101 loci related to heart disease in order to predict patients' risk, only to find that the resulting prediction was no better than that obtained by taking an adequate family history.[29] A variant allele that decreased the blood level of the antiretroviral nelfinavir paradoxically increased its efficacy. A patient who experienced the greatest clinical improvement in a tacrine trial turned out to be the one with the least favorable genotype. Clopidogrel, one of the most widely prescribed drugs in the United States, must be activated by P450-2C19 in order to have its antiplatelet effects. As a result, the FDA has recommended genotyping patients before its use. Two large randomized trials, however, found no benefit from doing so.[30] Hamburg and Collins admitted, "Genetic tests are not perfect, in part because most gene mutations do not perfectly predict outcomes." As long as clinical utility remains unclear, the future of pharmacogenomics will be at risk. As Weinshilboum's group warned, "In the absence of such evidence, payers will be unlikely to provide reimbursement for routine use of pharmacogenetic testing, and tests will remain inaccessible to the majority of patients."[31] The power of the purse can trump pharmacogenomic promises.

Collins's writings are particularly revealing. When he predicted the future of genomic medicine in 1999, he described his vision for what the management of John, a young man with a family history of heart disease, might look like in 2010. John's doctor would order a battery of genetic tests to specify his risk for heart disease: "Confronted with the reality of his own genetic data, he arrives at that crucial 'teachable moment' when a lifelong change in health-related behavior, focused on reducing specific risks, is possible. And there is much to offer. By 2010, the field of pharmacogenomics has blossomed, and a prophylactic drug regimen based on the knowledge of John's personal genetic data can be precisely prescribed to reduce his cholesterol level and the risk of coronary artery disease to normal levels."[32] As any young man or internist now knows, this moment has not come to fruition.

Even when Collins and others acknowledge such disappointments, they maintain their faith that the promise will soon be fulfilled. It is just a matter of time and continued research funding before routine pharmacogenomic testing becomes a reality and gives patients access to the right drug at the right dose at the right time. Are such persistent hopes realistic? Much can be learned by taking a close look at four cases: cancer chemotherapy, warfarin, codeine, and BiDil. These cases, which reflect different aspects of the field, reveal both the promise and the limitations of pharmacogenomics and help set realistic expectations for what the future might bring.

Cancer Therapeutics: The Poster Child of Pharmacogenomics?

After the initial discoveries of the 1950s and 1960s, cancer became one of the first areas where pharmacogenomics gained traction. Pharmacogenomics appealed to oncologists for two principal reasons. First, traditional chemotherapy agents, by their nature, are quite toxic. They have a narrow therapeutic index with a very fine line between expected side effects and intolerable complications. This requires careful monitoring of drug dosages and drug levels, something that makes even subtle genetic differences in drug metabolism relevant. Second, as became clear over the second half of the twentieth century, cancer is a genetic disease, offering the possibility that specific medications might target the genetic basis of each type of cancer or even of each individual's specific tumor. Cancer pharmacogenomics developed in three phases.

In the first phase cancer researchers and clinicians characterized genetic variations in responses to traditional chemotherapy agents. For instance, 5-fluorouracil is widely used to treat colon, ovary, and breast cancer. In the 1980s clinicians described patients who suffered fatal nervous system toxicity from routine doses of the drug. Investigation revealed that these patients had an inherited deficiency of dihydropyrimidine dehydrogenase, an enzyme that breaks down fluorouracil and endogenous pyrimidines. Subsequent work found that 1 percent of the population carries the deficient allele and, as a result, experiences enhanced drug effects. The rare homozygous patients face severe toxicity. Pharmacogenomic testing in advance can reveal who is at risk.[33] Other similar examples have been found. Mercaptopurine and thioguanine, both widely used against leukemia, are inactive pro-drugs that must be metabolized by one enzyme, HPRT, into their active forms. A second enzyme, TPMT, metabolizes them so that they may be excreted. Patients with low HPRT activity experience low efficacy, while patients with low TPMT activity develop dangerously high drug levels and even fatal bone-marrow toxicity. If the TPMT variant is recognized, then it is possible to treat patients successfully at 6 to 10 percent of the usual dose.[34]

In the second phase researchers used their increasing knowledge of genetics to characterize each individual's tumor in order to predict drug response and design optimal drug regimens. For instance, in the 1990s Evans and his team of researchers at St. Jude's Children's Research Hospital in Memphis showed that rates of drug metabolism in children with acute lymphoblastic leukemia (ALL) could vary tenfold. Patients with rapid drug metabolism received inadequate treatment effects. At first researchers measured drug levels to individualize methotrexate doses. This improved five-year survival from 66 percent to 76 percent.[35] The development of "gene chips" in the late 1990s allowed clinicians to test hundreds of genes and discover clues about tumor biology and treatment response before treatment: "Putting all of these molecular diagnostics on an 'ALL chip' would provide the basis for rapidly and objectively selecting therapy for each patient." By the early 2000s this approach began to produce results. Studies of ALL found that some tumors had hyperdiploid cells, with three to four copies of chromosome 21. Since this chromosome carried the gene for a transporter that moved methotrexate into cells, tumors with multiple copies had increased sensitivity to the drug. Similarly, tumors that overexpressed TEL-AML1 were unusually sensitive to asparaginase, while those with a translocation of a piece of chromosome 4 onto chromosome 11 were more sensitive to cytarabine.[36]

In the third phase researchers sought new drugs that would specifically target the genetic aberrations of individual tumors. As cancer genetics developed after World War II, researchers realized that specific tumors possessed characteristic mutations. Drugs that targeted the specific product of these mutant genes might function as "magic bullets," selectively killing cancer cells with little or no collateral toxicity. The dream of a specific agent was realized most dramatically with imatinib mesylate (marketed as Gleevec).[37] In 1960 researchers had described one of the first tumor mutations, the "Philadelphia chromosome," reciprocal translocation between the long arms of chromosomes 9 and 22 seen in chronic myelogenous leukemia (CML). By 1987 David Baltimore had identified the resulting aberrant fusion protein, BCR-ABL. This protein produces a cellular receptor, a tyrosine kinase, that is constantly active and drives cells to divide relentlessly. Meanwhile, researchers at Ciba-Geigy had been looking for drugs that could inhibit kinases. By the early 1990s they had found dozens of drugs that selectively targeted different kinases. When one of the Ciba researchers, Nick Lydon, went searching for possible disease targets, he met oncologist Brian Druker at the Dana Farber Cancer Institute. Druker studied one of these inhibitors, CGP57148, and found that it had dramatic effects against CML cells, first in cell cultures, then in animal models, and finally, starting in 1998, in patients with the disease. Approved in 2001, imatinib transformed the treatment of CML: the five-year survival rate is now nearly 90 percent. Subsequent research showed that the same drug also worked against gastrointestinal stromal tumors (GISTs), improving the treatment response rate for those patients from 5 percent to 50 percent. Seventeen other cancer drugs now have pharmacogenomic labeling.[38] Breast cancers that overexpress HER2 (a human epidermal growth factor) have a prognosis that is worse than usual, but they also have a better response to the targeted drug trastuzumab, while non-small-cell lung cancers with certain variants of EGFR respond to gefitinib, and colon cancers with EGFR variants respond to cetixumab.

Yet even amid these success stories there are seeds of concern. First, although pharmacogenomic testing can help predict which patients might have unusual responses to drugs, comparable information can often be obtained by giving a patient a test dose of the drug and measuring the drug levels in blood or urine. This can be more useful than a genetic test since it reflects the actual outcome of the many genes and other interactions that influence drug levels. Second, as impressive as imatinib has been, the drug barely came to fruition.[39] Druker's initial desire to test the molecule was stymied by the inability of lawyers at Ciba-Geigy and Dana Farber to reach consensus about a research contract. Druker then left Boston for Oregon Health and Sciences University, where he obtained a small sample and found the dramatic response in cell culture and mouse studies. In the meantime, however, Ciba-Geigy had merged with Sandoz to form Novartis. Executives there were unwilling to invest the $100 million to $200 million needed to take the drug through clinical trials, presumably because they considered the potential market—patients with CML—too small to justify the costs. Druker worked year after year, as patients continued to die, to convince Novartis to approve clinical trials. The company grudgingly consented to a small trial in 1998. Only the dramatic results of this pilot study convinced it to continue. Pharmacogenomics remains a vulnerable science.

A third problem appeared soon thereafter. Once Novartis realized the promise of imatinib, it priced the drug aggressively, charging patients $4,500 per month for a drug that they must take for the rest of their lives. Insurers have generally covered these costs, but patients without insurance have been less fortunate. Researchers estimate that 30 percent of patients with CML and GIST are noncompliant with imatinib because of its price. In a bitter irony, three of the patients from the original GIST trial had to discontinue the medication when they changed or lost their health insurance. All three had been in sustained remission while on imatinib; all three subsequently relapsed.[40] The high price of such wonder drugs can become an obstacle to their potential efficacy.

These problems may be soluble. The success of imatinib, and especially its expansion from CML to GIST, will inspire other drug companies to pursue potentially profitable compounds even if the patient population is small. Cancer researchers can now list twenty-four gene targets and potentially thirty new drugs that target them. This has fueled continuing optimism. One recent review argued that the "rapid development of next-generation sequencing technologies seems likely to be transformative. Within a few years, a complete cancer genome sequence will be obtainable for a few hundred dollars or less."[41] Researchers, however, have found more than they had bargained for. In November 2008 scientists in St. Louis published the first complete tumor genome, from a patient with acute myelogenous leukemia. Many others have followed, each holding the promise of new targets for drug therapies. This proved to be an embarrassment of riches. When a group at the Welcome Trust Sanger Institute's Cancer Genome Project sequenced a lung cancer and a melanoma, they found 22,910 and 33,345 mutations, respectively. Although this might provide "the foundation for prevention and treatment," doctors must first figure out which mutations are meaningful.[42] It will take back-breaking work to determine which of those tens of thousands of mutations hold the key to the next imatinib.

Warfarin: What Is the Value Added?

Many of the findings in cancer pharmacogenomics affect small numbers of patients. The situation is quite different with warfarin. Approved in 1954, warfarin is now the most widely used anticoagulant in the United States. Doctors prescribe it to prevent blood clots in people with clotting disorders, atrial fibrillation, and implanted vascular devices (e.g., pacemakers and heart valves). Each year 2 million Americans begin the drug, and doctors write 30 million prescriptions. Warfarin, however, is tricky to use. Like many cancer chemotherapies, it has a very narrow therapeutic index. If the drug levels are too low, patients do not receive the desired therapeutic benefit; if the levels are too high, they risk life-threatening bleeding after falls or other injuries. To make matters worse, individual responses are idiosyncratic. Effective doses vary by a factor of 100: patients require anywhere from 0.5 to 60 mg per day to achieve the desired degree of anticoagulation. Furthermore, drug levels are influenced by diet, especially by vegetables rich in vitamin K, and by the regularity with which patients take each dose. As a result of its wide use and many dangers, warfarin complications are the second-leading drug-related cause of emergency department visits.[43] To prevent these complications, hospitals and clinics have developed elaborate systems. Many have specialized "clotting clinics" that manage the initiation of warfarin therapy: patients receive strict counseling about diet and compliance, physicians use complex protocols to predict the most likely therapeutic dose, and patients submit to frequent blood tests to monitor their response until a stable regimen is found.

Pharmacologists have long looked to genetics to understand some of the person-to-person variability. Warfarin is largely metabolized by P450-2C9. Researchers in the 1990s showed that genotyping patients for 2C9 activity could help them predict treatment responses. Specifically, poor metabolizers of warfarin were at increased risk for bleeding complications during initiation of warfarin treatment. Homozygotes for one low-activity 2C9 variant can be treated at a fraction of the usual dose. Such discoveries made warfarin a prominent example of the promise of pharmacogenomics.[44]

This early phase of warfarin pharmacogenomics, however, ran into complications. Although genotyping patients for 2C9 can help predict response to warfarin, there is enormous variation even within the wild-type genotype, with drug clearance varying by more than an order of magnitude.[45] Further study provided some insight. Although 2C9 activity could predict certain complications, 2C9 genotyping did not explain most of the observed patient variation. Researchers in the mid-2000s identified two other relevant enzymes, Vitamin K epoxide reductase complex subunit 1 (VKORC1) and P450-4F2. Together, these three loci explained 30 to 40 percent of the observed variation in final warfarin dose. These findings led the FDA to revise the warfarin label to encourage physicians to genotype patients and use lower starting dosages when indicated.[46]

Clinicians, wondering whether the pharmacogenomic data would actually improve clinical outcomes, put the new knowledge to the test. The Warfarin Consortium developed two algorithms to predict final dose, one that incorporated genotyping and one that did not. It found that the addition of the pharmacogenomic data allowed it to detect a higher percentage of the patients who would end up with either unusually high or low warfarin requirements. However, as it admitted, these results did "not address the issue of whether a precise initial dose of warfarin translates into improved clinical end points, such as a reduction in the time needed to achieve a stable therapeutic INR [a measure of clotting activity], fewer

INRs that are out of range, and a reduced incidence of bleeding or thromboembolic events."[47] Such data forced the FDA to consider explicitly what level of evidence would be needed to justify new clinical guidelines. Would it require a randomized trial of traditional best practices versus enhanced genetic algorithms? Would it rely on surrogate markers (e.g., laboratory measures of clotting parameters) or clinical endpoints?[48] In May 2009 the Center for Medicare and Medicaid Services remained unconvinced: its analysis concluded that there was not yet enough evidence to justify reimbursement for routine genetic testing for warfarin. In February 2010, however, the FDA revised its label once again, providing more detail about genetic variants and predicted dosages. Support came from a comparative effectiveness trial that found that patients managed with a genetic-informed algorithm had 31 percent fewer hospital admissions (all causes) during warfarin initiation than those managed conventionally.[49]

Clinicians, however, remain unconvinced. Weinshilboum and other proponents of pharmacogenomics are frustrated that despite the growing knowledge base, "the clinical adoption of genotype-guided administration of warfarin has been slow, even though the evidence supporting such adoption is similar to the evidence supporting currently used clinical variables, such as age, drug interactions, and ancestral origin."[50] According to one review by experts at the Cleveland Clinic, neither the FDA nor expert panels endorsed routine testing "because there is currently insufficient evidence to recommend for or against it." Attempts to support warfarin genotyping through cost-effectiveness analyses have been undermined "because the benefit (clinical utility) is yet to be sufficiently characterized."[51]

The jury remains out. A sound biological rationale supports genetic testing for warfarin response, but testing has not entered routine clinical practice. Conventional warfarin management remains a viable alternative: careful dosing and close management, though administratively difficult, get patients to a safe and stable dose quickly. There is only a small window in which pharmacogenomics could demonstrate a marginal benefit. Furthermore, many of the warfarin complications occur not because of genetic factors, but because of disruptions caused by noncompliance or food interactions. No genetic algorithm will eliminate these effects, and this fact limits the possible impact of pharmacogenomics testing. The question may soon become moot. In October 2010 the FDA approved the first of a new generation of anticoagulants—rivaroxaban, debigatran, and apixaban. Operating by a different mechanism, these promise the same clinical benefit as warfarin but without the need for careful dose titration and close monitoring of clotting parameters.[52] If they really prove to be this safe, with a wide therapeutic index, then it is unlikely that pharmacogenomics testing will be relevant for their use. All the pharmacogenomics theory and practice of warfarin would fall by the wayside in favor of these new, simpler, and safer drugs. This suggests another potential response to pharmacogenomics: in cases where genetic testing becomes relevant because of the potential dangers of a drug, a better solution might be to develop an alternative drug.

Codeine: The Forgotten Exemplar

Despite all the advances of modern medicine, one of the most common presenting complaints of patients remains the same as it has been for millennia: pain. One of the most commonly prescribed types of drugs also remains one of the oldest: opioids. Codeine, for example, has long been a mainstay of pain treatment after tooth extractions, surgery, and fractures. Doctors know that patients vary in their response to codeine and other painkillers, although it is not clear why. Pain varies with the severity of the disease, the pain threshold of the patient, and the patient's response to medicines. Many patients do fine. They take one or two codeine pills after having their wisdom teeth extracted and recover without complication. Others experience less relief and quickly consume their full allotment of pills. When patients do consume a prescription more quickly than their physicians expect, this raises red flags. Is the patient a drug seeker, an opioid addict feigning symptoms in order to get more drugs? Is the patient selling the pills on the street for money? Disagreement about the need for opioid refills is a frequent source of patient-doctor conflict.

Developments in pharmacogenomics in the 1980s cast this problem in a new light. Codeine, it turns out, is a pro-drug: 3-methylmorphine. It must be metabolized by P450-2D6 into its active form, morphine. Researchers realized that poor metabolizers, unable to metabolize codeine adequately, experience only a weak analgesic effect from the drug. This affects 2 to 10 percent of whites in the United States. As Evans pointed out in 1999, "It is thus not surprising that there is remarkable interindividual variability in the adequacy of pain relief when uniform doses of codeine are widely prescribed."[53] Some patients suffer in silence from undertreated pain, while others protest, only to be labeled drug seekers or addicts.

Still other patients have the opposite problem. Ultrarapid metabolizers activate codeine much more quickly than usual and can experience life-threatening opioid intoxication. One patient became comatose after a routine dose given for cough suppression. Laboratory testing revealed that the resulting serum morphine level was twenty to eighty times what had been expected. This genotype has been reported in 1 to 7 percent of white patients and 25 percent of Ethiopians.[54]

Tests of P450-2D6 activity were among the first pharmacogenomic tests to become commercially available. Advocates worked assiduously to educate their readers about the potential of gene testing for codeine response. Many prominent reviews cited codeine as an example of the relevance of pharmacogenomics.[55] Yet the vast majority of prescriptions for codeine were given without pharmacogenomic testing. Why did physicians not bother to implement the potential of pharmacogenomic science? There are many possible reasons. First, codeine is often prescribed in acute settings, in dentists' offices or emergency departments where patients need immediate relief. No one wants to wait for the results of a pharmacogenomic test. Second, most physicians allow their patients to titrate their own dose, telling them to take one or two pills every few hours as needed for pain, an approach that works well enough most of the time. Third, doctors have many other options. If patients report that the codeine does not work, it is not necessary to do genetic testing: doctors can simply try other opioid or nonopioid analgesics. Fourth, doctors may not consider pain a problem important enough to justify the investment of time, money, and mental energy needed for genetic testing. As a result, as other exemplars of pharmacogenomics rose to prominence, codeine slowly disappeared from the reviews.[56]

A 2005 tragedy, however, turned attention back to codeine pharmacogenomics. A baby in Toronto died unexpectedly. Investigation revealed that the baby's mother had been taking codeine for two weeks—the drug is used widely to treat pain from episiotomy or cesarean incisions. Unfortunately, both the mother and the baby were ultrarapid metabolizers. The mother's breast milk had a morphine concentration of 87 ng/ml (1.9 to 20.5 would have been expected), and the baby had a level of 70 ng/ml (0.2 to 2.0 would have been expected on the basis of the mother's prescription). Other ultrarapid-metabolizing children have died from routine prescriptions after tonsillectomy. The FDA considered recommending 2D6 testing before prescribing codeine to breast-feeding mothers but, in the end, chose not to. A conjunction of several rare variants in other enzymes was usually needed for codeine toxicity to reach lethal levels.[57]

Similar ambivalence has emerged about other pain medications. Many enzymes have been implicated in determining how patients respond to opioid and nonopioid pain medicines. Results, however, have been inconsistent because of poor study design, the complexity of pain as a symptom, and the impact of multiple, interacting enzymes. This leaves experts unwilling to endorse routine pharmacogenomic testing for analgesics: "Each of the genetic variants investigated up to now seems to contribute in a modest way to the modulation of analgesic response . . . but it is premature to recommend the clinical application of a single genetic test." The potential remains unrealized. For testing to become commonplace, researchers will need to demonstrate "the superiority of analgesic therapy guided by pharmacogenetics to improve response rates, decrease adverse drug reactions and consequently increase cost efficacy."[58] Clinicians remain unconvinced.

BiDil: A Genetic Drug That Wasn't

No drug in the domain of genetic medicine has attracted more strident and sustained controversy than BiDil. Its story began in the 1980s.[59] Physicians then had several drugs to treat high blood pressure. Two of them—hydralazine and isosorbide dinitrate—both work as vasodilators. Cardiologist Jay Cohn speculated that because of their common mode of action, they might have synergistic effects if they were used in combination. He led one trial, Vasodilator Heart Failure Trial I (V-HeFT I), that demonstrated the value of the combination therapy. A follow-up trial, V-HeFT II, however, found that the combination was less effective than another new class of drugs, angiotensin-converting-enzyme inhibitors. Despite this setback, Cohn pursued and won a patent for the method of using the drugs as a combination treatment in 1989. He trademarked the combination as BiDil in 1992. The FDA, however, rejected his application for approval because of insufficient evidence of efficacy. Cohn returned to his original data and found a hint that the drug had greater efficacy in the forty-nine African American patients in V-HeFT I than it did in the white patients, a result he published in 1999. The FDA encouraged him to seek more evidence for this theory. In 2001 he launched the African American Heart Failure Trial (A-HeFT), a randomized study of the efficacy of BiDil involving over 1,000 heart-failure patients who self-identified as black. He received another patent for the use of BiDil specifically in African Americans in 2002. The A-HeFT trial monitoring committee stopped the trial prematurely in 2004 because of promising early results: a 43 percent reduction in heart-failure mortality in the patients on BiDil. Experts estimated that 750,000 patients might switch to BiDil, generating $825 million in profits. The stock of NitroMed, the company that owned BiDil, soared. The FDA followed through on this enthusiasm and approved BiDil for use in patients who self-identify as black in June 2005.

An astute observer might ask at this point, where is pharmacogenomics? Genetics played little explicit role in this early phase of BiDil history. The official drug label approved by the FDA in 2005 did include pharmacogenomic information, specifically, a warning that roughly half of all patients are fast acetylators of hydralazine and, as a result, will experience lower exposure to the active drug.[60] But the developers of BiDil never made this fact part of the BiDil discourse. Acetylation variants did involve pharmacogenomics, but they had nothing to do with the race specificity on which the 2005 incarnation of BiDil depended. Instead, the logic of BiDil depended on an implicit assumption that blacks and whites were different, and that this difference—at least as it related to heart-failure pathophysiology and treatment—was rooted in genetics. The superior efficacy of BiDil in blacks versus whites seemed to validate both the basic assumptions of pharmacogenomics and the widespread faith that race could be used as a proxy for human genetic variation.

Such claims produced a furious outcry, especially from social scientists attuned to the arguments about race and genetics in the United States. They argued that the appeal of racial therapeutics led the FDA to approve the drug despite obvious problems with the evidence presented by NitroMed and A-HeFT. They described how the legacy of the Tuskegee syphilis study loomed in the background of the FDA debates, with the federal government wanting to atone for past sins by approving a drug specifically to help African Americans. They traced how NitroMed funded the Congressional Black Caucus, the National Medical Association, the Association of Black Cardiologists, and the National Association for the Advancement of Colored People, all of whom encouraged the FDA to approve the drug. They described BiDil as a cynical effort to exploit race and loopholes in patent law and FDA policy in order to extend patent protection on an old drug. After all, even as proponents of BiDil sought a race-specific approval—the only thing that would extend their patient protection—they argued that the drug would likely work well in all races.[61] Defenders of the drug went so far as to accuse social scientists of trying to kill black people by sowing controversy about BiDil and misrepresenting it as a racial drug, even though it was Cohn and NitroMed who had pursued the race-specific approvals in the first place.[62]

Amid the claims and counterclaims, some BiDil proponents attempted to stake out a middle ground. They argued that the racialization of the drug and racial therapeutics in general were necessary evils, a passing phase on the road to fully individualized treatments. Race provided enough information to motivate race-specific use of a drug like BiDil, but ideally, it would be abandoned once further research identified the underlying genetic basis of the variations in BiDil response. As Gary Puckrein, an advocate for minority health, argued: "Both the drug's manufacturer and its critics advocate continued investigation into the genetic and other possible reasons for the difference in response to the drug, and both anticipate that the social concept of race will be superseded by more objective and precise criteria, narrowing the focus among African Americans to specific individuals who will benefit from the drug while also identifying non–African Americans who will benefit—probably on the basis of a genetic feature that is more common among African Americans than among other racial or ethnic groups." Doctors could use BiDil racially and, in the meantime, figure out how to use it individually.[63]

BiDil did not turn out as expected. Despite projections of a billion-dollar bonanza, the drug proved to be a commercial failure. NitroMed lost $108 million in the first year after FDA approval. It is not clear which of several factors contributed most to its demise. NitroMed priced the drug high—$1.80 per pill, or $10.60 per day for a common dose. Since BiDil was simply a fixed-dose combination of two existing drugs, each of which was available as a cheaper generic, many insurers simply substituted the generics whenever physicians did prescribe BiDil. Moreover, the controversy over the "black drug" was read in many ways by patients and doctors. While some saw it as something valuable, others saw the "special treatment" as uncomfortably reminiscent of Tuskegee. Whatever the causes of its failure, NitroMed laid off most of its workforce and stopped marketing BiDil in January 2008.[64]

Even as the drug failed—and perhaps inspired by its failure—NitroMed did follow through on one initial promise: it searched for genetic predictors of BiDil response. Finding them would have accomplished several things. First, the company might have been able to obtain new patents for the use of BiDil in patients with specific genetic markers. Second, it would have been able to market BiDil not just to people who self-identify as black, but to anyone who had those markers. This would have both specified and broadened the future market for the drug. NitroMed's research team quickly set out to analyze the A-HeFT patients, looking for genetic markers that predicted response rates. By 2007 it had some preliminary but enticing results. As NitroMed's Jane Kramer explained, "We often say that our indication for self-identified blacks is a very uncomfortable and a very uncertain proxy for patients with heart failure who could benefit from our drug. . . . We do hope that further genomic studies are going to help focus that indication much better, but at the moment this is all we have."[65]

One trial looked at polymorphisms in endothelial nitric oxide synthase, an enzyme relevant because of BiDil's putative role in nitric oxide metabolism. As the authors explained, "Differential responsiveness to medical therapeutics may occur as a function of race and ethnicity and these differences at least in part may reflect differences in genetic background or gene-environment interactions." Since past studies had found that NOS3 variants influence heart-failure outcomes, the researchers examined the impact of different NOS3 genotypes in the A-HeFT data. The study found that most patients who benefited from BiDil did indeed have a particular NOS3 variant that was common in blacks but was also found in 40 percent of whites. Moreover, this finding was mostly due to differences in the quality-of-life measure (a comparison of the Minnesota Living with Heart Failure Questionnaire at baseline and at six months). There was no difference in event-free survival (i.e., deaths or hospitalization). As a result, the authors admitted that "this finding must be interpreted with caution." The authors, as authors often do, concluded with the hope that future research would clarify these results.[66] Nothing further, however, has been published on the genetic basis of BiDil's efficacy. The FDA has taken no action on BiDil's putative pharmacogenomic biomarkers.

BiDil offers a cautionary tale about the potential of race-based therapeutics and the use of race as a passing phase on the road to fully individualized therapeutics, but it is unlikely that it will be the only example of such a tale. Race has been an important feature of the field ever since the first emergence of pharmacogenetics in the 1950s.[67] It appears repeatedly in prominent reviews. Writing in *Science* in 1999, for instance, Evans and Relling asserted that "the marked racial and ethnic diversity in the frequency of functional polymorphisms in drug- and xenobiotic-metabolizing enzymes dictates that race be considered in studies aimed at discovering whether specific genotypes or phenotypes are associated with disease risk or drug toxicity." Other reviews typically mention race, at least in passing, as a relevant pharmacogenomic concern.[68]

However, as happened with BiDil, the details of race-based prescribing have not turned out as expected. For instance, in 2001 Alastair Wood defended the use of race in medicine in the *New England Journal of Medicine*. He pointed out that the frequency of two poor metabolizer alleles of 2C9—relevant for warfarin—were higher in whites (11 percent and 8 percent) than in blacks (3 percent and 0.8 percent). The frequencies of these alleles were also low in Asian populations. So are whites, on average, more likely to be poor metabolizers and thus poor responders? The opposite is true. As Wood admitted elsewhere that same year, "In contrast to what would be predicted from ethnic differences in the frequency of the CYP2C9*3 allele, white patients require higher warfarin doses than Asians to attain a comparable anticoagulant effect."[69] The promising exemplar of racial pharmacogenomics proved to be yet another example of the complexity of the genetics of drug response. One of the best reviews of this problem looked at race-based differences in response rates to antihypertensives.[70] Many clinical trials have described racial differences in drug response. Captopril, for instance, lowered diastolic blood pressure more in white patients than in black patients. The racial differences have been consistent and significant. In every study, however, the amount of variation within each racial group was far larger than the differences between the groups: each group had a broad bell curve of drug responses, and the curves of the two groups overlapped substantially. As a result, 80 to 95 percent of all black and white patients will likely have indistinguishable responses to each medication. Although racial differences might exist, they are irrelevant for the majority of patients.

In the end, BiDil provides a telling example of the dangerous allure of race and genetics in medicine. No other pharmaceutical company has followed NitroMed's lead and applied for race-specific patents. However, a move toward racial patenting and marketing has appeared in less regulated domains. The number of race-based patent applications and awards has increased significantly over the past twenty years. Companies now market race-based vitamins. Nike has even developed a specific line of walking shoes, the Nike Air Native, tailored for the supposedly unique thickness of Native American feet. Such products stand less as a testimony of science than as a testimony of the identity politics and the "durable preoccupation" with race that made BiDil possible.[71] The adverse consequences that follow this reintroduction of race thus far outweigh whatever benefits might accrue.

The Promise and Perils of Pharmacogenomics

No one doubts that pharmacogenomics makes good sense. Of course there is genetic variability between people. Of course it can be detected. Of course some of it influences how individuals respond to drugs. But does this make it a relevant guide for clinical practice? Does it merit a substantial investment of health-care resources? The answer in some cases will be yes—but in which ones? In 2010 Collins and colleagues launched a series of articles on genomic medicine in the *New England Journal of Medicine.* Their introductory essay began with another clinical vignette, but this one captured the mixed legacy of pharmacogenomics. A 40-year-old woman, concerned about her risk for breast cancer, obtained a commercially available gene test kit. Her doctor dismissed the results as unreliable. Because of her family history of breast cancer and her Ashkenazi ancestry, she was referred for genetic testing (of *BRCA1* and *BRCA2*). Although both tests were negative, a screening mammogram found breast cancer. A biopsy allowed the tumor to be tested for HER2 expression; since this test also was negative, the patient received the conventional treatment, tamoxifen.[72] Genetic medicine appeared repeatedly in the vignette but never made a useful contribution.

It is likely that the clinical prospects of pharmacogenomics will be disease specific, depending on the particular pathophysiology. Some diseases have especially good prospects. Cancer pharmacogenomics makes some sense. If treatment is influenced by, or even depends on, genetic polymorphisms of specific tumors, then it would be wise to take these into consideration. Infectious diseases offer a parallel case. The genetic diversity of viral and bacterial pathogens, and especially of their patterns of drug resistance, justifies efforts to genotype them. Other diseases, however, lend themselves less well, typically because the genetics of the disease and its treatment responses are less well understood. Much more needs to be known about the science of depression or coronary artery disease before pharmacogenomic practice becomes a compelling clinical reality.

As promising cases for pharmacogenomic medicine emerge, financial considerations will become more significant. Cost will be an issue not just for the genetic testing but for the targeted treatments as well. It is possible that physicians and patients will accept these costs for some diseases but not others. Cancer, for instance, motivates a no-expense-spared approach to a greater extent than other areas of medicine. Novartis can charge $4,500 per month for imatinib, and insurers will pay. If they did not, patient advocacy groups would rise up in protest.[73] It is not hard to imagine ranking diseases on a priority scale and guessing where insurers and society would draw the line for expensive therapies. Would they be tolerated for diabetes (and would a distinction be made between type 1 and type 2?), asthma, obesity, depression, arthritis, or erectile dysfunction? Pharmacogenomic pioneers will have to remain closely attuned to what they think the market will bear as they pursue new and complex treatments.

A closely related question is exactly what value is added. For drugs like imatinib, the payoff seems clear. Patients with CML or GIST had a terrible prognosis before imatinib. Now patients with CML can expect to live thirty years after diagnosis. Long-term survival for GIST has improved tenfold.[74] But in other cases the benefit is less clear. Will genotyping for warfarin or clopidogrel reduce complications? How useful is a new cancer therapy that extends survival by only a few months? In some cases other options exist. If trial and error can get a patient to a stable dose nearly as quickly and safely, then is this traditional approach good enough? Trial and error has an added advantage: by considering the most downstream variable—the actual clinical endpoint—it incorporates other variables, especially environmental exposures, polygenic effects, and patient compliance, that are essential to the ultimate treatment outcome. Gene tests will never be a replacement for careful monitoring of treatment response. At best they will be one tool among many to optimize outcomes.

A wholehearted commitment to pharmacogenomics would also bring certain downsides. One is the question of attention in both research and clinical care. Because so much effort is now invested in finding pharmacogenomic treatments, less effort is invested in other avenues. The race to develop drugs for cancer patients who happen to have identified mutations does little for the other patients whose tumors lack those mutations (these patients also have little commercial appeal for pharmaceutical companies). As the Nuffield Council on Bioethics wrote in 2003, "Pharmacogenetics may significantly improve medical treatment for some people, but it may also result in more people falling into categories for which effective drugs are not developed, because of inadequate financial incentives to bring to market a drug that may be very effective but only for a small population or for a large but poor population."[75]

The Nuffield Council expressed similar concerns about clinical care: "The application of pharmacogenetics might impede healthcare delivery, by taking up too much of clinicians' time."[76] This is especially relevant in light of uncertainty about the most important determinants of treatment failure. Advocates of pharmacogenomics assume that genetic factors play a substantial role and that attention to pharmacogenomics will improve clinical outcomes. Could a similar benefit be obtained by attending to diet and other environmental factors that influence drug metabolism? Could an even greater benefit be achieved by rooting out noncompliance? Physicians have worried about the regularity with which patients take medicines since the time of Hippocrates. Clinicians have known since the 1950s that patients do not always take medications as directed. Estimates suggest that patients take only half of the prescribed doses.[77] If a patient misses half of all doses, then drug levels will be half of what the doctor intended.

Another issue is race. Race rode the coattails of personalized medicine and has achieved prominence within clinical practice once again. Some proponents of pharmacogenomics have encouraged the use of race as a proxy for individual genetic variation in order to guide treatment decisions, especially when relevant drug-metabolizing alleles have been shown to vary among conventionally described racial groups. Is there a clear link between genetics and race? Racial categories are malleable: different countries parse similar substrates into different categories. Some categories make little sense at all. Someone who self-identifies as Hispanic in the United States could have ancestry that is 100 percent European, 100 percent African, or 100 percent Mayan. Such ambiguities have led some experts to urge caution. As the Nuffield Council concluded, "Since clear-cut divisions between racial or ethnic groups are highly unlikely, we take the view that membership of a particular racial group should not be used as a substitute for a pharmacogenetic test, even if it is the case that the genetic variant being tested for is known to be more or less prevalent in particular groups."[78] Because of the complexity of human genetic variation, genetic medicine will need to be individualized and cannot stop at the level of race, ethnic group, or other label.

Last of all come the questions of cost and equity. If pharmacogenomics produces highly specific and highly expensive new drugs, as has happened in cancer therapeutics, it will exacerbate disparities between those who do and those who do not have access to them. This is true within countries such as the United States as well as between the United States and middle- or low-income countries. Pharmacogenomics would thus exacerbate existing health inequalities between rich and poor.[79]

So what could personalized medicine involve, beyond Hamburg and Collins's narrow vision? Pharmacogenomics has probably earned a role as one of several approaches. Doctors should also attend to diet, smoking, and other environmental exposures. They should explore the social or economic obstacles that prevent patients from following prescriptions closely. They should develop their relationships with patients in order to create the strongest possible treatment alliance. Some discussions of pharmacogenomics appreciate this. Varmus, for instance, emphasized the broad possibilities implied by "personalized medicine": "Both genetic and nongenetic information are important; the more we know about a patient—genes and physiology, character and context—the better we will be as physicians."[80] For medicine to be fully personalized, clinicians, funders, and regulators will need to look beyond genomics. What is needed is not a national highway system but a network of country roads. The more intimately physicians know their patients' worlds, the more successful their treatments will be.

PART THREE

Genetics in Human Behavior and Culture

CHAPTER TWELVE

The Persistent Influence of Failed Scientific Ideas

JONATHAN BECKWITH

"THEY SOUND LIKE the Jukes and the Kallikaks." When I was a kid in the 1940s, my mother would occasionally blurt this out while talking about some problematic family. She apparently accepted the claims about these two icons of family genetic inferiority. Two books about the Jukes and the Kallikaks, one appearing in the late nineteenth century and the other in the early twentieth century, made these families famous and infamous. The Kallikaks, whose notoriety lasted well past the middle of the twentieth century, were described by psychologist Henry Herbert Goddard in his 1912 book *The Kallikak Family: A Study in the Heredity of Feeble-Mindedness.*[1] According to Goddard, the pseudonymous Martin Kallikak had fathered several children with two women, one a respectable Quaker and the other a supposedly "feeble-minded" woman. According to Goddard, all of the children from the former relationship were "wholesome" and had no signs of "retardation," while the latter union produced what he called "a race of defective degenerates." Appearing in the early days of the eugenics movement in the United States, Goddard's work, including his 1912 book, was taken to justify much of what the eugenicists proposed. Yet even by the scientific standards of that time, the study of the Kallikaks was seriously flawed. He and his colleague used extraordinarily subjective evaluations of the mental capacities and character of its subjects. Estimates of the qualities of deceased Kallikaks were deduced from their "reputation." Estimates of those who were still alive were based on a research assistant's personal observations. As a result, some scientists, including most notably James McKeen Cattell, psychologist, eugenics fan, and editor of *Popular Science Monthly*, were critical of Goddard from a scientific perspective. These critical voices from the scientific community increased as the years went by. By 1940, one leading psychologist stated that the Kallikak study had been "laughed out of psychology."[2]

Despite the disdain expressed by members of the scientific community, the myth of the Kallikaks retained its hold on the public imagination. In 1939 James Joyce alluded to the Kallikaks in a passage of *Finnegan's Wake* when he was referring to hereditary feeblemindedness. And in a 1961 psychology textbook by Garrett and Bonner, a section on heredity included a dramatic cartoon showing the two "contrasting" Kallikak family lines.[3] Thus, from the publication of Goddard's book in 1912 to Garrett's textbook, which was used in many psychology courses for years after 1961, the myth of the Kallikaks was perpetuated for more than fifty years. Although scientists had rejected the conclusions of Goddard's report at an early stage, they failed—or did not try—to reach the public to discredit it sufficiently and widely enough to interfere with continued adherence to the myth.

I use this history to introduce a number of examples from the research field of behavioral genetics where faulty ideas have had significant social consequences. Poorly done or overinterpreted and biased studies in this area have often been published in leading scientific journals, have received widespread publicity, and have eventually been dismissed and rejected by much of the scientific community, but nevertheless have taken on a life of their own. Some of these ideas have had a surprisingly long shelf life after their rejection. The length of the shelf life and its attendant social consequences are in part due to the failure of the scientific community to take more responsibility for informing the public of the flaws of such socially consequential science.

XYY and Crime

In 1965 an Edinburgh research group led by Patricia Jacobs reported in *Nature* magazine the discovery of a possible genetic link to criminality and aggressive behavior.[4] The article, "Aggressive Behavior, Mental Subnormality and the XYY Male," described the finding of an unusually high proportion of males with an extra Y chromosome in a state hospital for the criminally insane. Although this study attracted considerable attention when it was published, the idea of an association between the extra Y chromosome and criminality was greatly amplified when a mass murderer in Chicago was identified in the media as an XYY male.

One night in 1966, Richard Speck entered a residence for nursing students and murdered eight of the sleeping students. He was quickly arrested by the police. A geneticist was quoted by newspapers as speculating that Speck might be an XYY male because he was quite tall and suffered from severe acne, two characteristics considered typical of XYY males. This speculation plus a miscommunication between a pathologist who examined Speck's chromosomes and a *New York Times* reporter led to widespread reporting that Speck was indeed an XYY male. The media leaped on this story with headlines such as "Chromosomes and Crime" and "Born Bad."

But Speck was an XY male. Vanderbilt University cytologist Eric Engel had asked in 1966 if he could look at Speck's chromosomes and had found that Speck had only one Y chromosome. When the newspapers reported in 1967 that Speck was an XYY male, Engel repeated his analysis and again found him to be XY. Although the Associated Press distributed a report of Engel's finding, it was never covered by major newspapers or news magazines. Engel later published an article in the *American Journal of Mental Deficiency Research* where he reported his finding and expressed being "amazed" that the misinformation about Speck had persisted despite the chromosome analysis.[5]

In the years that followed, scientific reports about XYY males appeared, although the links to aggression and criminality were few. XYY males in institutions (including those of the original Jacobs study) were the least aggressive inmates and had committed crimes against property, not people. Very few studies could replicate Jacobs's findings; in particular, there was no higher frequency of XYY males in the prisons. A 1974 scientific review concluded that "the frequency of anti-social behavior of the XYY male is probably not very different from that of non-XYY persons."[6] In 1975 a major study of XYY males supported the conclusions of the 1974 review.[7] In 1982 Jacobs, dismayed by the media coverage of her study, stated, "In retrospect, I should not have used the words 'aggressive behavior' in the title of my paper and should not have described the institution as a place for 'the treatment of individuals with dangerous, violent or criminal propensities.' "[8]

The research done since the 1970s indicates that the only characteristics that may be applied to XYY males generally are that they are taller and tend to exhibit slower learning curves in certain subjects. One report studied the development of XYY males from birth along with XY and XXY males as controls. The authors present "evidence for a slightly increased liability to antisocial behavior in XYY men." However, controlling for the various factors that might have influenced behavior of the

XYY males, the authors conclude that "IQ contributes significantly to the likelihood of having a conviction, but that social class and karyotype [the extra Y chromosome] does not."[9]

Despite the criticisms, failed replications, and eventual dismissal of the idea of the criminal chromosome among scientists, the myth of the XYY male persisted and spread through the culture as an indication of genetic roots for criminality, much as the Kallikaks had as an exemplar of intelligence's roots in heredity. In high-school biology and medical-school psychiatry textbooks, sometimes illustrated with Richard Speck's picture, students continued to learn well into the 1970s that XYY males were superaggressive and prone to criminality. A series of British novels about "the XYY man," a superspy for MI5, was turned into a Granada TV series in 1976.[10] Two movies "featured" XYY males, Italian horror-movie director Dario Argento's *The Cat o'Nine Tails* in 1971 and *Alien*³ in 1993. In the latter, Sigourney Weaver lands her damaged spaceship on a planet inhabited by XYY males who have been exiled from Earth. Their leader tells Weaver that they are a bunch of "thieves, murderers, rapists and child molesters . . . all scum." In the same year in which *Alien*³ appeared, an episode of the television show *Law and Order* featured a story of a 14-year-old boy with an extra Y chromosome who beat his friend to death. In 1985 Harvard professors James Q. Wilson and Richard J. Herrnstein used the supposed link of XYY to criminality to support arguments in their book *Crime and Human Nature*.[11] In a class I teach, I still hear from some graduate students that they were taught in high-school biology courses the validity of the XYY-criminality connection.

The wide acceptance of claims about the extra Y chromosome has also had a direct social impact. In the 1970s, in at least two states, juveniles involved in crimes were screened for the XYY genetic makeup.[12] Studies in the United States and other countries screened newborn baby boys for the extra Y chromosome in research that posed significant ethical problems. Broader claims that criminality was genetically determined, not the result of poor environments, were bolstered by the reports on XYY males.[13]

Like the Kallikaks, the XYY story has had an extraordinarily long shelf life, having been perpetuated for at least forty years after the original report by Jacobs and her colleagues. Both Eric Engel, the cytologist, and Patricia Jacobs, the geneticist, stated disappointment and regret that the story had been blown up to such proportions, but their subsequent statements in scientific journals never received any attention, while the original "criminal chromosome" version of the story persisted. One possible sign that after all these years the tide of public misinformation may have begun to turn is a 2007 episode of *CSI Miami* in which the crime investigators state that the original claim for a determinative role of the extra Y chromosome in criminal behavior has been disproved.

MAOA and Crime

In 1993, twenty-eight years after the XYY story had begun (also the year of *Alien*³), a new stimulus of scientific and media interest in genetics and criminality appeared. A research group in the Netherlands published two articles describing studies of a family in which many of the males exhibited incidents of aggressive or antisocial behavior.[14] In *Science* magazine the researchers reported that all the male members of the family who showed this behavior carried an altered version of a gene involved in brain function. The gene encoded the enzyme monoamine oxidase A (MAOA), an enzyme involved in the metabolism of serotonin. The MAOA gene mutation in these male subjects eliminated all MAOA enzyme activity. The article ended with the proposal that "given the wide range of variation of MAOA activity in the normal population, one could ask whether aggressive behavior is confined to complete MAOA deficiency." In this relatively moderate language the authors were suggesting that their finding might lead to a much broader understanding of aggressive behavior in society. It is the norm rather than the exception for scientists to speculate about the potential broader implications of research. In that sense, this last sentence does not seem any different from many mainstream scientific presentations. The authors apparently had no idea how this suggestion would be taken.

When these two articles appeared in 1993, *Science* chose to highlight the research findings with a news article titled "Evidence Found for a Possible 'Aggression Gene.'"[15] The author of the news article reported that "it might be possible to identify people who are prone to violent acts by screening for MAOA gene mutations." It is likely that this highlighting and the reference to an "aggression gene" helped initiate the media frenzy that followed. *Newsweek* reported on the study in an article titled "The Genetics of Bad Behavior: A Study Links Violence to Heredity."[16] Accompanying the report was a photograph of a bloody confrontation between Israelis and Palestinians. A local Boston television station reported the story with film footage showing actors re-creating a mugging. *U.S. News and World Report* showed a baby dressed in a black-and-white-striped prison outfit on its cover designed to highlight what was actually a somewhat critical article.[17] In 1994 lawyers for a man convicted of murder argued for leniency in sentencing by suggesting that the man's actions might be due to MAOA deficiency or some similar defect on the basis of generally aggressive behavior among the man's relatives.[18] The discussions of criminal responsibility and genetics heightened by these stories led to discussions two years later by judges at a summer conference who asked geneticists whether genetics was now saying that many criminals had no free will[19]—an extraordinary voyage from the study of one family to potential influences on courtroom decisions.

Like Patricia Jacobs and Eric Engel in the XYY case, two of the researchers involved in the MAOA studies expressed concern and regret over the public impact of reports of their work. The leader of the project, Hans Brunner, stated at a symposium on genetics and criminality, "The notion of an 'aggression gene' does not make sense." And Xandra Breakefield, who had determined the nature of the mutation in the MAOA gene, quit working on the subject out of dismay at the way it was presented by the media.

In the years since the MAOA publications from Brunner's group, no other family has been reported with the mutation identified in the Brunner study. However, an article published in 2002 by Caspi and colleagues on common polymorphisms of the MAOA gene suggested that variation in the levels of the MAOA enzyme could be associated with antisocial behavior under certain cirumstances.[20] The authors presented evidence that children who carried a polymorphism thought to lower MAOA levels and who had been subjected to child abuse were more likely to exhibit antisocial behavior than those children who carried a different polymorphism. However, attempts to replicate this finding have given mixed results, leaving the validity of this finding in question.[21] Although some more significantly direct deterministic genetic variants for some fraction of antisocial behavior may eventually be established, neither XYY nor the common MAOA polymorphisms represent such variants.

Boys, Girls, and Mathematical Ability

In 1980 researchers from Johns Hopkins University published in *Science* magazine an article on mathematical test performance differences between boys and girls. The impact of this study continued to reverberate in subsequent decades. In their article titled "Sex Differences in Mathematical Ability: Fact or Artifact?,"[22] Camilla Benbow and Julian Stanley reported that they had administered the math component of the Scholastic Aptitude Test to seventh- and eighth-grade students who were mathematically precocious and had analyzed differences between performances of boys and girls. The most striking finding was that at the highest-level scores on the test (700–800), boys outnumbered girls thirteen to one. Attempting to distinguish between explanations that invoked social factors and those that invoked biological explanations to account for this dramatic difference, the authors quantified the number of courses that the students had taken and the students' expressed interest in mathematics. Finding no differences between boys and girls in these criteria, the authors concluded their *Science* article: "It seems likely that putting one's faith in boy-versus-girl socialization processes as the only permissible explanation of the sex differences in mathematics is premature."

Again, *Science* featured this study in its "News and Views" section with an article titled "Math and Sex: Are Girls Born with Less Ability?" and subtitled "A Johns Hopkins Group Says 'Probably': Others Are Not So Sure."[23] Statements from the authors themselves heightened the impression that social explanations for the differences should be downplayed. For instance, Benbow stated in the *Science* news article, "Women would be better off accepting their differences and working to encourage girls to achieve as much as they can than to constantly blame their lesser achievements in mathematics solely on social factors."

A media blitz followed. *Time* magazine's article "The Gender Factor in Math: A New Study Says Males May Be Naturally Abler than Females" was accompanied by a cartoon in which a schoolgirl is unable to multiply two single-digit numbers while a boy completes the multiplication of a four- and a three-digit number.[24] *Newsweek* titled its report "Do Males Have a Math Gene?"[25] Articles in both the weekly and Sunday *New York Times* also reported on the study immediately upon its publication. This story in its most determinist form spread very rapidly to magazines such as *Playboy, Reader's Digest,* and the popular science magazines that were being published in that era.[26]

The "others not so sure" of the subheadline in the *Science* article were many. Articles published in educational, psychology, and other academic journals pointed to the many social factors that Benbow and Stanley had ignored.[27] These articles referred to studies showing the impact of the kinds of toys boys and girls played with, social attitudes that led to boys' disdain of girls who were "science whizzes," and the influence of parents' and schoolteachers' preconceptions of differences in girls' and boys' aptitudes in math, as well as prejudices that girls interested in mathematics faced in college and beyond. In fact, the number of articles that had explored social factors influencing math ability vastly outnumbered any that leaned toward a biological explanation. Benbow and

Stanley had hardly exhausted possibly explanatory social factors by comparing the number of courses taken by boys and girls and assessing expressed interest in math of boys and girls. Despite these critical responses, restricted largely to academic journals, it was the idea of the "male math gene" that persisted in the public's mind. Studies done to estimate the impact of the extensive publicity reported a negative influence on parental attitudes and on the attitudes of school students themselves toward girls' math abilities.[28]

A further indication of the persistent impact of this study came twenty-five years later, in 2005, when Lawrence Summers, president of Harvard University, in a widely reported speech offered possible explanations for the low number of women in mathematics and related fields. For one of these explanations, he alluded to scientific evidence supporting a biological explanation for this difference, stating that "in the special case of science and engineering, there are issues of intrinsic aptitude."[29] Elsewhere Summers cited as backing for this statement psychology professor Steven Pinker's book *The Blank Slate,* in which the Benbow-Stanley study is referred to as evidence for the "intrinsic aptitude" explanation.[30] The negative response of many Harvard faculty to Summers's speech on this subject may have contributed, along with other issues, to his resignation in 2006.

But what is surprising about this scientific myth's lasting power is that continuation of the Johns Hopkins University study of mathematically precocious youth in the years since 1980 has shown a progressive reduction in the difference between boys' and girls' performance on the math tests. By 1997 the 13:1 ratio had dropped to 4:1 and in 2005 to 2.8:1.[31] Clearly this improvement in girls' performance is not due to changes in girls' genes in the intervening years. One possible explanation for these changes is the altered social environment beginning in the 1970s in which women pushed for improved educational environments for girls studying math and science and increased understanding of the social factors that prevented their advancement. Recent cross-cultural studies have presented evidence that women's achievements in math and science are correlated with attitudes toward gender of the society in which they live.[32]

What is perhaps not startling, but quite disappointing, is the absence of any media coverage of these recent findings and others that effectively undercut the basis for the conclusions of Benbow and Stanley in 1980. There have been no "News and Views" stories in *Science,* no newspaper reports, and no headlines in *Time, Newsweek,* or *Playboy.* It seems that refutations of what were once highly newsworthy stories and, in cases such as this, ones that influenced public attitudes are not newsworthy themselves. One journalist related to me that when she proposed to write an article for her newspaper about the controversy over the boys-girls-math study, her editor's response was that he was not interested in such an article unless fraud in the original study could be established.

Smart Genes, Stupid Genes

In 2005 *Science* published an article reporting the discovery of a human gene variant that was proposed to be important for the ongoing evolution of the human brain.[33] The researchers, led by Bruce Lahn of the University of Chicago, found that most humans have one of two versions (polymorphisms) of a gene called microcephalin. The microcephalin gene had previously attracted attention when researchers found that complete deletion of this gene led to a rare condition in which babies are born with very small brains. However, the specific function of the protein encoded by the microcephalin gene is unknown.

Lahn and his coworkers investigated the frequency of these two polymorphisms in peoples from different parts of the world. They found that one of these variants was quite common in peoples of sub-Saharan Africa, and the other was found more often in populations in the rest of the world. They then asked whether one or the other polymorphism had been evolutionarily selected for in the populations where it was common, thus conferring some advantage on that population. To answer this question, they used a computer program to tell them whether the prevalence of one of the polymorphisms in particular populations was due to genetic selection in the evolution of modern humans. They reported that results from the computer analysis indicated that the polymorphism found more commonly outside Africa had been selected for sometime between 14,000 and 60,000 years ago. They then went on to point out that this period was one in which "modern human behavior, such as art and the use of symbolism," arose. As a result, in their conclusion they speculated that the polymorphism found more commonly in Europeans and other populations and not in sub-Saharan Africans may have enhanced one or more of several traits, including "brain size" and "cognition." These speculations on their findings were readily interpreted by others to suggest the possibility that sub-Saharan Africans were not as culturally advanced as other peoples for genetic reasons.

In the same issue of *Science,* Lahn's article was highlighted with a "News of the Week" article titled "Are Human Brains Still Evolving? Brain Genes Show Signs of Selection."[34] Despite the cautious wording of the "News" article, Lahn's study was picked up by many newspapers with reports that included Nicholas Wade's *New York Times* article "Brains May Still be Evolving, Studies Hint" and Ronald Kotulak's *Baltimore Sun* article "Two Evolving Genes May Allow Humans to Become Smarter."[35] *Discover* magazine deemed the report one of the "top discoveries" of 2005. Without waiting for confirmation of these speculations, those with a political ax to grind joined in with their comments. John Derbyshire in *National Review Online* stated, "Our cherished national dream of a well-mixed and harmonious meritocracy . . . may be unattainable."[36] To the *New Scientist,* University of Wisconsin anthropologist John Hawkes claimed, "Whatever advantage these genes give, some groups have it and some don't. This has to be the worst nightmare of people who believe strongly there are no differences in brain function between groups."[37]

This rapid extension from the scientists' preliminary results and expansive speculation to commentators' claims of relevance to issues of equality is disturbing, but not unusual, given the other examples I have discussed. Nevertheless, it is hard to see how one is justified in making such leaps of logic in the absence of any knowledge of what the microcephalin gene does other than the extreme effects of its complete deletion. Nothing is known about the protein product of the gene, about its function in the brain or in other parts of the body, and about whether there are any effects of the two polymorphisms on the expression of the gene. At the time of publication, the authors had not sought evidence for differences in brain size or brain function in individuals carrying one or the other polymorphism. Moreover, during the vast time range of 14,000 to 60,000 years ago, which the authors concluded was when one of the polymorphisms had been selected for, many other changes in human living conditions and culture were occurring in addition to the rise of art and symbolism. A focus on those specific features of the evolution of human culture seemed a particular stretch. Finally, other human geneticists questioned the validity of the specific computer program used by Lahn's group as a means of concluding anything about the evolutionary selection of one of the polymorphisms.[38] Using what they considered to be a more sophisticated computational approach, these scientists concluded that this polymorphism had not been selected for. However, within a year all these criticisms became irrelevant when other scientists, trying to test the speculative suggestions of Lahn and coworkers on brain size and cognition, showed that they could not be correct. Specifically, these groups reported that there was no difference in brain size or in IQ test performance between people with one or the other polymorphism.[39] Moreover, Lahn stated that he had tried determining the version of the microcephalin gene he himself had, that the experiment's outcome was "blurry," and that "it wasn't looking good." All of this was reported in the second news article in *Science*.[40]

In this case it seemed that the corrective features of the scientific process had worked. Scientists report, others test, and proposals are shown to be incorrect. Yet, as I described in previous examples, there is a problem in the absence of media coverage of new results that falsify aspects of the original reporting. Other than the *Science* article, the only media report I could find on the refutations of the suggestion of a cognition connection was from the Australian Broadcasting Company, which proudly pointed out that some of the researchers who were coauthors of a refutational article were Australian.

Where Do We Go from Here?

The examples presented here follow a fairly common pattern. Scientists carry out a research project producing results, conclusions, and speculations that appear relevant to questions of social importance. The authors submit a paper for publication on the work that includes speculations going substantially beyond their evidence, a reasonable practice in scientific publications. However, in these particular cases the science is weak or faulty, and the ideas presented may be used by others in socially harmful ways. The paper is submitted to a prestigious scientific journal, where it presumably undergoes serious peer review. The editors of the journal may be influenced to favor publication of such reports because of the wide attention they may attract. Upon publication, the journal may accompany the article with its own news article, often sporting a headline that presents in more popularized language the academically worded speculations of the scientists. Several days before the article appears, a copy is sent to the media to emphasize, along with the news article, the potential public interest of the story. From that point on, there is a cascading amplification of the message as it passes from newspapers, television, and the Internet news providers to more popular magazines and blogs. In many instances the claims are incorporated into popular culture and school textbooks and are even used by the scientists themselves, policy makers, or commentators to promote the presumed social utility of the work. Public attitudes and public policy are influenced, and, at least for a while, social norms may be changed. In the cases I have described, a view of society is reinforced in which human behavior and aptitudes are largely a product of our genes. This process takes place even though serious criticisms are made at the time of the publication and further studies refuting or failing to replicate the original claims appear in the scientific journals. As the examples given illustrate, refutations of a research study are only infrequently considered to be worthwhile news stories by the media (unless the original study is shown to be a "fraud").

With regard to the scientists who originally published the work, some may be appalled at the course the story has taken after its publication. Sooner or later that regret may be expressed, but usually only within academic circles and with little impact on public knowledge. In contrast, other scientists who publish such studies may be content with the publicity and may even amplify the speculations by their public statements. These scientists may have strong prior commitments to a more genetic-determinist view of the world. Any attempts to publicly criticize the work or the media coverage often come from other scientists who are concerned about the social impact of questionable scientific reports (and who may have a very different worldview).

Some scientists I have referred to were "amazed" (Engel), expressed "regrets" (Jacobs), or stopped research on the subject (Breakefield) after seeing the social fallout related to their scientific work. Could they have anticipated that fallout? They very well might have if they had been aware of how such claims from behavior genetics have been used throughout the history of the field. But ordinarily there is nothing in the education of scientists that alerts them to the connections between the scientist's work and social impacts of this sort. I argue that a broader education of scientists than they currently receive is needed to remedy such lack of knowledge and its potential consequences. There are courses taught for science students in a few university or college science curricula that cover such issues and present the concept of social responsibility in science. Such courses often include analyses from history, philosophy, sociology, and other disciplines. They can better prepare students to anticipate times in their scientific careers when there may be "unintended consequences" of their work and to consider in what ways they might act to prevent any social harm. If caution in presentation of their work is not sufficient, they should actively engage with the public to correct misimpressions. Such courses can also teach future scientists to become aware of the assumptions they bring to their research and to realize how these preconceptions frame the questions they ask or the speculations they make.

Second, the scientific journals and their referees should be particularly watchful in evaluating the scientific validity of work that has the potential to cause social harm. Referees and editors could suggest moderation of speculations of a paper's authors when those speculations far outreach the paper's content and may generate unwarranted and destructive social interpretations. Among the referees who review articles that fall into this category should be ones who understand the problems that can arise from the miscommunication of scientific ideas to the public. I am not suggesting censoring publications. Rather, caution is called for, given the many mistakes prestigious scientific journals have made in publishing and featuring weak or overinterpreted scientific stories the impacts of which have had extraordinarily long shelf lives. Heretofore, journals have conceded to the "breakthrough" mentality by promoting unreplicated or weak studies, highlighting them to the media, and featuring them with news stories. (I will not go here into the broader question of the tensions between science journalism and science.)

Finally, there is also a need for scientists to consider a responsibility that goes beyond being careful in the way their own work is presented and used. When scientists see faulty reports by others in their field that may cause social harm, who better to present an alternative view to the media and the public than those who are knowledgeable in that field? I consider such efforts to be comparable to those of scientists such as Kenneth Miller and Stephen Jay Gould who have entered the public fray to defend the teaching of evolution and debunk theories of "creationism" and "intelligent design." Gould argued in his book *The Mismeasure of Man* for "debunking as positive science" and stated that "sound debunking must do more than replace one prejudice with another. It must use more adequate biology to drive out fallacious ideas."[41] Scientists will have to be creative in trying to ensure that when a study of the sort described here or its speculations are falsified, the public learns about it and the media cover it. There is much concern today about the sorry state of science education and the need for scientists to remedy this situation by engaging more with the public.[42] But included in this engagement should be an effort to point out when flawed or overinterpreted science with significant potential to generate harmful social consequences has been misrepresented to the public.

CHAPTER THIRTEEN

Map Your Own Genes!

The DNA Experience

SUSAN LINDEE

IN THE 1960s, in response to the truly strange climate of creationism in the United States, the Russian-born evolutionary biologist Theodosius Dobzhansky proclaimed that "nothing in biology makes sense except in the light of evolution."[1] I want to modify Dobzhansky's claim for the new century: increasingly, nothing in biology makes sense except in the light of the marketplace. Markets justify what would be inexplicable to the generation of geneticists, including Dobzhansky, who founded the American Society of Human Genetics and worried about radiation effects on heredity in the 1950s and 1960s.[2] That group of genetic pioneers would presumably be puzzled by the commercial promotion of a geneticized race that geneticists have been questioning for the last six decades; the investment of many millions of dollars in research programs, like deCODE's, that do not generate scholarly papers;[3] the application of nonconsensus, varying standards for what constitutes a reportable association between a gene and a disease or a marker and a population; and the acceptance of a customer's own decision about his or her race, based on a short drop-down menu of options that are reminiscent of those of the eighteenth-century physician and race theorist Johann Friedrich Blumenbach (European, Asian, African) and that have no relationship to the population groups for which gene frequencies are known to vary.[4] These customers who are always right can then guide biological conclusions about human population genetics, or so the companies selling genetic tests directly to consumers claim. Ancestral Origins, a company marketing DNA tests to the public, even sells glossy individualized maps "suitable for framing and displaying with pride in your home" depicting the geographic origins of identity.[5]

In recent years a bewildering world of biomarkets has reconfigured what it means to be a geneticist, a consumer, and a racially marked person.[6] The most prominent direct-to-consumer (DTC) personal genome-testing companies are 23andMe, Navigenics, and deCODEme, but in the more astrological (or what Hall and Gartner have called "astrologicogenomics") categories are a large number of ancestry-testing companies that vary in the caution of their claims and the nature of their promises. States are not sure what to make of these consumer products; New York and California have launched investigations.[7] Nor are the federal regulatory agencies or congressional oversight committees sure about what to do. The U.S. Federal Trade Commission warned as early as 2006 that skepticism was warranted.[8] In the spring of 2010, when the Walgreens drugstore chain announced plans to sell a saliva-collection kit marketed by Pathway Genomics, the Food and Drug Administration (FDA) sent a formal query to all parties that deferred the deal.[9] A few days later the chairman of the U.S. House of Representatives Committee on Energy and Commerce, Henry A. Waxman, announced a formal investigation into the DTC genetic-testing industry.[10] In July 2010 the FDA held a very well-attended public meeting in Maryland that asked for input on its plans to regulate laboratory-developed tests, which include genetic tests.[11] The overwhelming conclusion was that regulation was necessary.[12] Meanwhile, the new biomarkets have attracted relatively high scholarly interest among physicians, geneticists, ethicists, sociologists, and science studies scholars. Some observers have approached the problem in economic terms, as a question of both health-care costs and consumer fraud. Others have promoted a "deficit model" in which the problem is that consumers do not know enough about genetic testing and therefore are not able to decipher/interpret the results or figure out whether a particular test is worth the money.[13] Still others worry that doctors are the ignorant ones, unable to counsel their patients who arrive with such test results seeking guidance.[14] There does seem to be a rough consensus in this literature that DTC genetic testing constitutes "a major challenge to health-care systems because it distorts and confuses notions of diagnosis and screening and capitalizes on ignorance about causation for commercial ends."[15]

In March 2010, in one response to this intersection of technology and marketing, the National Institutes of Health (NIH) announced that it was creating "a public database that researchers, consumers, health care providers, and others can search for information submitted voluntarily by genetic test providers. The Genetic Testing Registry (GTR), now operational, aims to enhance access to information about the availability, validity, and usefulness of genetic tests."[16] A fact-checking database, this registry in theory provides physicians and consumers with a trustworthy online source of information about particular tests and resources for finding more information. The site notes that the "NIH does not independently verify information submitted to the GTR; it relies on submitters to provide information that is accurate and not misleading. NIH makes no endorsements of tests or laboratories listed in the GTR. GTR is not a substitute for medical advice." Information is organized by disease—so all providers who can test for cystic fibrosis show up together—and eventually, after several layers of information, a link to the company website appears. When it was first announced, one of the founders of 23andMe instantly endorsed the plan.[17] The NIH has thus made a move to bring this industry into some kind of alignment, consensus, or logic—a daunting task.

From a different perspective, the Coriell Personalized Medicine Collaborative has now begun a study of whether genomic information actually matters to health. Coriell is a not-for-profit biobank in Camden, New Jersey, established in 1953 and holding an astonishing collection of cell lines from populations as diverse as Georgian centenarians, the Amish, the Yerkes nonhuman primate research center, the Wistar Institute, Huntington disease families, and many other sources. Coriell has at least 42,000 samples from many institutions and populations and relevant to many kinds of research.[18] The Personalized Medicine Collaborative, launched in 2008, requires that each consumer-participant (I became one of the consumer-participants in 2010) authorize the continued use of his or her DNA in a large-scale research program aimed at determining the impact of genomic information on behavior and ultimately on health. It is not simply a matter of telling individuals about their own genes. Rather, the program is aimed at assessing the consequences of information for health care. Instead of following the DNA, the research will be following the people who learn about the DNA and seeing what they do with what they know.

One possible finding of this study could be that providing consumer-patients with genomic information has no impact at all on long-term health. The Personalized Medicine project thus engages directly with some of the most powerful ideas that have shaped the rise of genomic medicine generally over the past thirty years: that genetic information will become medically relevant at a detailed and specific level; that DNA tests can and will be democratized and made generally available to a broad population; that complex diseases will be understood in genetic terms; and that this understanding will be of benefit to patients. The project is furthermore situated at contentious intersections that have a broad relevance to contemporary biomedicine. It involves genetic data, the quotidian use of the Internet as a medical technology (all of my Coriell results have been available, password protected, online), and the exploration of complex relationships between research and clinical care, between medical professionals and research scientists, and among privacy and efficient information management, data collection, and (long-term) profitability. The Coriell Personalized Medicine Collaborative is one of the new forums for biological citizenship in the genomic age.[19]

In a wonderful essay that concludes "We are all geneticists now," Misha Angrist wonders "if we are to venture further down this road, who will handle the onslaught? A cohort of geneticists whose numbers are roughly the same as the number of astronauts?"[20] He describes a "gathering storm of personal genomics"[21]—a storm that, according to the 2009 consensus statement of the American College of Clinical Pharmacology, threatens to undermine consumer trust in genetic testing. "This negative branding can be an unintended marketing consequence of premature promotion and uptake of DTC/P (Direct to Consumer/Personalized) genetic testing. This outcome could deter the future utilization of pharmacogenetic testing to inform choices about medication use, which is a long-awaited scientific advance in our discipline."[22]

I suggest that this storm is an unwanted climate event produced by a certain historical lineage—an ancestor story, if you will. It is the progeny born of James Watson and Walter Gilbert, instantiated in the Human Genome Project (HGP), and now returning with a vengeance. Jennifer Reardon has invoked Frankenstein's monster, cobbled together in a fit of hubris and now roaming the melting ice caps.[23] Intense consumer desire for what I have come to think of as "the DNA experience" was probably a predictable outcome of early promotion of the HGP. The tone of DNA marketing today—which might indeed threaten genomic medicine—reflects DNA marketing at the time when the HGP began. What the fathers of the HGP perhaps did not anticipate is that this marketing would be taken away from the amateurs and become the business of the professionals on Madison Avenue.

When Dorothy Nelkin and I first started thinking twenty years ago about what we later came to call "the DNA mystique," we expected it to be one of those temporary, shimmering sociocultural phenomena: a brief, odd, popular wave that would soon be gone. Indeed, at one point, during one of those times when writing the book was going slowly, we feared that if we did not hurry up and turn around that painful chapter, everything we were writing about would be irrelevant, the moment would have passed, and our book would be about history. Nelkin died in 2003, but she lived long enough to realize that we were in fact chronicling the early stirrings of something that has only grown more intense and more powerful over the years. Popular and scientific claims about the deep, profound powers of genes have escalated, sharpened, and gained credibility and resonance, and the institutions and professional groups promoting these claims have expanded.

That earlier efflorescence reflected the interests of scientists in the United States and some other industrialized nations who sought public attention and national support that would facilitate funding for genomics research.[24] Now the DNA mystique is tuned to a high pitch by teams of marketing professionals, public relations firms, and highly creative writers and image experts, and DNA is the hook for products like living-room art (your own ancestry map in full color, with frame) and for moving experiences of historical connection (often with famous historical individuals like Genghis Khan or well-known, historically interesting populations like the Phoenicians) recounted in emotional first-person narratives on websites. DNA has become an intimate experience and an actor in a network of kinship, identity, and meaning in industrialized and prosperous nations in which consumers can afford to purchase this kind of citizenship. The DNA experience is the central product of an industry that promises consumers various kinds of truth, generally for $79 to $399. The DNA mystique is what they are selling, along with "stunning" personalized ancestry maps.

Consumers buy a lot of experiences (the movie *Avatar,* for example) and there is nothing wrong with selling an experience. The marketing of the DNA experience, however, is interesting and complicated, and it is not trivial for the scientific community or for our understanding of the implications of genomic medicine. It could play a role in the long-term consequences of increasing access to genetic information by individuals, employers, health insurers, and research institutions—and the medical efficacy with which that information is applied. The 2009 American College of Clinical Pharmacology consensus that "the response of consumers to such advertising can have both immediate and long-term effects on public health and the future adoption of pharmacogenetic/genomic testing" seems warranted.[25]

I am interested in the roles that DNA has come to play in the stories we tell ourselves and in the deep roots these stories have in the development of genetics as a science over the past century. The official skepticism about DTC genetic testing has focused more on disease testing than on ancestry testing, perhaps because ancestry testing is so clearly an imaginative, vague realm of limited medical relevance. But I am generally biased in favor of noticing the importance of what is popular, common, and possibly false in technoscientific culture.[26] The casual, messy domains of ancestry genomics—unmoored from the serious business of disease and risk—may be more revealing by virtue of their undisciplined commercialism. They may tell us more than serious medical testing does about the underlying expectations and beliefs that facilitate the DNA experience.

I here consider some of the marketing narratives posted to attract buyers on ancestry websites, where stories mix vague family lore and haplotypes, deep human suffering and magical feelings of connection. It may be that DNA as an online relational product does not work without these emotive threads, which in many cases emphasize the moving personal relevance of the technical data. As the ETC Group noted in a trenchant critique in 2008, DTC genetic-testing sites ironically emphasize the opportunity for a warm personal connection that comes about in social isolation: there is no contact with a doctor, counselor, or other healthcare professional in this testing system, so consumers connect all by themselves to history, a migratory group in the distant past, or relatives they do not know.[27] These first-person accounts of DTC ancestry testing may express powerful social beliefs about a biological material that has been freighted with meanings that are unrelated to what we know of its technical roles in the cell and the body and of its history. Ancestry testing is a particularly blank screen because the quantitative reality is that if you go back far enough, all humans are related to all other humans—at fourteen generations back, or only about 300 years, each person has more than 30,000 direct ancestors. A few more generations, and the numbers get spooky. Why would DNA possibly associated in a vague geographic, statistical, populational way with any one of these 30,000 or more ancestors be meaningful to an individual buying an ancestry test? It is not the technical detail that generates the meaning, I suggest. It is the DNA experience.

"DNA Genealogy Has Never Been So Easy to Understand!"

The claim in the preceding heading is on the marketing website of Ancestral Origins DNA.[28] The hobbyists in genetic genealogy buy tests that advertise the ability to provide new kinds of family history. Although obtaining the DNA results is relatively straightforward—involving a spit test, a mailed package, and online password-protected results—their interpretation has proven to be complicated. My PhD student Joanna Radin has looked at how these hobbyists new to genetic genealogy have begun to seek assistance from more experienced hobbyists through web-based discussion boards with names such as DNA Newbie as they try to interpret the results. In 2005, the same year in which *National Geographic* launched a five-year initiative known as the Genographic Project to sample and archive human migratory history, the International Society of Genetic Genealogists (ISOGG) was formed to serve as a clearinghouse of information for members of the nascent but rapidly growing genetic genealogy community.[29] The "consultants" page on ISOGG describes "Y-chromosome consultants" and "autosomal DNA consultants" who can help hobbyists interpret their test results. The site also includes "famous DNA," "famous haplotype," and "presidential DNA" pages and moving success stories of family connections made and recovered through DNA. As Radin observes, this society produces user-generated guides to address complex matters of ethics and kinship that emerge through testing, including the possible accidental discovery of "nonpaternity."[30]

Similarly, the anthropologist Gísli Pálsson has looked at the blurring boundaries between experts and nonexperts and has considered how personal genomics has in some ways "democratized" genomics discourse. "The boundary between experts and lay persons has been blurred and refashioned," he notes, and the DTC companies "'democratize' genomics both in the sense that they offer test kits for a low price (ranging from $250 to $2,500)—within the reach of the public, at least not just the research elite and the wealthy—and in the sense that analyses and interpretations of genome scans are now a matter of intense public discussion through all kinds of media, including Web browsers and blog sites."[31] Pseudoscience appears in his story in the form of the old, rejected claims about race and human evolution that modern molecular genomics victoriously leaves behind and simultaneously keeps intact. Pálsson is nicely attuned to this confluence of what is both explicitly rejected and quietly retained. In purchasing his own genetic information from deCODEme, he became (self-consciously) one of the citizens of the brave new world of genomic medicine, and he can now compare his genome with that of James Watson (among others). The imagined community of genomic testing involves people who are close and intimate (the family), people who are famous (Nobelist James Watson, gene hunter Craig Venter, or deCODE founder Kári Stefánsson), and people who collectively stand for racialized groups that have been critically important in global history.[32]

African Americans constitute a crucial demographic for these tests, partly because marketers assume that DNA testing may be particularly meaningful for people who are, as Eric Wolf once described indigenous groups, "without history."[33] Rick Kittles, a biologist who is one of the pioneers in the ancestry-testing enterprise and one of the founders in 2003 of African Ancestry Inc., makes the critical point in company promotional material on the Internet that "unlike most Americans, tracing one's roots through traditional means can be challenging for African Americans in that they usually hit a brick wall"—the Transatlantic Slave Trade, which disrupted families and erased identities. DNA ancestry testing, he proposes, provides a way around this wall, this point in history at which their ancestors came through slavery, beyond which no text can take them except the "text" of DNA. Kittles construes the technical details as reliable enough and says that some of the public controversy about the legitimacy of this testing has done more to confuse the public than to enlighten it.[34] But the technical details themselves are confusing. DNA data available for these population studies come overwhelmingly from European or European American populations—from a relatively small collection of Utah Mormons and from the "1050 individuals in 52 world populations" banked at the Foundation Jean Dausset–Centre d'Étude du Polymorphisme Humain (CEPH) in Paris. The samples at Paris were collected from various laboratories as part of the Human Genome Diversity Project (HGDP) and the CEPH "in order to provide unlimited supplies of DNA and RNA for studies of sequence diversity and history of modern human populations." Even if all 1,050 individuals in the CEPH collection were from Africa, it would be a vexing business to link shared DNA to particular populations or historically recognized tribal groups.[35] The numbers for all the populations are small, and we might wonder whether the feeling of connection that such tests can produce comes at the expense of accuracy and technical legitimacy.

What is clear today is that race remains profitable. It is useful for product placement, for identifying proper consumers, and for telling people that they are the right kind of customers by tagging them with properties they will recognize and understand. Often the supposed details that make a test suitable for a given group are trivial—a tiny difference in presumed risk, a small, suggestive study—but these numbers are good enough for marketing.

The online DTC testing company Ancestry by DNA offers a racial test that "makes a great gift!" but warns that its test "does not predict or establish a person's race; it only gives an estimate of genetic ancestry or heritage, for example."[36] Race, the company's website says, "includes both a cultural and biological feature of a person or group of people. Given the fact that physical differences between populations are often accompanied by cultural differences, it has been difficult to separate these two elements of race." Its test promises to provide consumers with "insights on possible geographic regions of origin; however, it will not definitively state your nationality. Nationality and race are both determined by social and political factors that are independent of genetics."[37] The tests will reveal, the company says, your deep ancestry along a single line of direct descent (paternal or maternal) and show the migration paths they followed thousands of years ago. Most of the results do seem to come with a migration path, a wide ribbon of movement out of Africa in various directions. Similarly, the Genographic Project proposes that "your results will reveal the anthropological story of your direct maternal or paternal ancestors—where they lived and how they migrated around the world many thousands of years ago."[38] Genetic tests, then, are understood to provide evidence of historical migrations and ties to possibly distant geographic regions.

What do customers make of such claims? The personal testimonials produced to market the tests blend family lore, unsolved mysteries, and haplotypes. Presumably concocted from both consumer testimonials and marketers' suggestions and amendments, these stories seamlessly combine technical and emotional details in the same frame, and in their structure they almost perform a science studies analysis. I suggest that these narratives are communal products. They are not the isolated words of individuals but performances of an expected habitus, a way of being in the world, in which science reveals personal truths. They are also carefully crafted market products, the downstream results of the promotion of DNA. In our book Dorothy Nelkin and I quoted critic Robert Warshaw, who said that mass culture "is the screen through which we see reality, and the mirror in which we see ourselves."[39] DNA is now a player in mass culture. It has left the lab behind and moved into worlds that experts cannot control. This is not particularly good for genomic science.

Now I want to turn to some brief summaries of a few personal stories—testimonials—and consider them as literary products that can be read for structure, allusion, imagery, and narrative force. Some of these testimonials have named authors—sometimes a geographic location is even provided for the author—and I use these names even though I do not know whether they are real or are pseudonyms developed for the campaigns. It has occurred to me that the right kind of swindler might be able to use haplotypes in some creative way to gain a family's confidence, so I hope that the names on these sites are fictitious.

One testimonial describes a discovery that the family was Phoenician instead of Viking, and the discovery explained an apparent racial marker—"darker complexions"—and contradicted a father's convictions. ChrisB123's message, from 2009, stated that his father "was the Nordic type, blond hair and blue eyes. He just KNEW all of his ancestors were Vikings." The fact that some of his children had "darker complexions and green eyes" did not affect this conviction. But the test results "finding that his haplogroup was J2 shows just how off-base his father's line was from the typical Viking path some of his family from his mom's side took. J2, the haplogroup of the Phoenicians. WOW." ChrisB123 reported that the news affected his children. "Two of my boys have always had a deep interest in the Phoenicians, and now that interest has skyrocketed. What a gift. My brother can't wait to take another scraping to gather even more information. Thank you for this project. We really are one big family."[40]

In this single testimonial the properties of DNA that matter unfold: its authority and status as more persuasive than personal conviction or family lore, its relevance and force across a kinship network (father, brother, sons), and its extreme (masculine) potency across many, many generations from the Phoenicians (1550–300 BC) to the present, conferring on some members of this living twenty-first-century family "darker complexions and green eyes" through more than 2,000 years (an estimated eighty generations). It is perhaps too obvious to point out that the total number of ancestors at such a distance would be very high. One online calculation of such trees shows that twenty-four generations (back to the year 1170) would yield 134,217,727 ancestors.[41] In these calculations, of course, the numbers very quickly exceed the total estimated human population in the historical past, but this is because of what is called "pedigree collapse," the fact that human beings are all so closely related that many of those individual ancestors would appear more than once in any tree.[42] The imagined Phoenician skin color, then, has persisted with tremendous force (there is an ongoing scientific debate about the appearance of the Phoenicians), and this skin color undermines a family identity as "Viking." The entire brief narrative is dramatic, emotive, and mysterious, with DNA results upsetting the order of things, reconfiguring identities, and inspiring passionate interests.

Many other testimonials describe the creation of new or unexpected family relationships and the retrieval of lost connections, providing social networks to those whom I will call, after Eric Wolf, "people without relatives." For example, in one account on the site of the ISOGG, a "sad story" of early parental death and no knowledge of "his ancestral Stewart family" led a South Carolina man who envied his friends and his wife's family, who "seemed to know their ancestors for many generations back," to seek ancestry testing. Through a test from Family Tree DNA and a connection with a Stewart family genealogist, "we pinpointed his father, his grandfather, and several generations back. We located an old friend of his grandfather's family, age 99, and he had the intense pleasure of meeting her and hearing her say 'You're a Stewart. I can tell by looking at you!' " Now "he plans to join us at our 100th anniversary Stewart reunion in South Carolina. He has found his family at last."[43]

Another testimonial reported that the DNA results created new family members, with the data responsible for "the immense pleasure of finding out that we did have family we thought we had lost along 3 generations, 6000 miles and 90 years." People whom no one had ever met, had coffee with, or spoken to brought "immense pleasure" as new members of the family. This new bond was the result of a campaign by someone in California, a "possible cousin," who had been "trying to find family through the internet." "In no time" the results came back on this seeker and a "male cousin of ours" that showed a connection. "Needless to say, we couldn't feel happier now."[44]

What kind of happiness is this? It seems to be a kind of happiness that it is needless to explain. It is the return of ties and relationships to people "we thought we had lost" now living in California. DNA results make whole what had been incomplete and fragmented. They compensate for the vagaries of history, the lost connections, and the social distances. Patricia Mathes recounted her story on Family Tree DNA: "I have to admit I was a bit overwhelmed by all of this. Because I was abandoned at birth I went into this hoping only to find a clue to my nationality. Instead I not only found out that my ancestors were Hungarian Jews, but I found two distant cousins. First relatives I've ever known other than my children. This really is a very rewarding experience, well worth the wait and the cost."[45] The new relatives of DNA testing enter the industrialized kinship system through the laboratory, and their stories are cheerful, enthusiastic, and bright. They come not with tragic revelations of a grandfather's secret second family or traumatic breaking points at which the family was separated by war, genocide, or famine, but with entertaining and alluring exoticism. The new DNA relatives are relentlessly desired, the very wanted cousins of biotechnology.

In another account, by Joel Cherniss of San Francisco, a shared last name was revealed to be a shared biological history. Cherniss's father had origins in the former Soviet Union, and when he learned of a family in Houston with the same name, he arranged for both his own son and the son of that family to be tested. "I was overjoyed when your company said that both samples matched at all 11 places on the Y-Chromosome exactly. To add to my excitement was the fact that although I was brought up Reform Jewish, and thought of myself as Yisrael you were able to tell me that our genes matched, exactly, the Cohanim gene that I had read about a few years ago. My dad had told me that his grandfather was a cantor in his Ukrainian village synagogue for 47 years. I thought there might be a reason why, now I think I know that reason."[46] A famous Jewish line, the Cohanim, a Ukrainian cantor, and the collapsed Soviet Union are folded together in this family story, which makes the purchaser "overjoyed." What kind of happiness is this?

Finally, there is the story of the European African American. A participant in the Genographic Project, he reported that he expected "pretty straightforward" DNA results. "I figured, OK, I'm black." Instead, "It actually turned out quite interesting: EUROPE! My paternal DNA shows that my ancestors were Europeans, which I had no clue about until becoming part of this project. It does change the way I view myself because it gives me a richer and deeper appreciation of who I am and where I came from, because you think you know, but you have no idea! That's the most amazing part of the Genographic Project." The DNA thus provokes possible evidence of acts that in strict historical terms, given the practices of the slave trade, might have involved rape or coercion—sexual violence of some kind—now transmutated into a thrilling story of self-discovery. Even if the stories could be a biological record of human tragedy, in the world of DNA relationships they are positive, upbeat, and encouraging. "It's also brought me and my family closer together since we've been digging up old photos and discovering long lost family history that we physically own but has been lost in time, tucked away in attics and basements. Being a part of this project has been truly an amazing adventure!"[47] This is a world without sorrow. The stories on ancestry DNA sites fuse history, biology, emotion, and kinship in ways that capture the public meanings of genetics and genomics in the twenty-first century.

The Unnatural Growth of the Natural

In 2009, in the United States, the last few states to take action began mandatory newborn screening for the genetic disease cystic fibrosis (CF). Since 2007 CF has been added state by state to the newborn panels that now screen for an average of about thirty genetic diseases in every child born in the United States. This pattern began in the 1960s with screening for phenylketonuria, a condition for which a neonatal diagnosis has direct health benefits. In many cases now, newborn screening tests produce uncertain or ambiguous information—certainly this is so for CF—the timing of which may or may not help a newborn. To complicate the picture, in

June 2009, just as the last states geared up to begin newborn screening for CF, a major French survey showed that one of the CF mutations, R117H, included in the newborn screening panels had no reliable correlation with CF symptions. The group concluded: "These results suggest that R117H should be withdrawn from CF mutation panels used for screening programmes. The real impact of so-called disease mutations should be assessed before including them in newborn or preconceptional carrier screening programmes." This article was the fifth most read article in the *Journal of Medical Genetics* in December 2009.[48]

Screening is technocratic reason, surveillance, and the careful management of populations and resources. Yet even in the world of genetic screening, in the case of a disease that has been the focus of significant scientific attention (there are more scientific articles published about cystic fibrosis than about any other genetic disease), uncertainties and technical confusion persist. A mutation approved by consensus, built into panels all over the world, is found to be an unreliable sign of a very important genetic disease. How much more uncertainty and confusion prevail in the messy worlds of marketing DTC genetic tests?

In his much-cited essay "Making Up People," published in 2006 in the *London Review of Books,* Ian Hacking quotes Nietzsche: "There is something that causes me the greatest difficulty, and continues to do so without relief: unspeakably more depends on what things are called than on what they are."[49] Hacking invokes the quote to underline his point, that creating new names is enough to create new "things," new ways of being a person, and new forms of interaction with friends, families, employers, physicians, and counselors. In this process of creating new ways of being a person, looping effects are very important. People opt themselves into medical regimes and build identities around technical categories that shape how the categories themselves then stabilize or grow or change. DNA relatives are perhaps an example of new kinds of people.

In her complex examination of human labor and existence in *The Human Condition,* Hannah Arendt explored what she called "the unnatural growth of the natural." She proposed that in the modern scientific age humans have become capable of "unchaining" natural events that would never occur in nature. They "provoke" these events and initiate "natural processes which without men would never exist." They "import" cosmic processes into made things, into technical systems, societies, and governments—both literally and figuratively—so that world history has become increasingly naturalized, made to appear not as the product of conscious human effort but as an unfolding biological development.[50]

As scholars in science studies think about spit parties, patient blogs, biobanks, and democracies of consumer marketing that have incoherent politics, both Arendt and Hacking are perhaps good philosophical collaborators. Genomics is growing like herbicide-resistant weeds in a corn field, sprouting new kinds of experts, new relationships, and new forms of identity. Individuals enroll themselves and their family members in a story that gains credibility just by being told. DTC genetic testing appears in this broad narrative not as the product of human effort but as an unfolding biological guide to intimate personal knowledge. It provides access to new bonds, new identities, and new families. The effervescent first-person testimonials—the relentlessly happy revelations of warm connection—hint at the stakes involved.

Genomic medicine today seems to involve deep inconsistencies. The genome is construed as individual, highly revealing of particular, specific risk, and also as easily categorized and linked to mass group identities—literally based on something as indefinite as a continent. "Europe" is a category now under siege in history departments, where it is seen as a particular invention, a political claim rather than a geographic space. But this ephemeral and charged entity—Europe—is also now a solid biological category, a way of understanding heredity at the very moment when its political legitimacy is in a downward spiral. Buyers of genetic tests can "choose" to be European; their choice affects what their risks will be reported to be (their own experience of risk); and then the category "European" can in the aggregate be modulated by their self-identity—their DNA becomes a part of the European category after the fact. The social choice ("I am European") gets written into the biological data.

Hacking's looping effects here are profound, and as Arendt suggests, nature is constantly being "provoked" into culture, growing and expanding in ways that call us to attention. Consumers drawn to town-hall meetings about an NIH database by their interest in individual genomic data are disappointed to learn that their participation will not result in their acquisition of this data. Karen Sue Taussig suggests that promoters of the power of genomic medicine have created an awkward groundswell of public interest—the genome is the focus of desires and longings that will be difficult to fulfill.[51]

A 2002 marketing report by Patent Insights, Inc., noted that "marketing studies predict that genetic testing will be a multi-billion dollar industry in just a few years."[52] A 2001 report by Frost and Sullivan observed that "genetic testing is the highest growth segment of the diagnostics industry, with the number of new products and services far exceeding other market sectors."[53] In April 2006 BCC Research reported that the U.S. market for therapeutics and diagnostics for genetic diseases reached $4.8 billion in 2005 and is expected to grow at an average annual rate of 8.7 percent, reaching $7.3 billion by 2010.[54] And in early 2008 Fuji-Keizai USA reported that "DNA-based testing is moving into a new phase. Transformational technologies are allowing complex genetic (specific gene) and genomic (large numbers of genes) tests to move from research-only labs into medical and clinical labs."[55] These quotes are all taken from the teasers online for these marketing reports, each of which costs thousands of dollars to buy. The marketing experts are on the job.

In the world of DNA ancestry testing, fantastic historical stories of Vikings and African royalty provide a vivid counterpoint to haplotype names, and individual, personal true selves intersect in databases with promiscuous meanings, open to a wide range of potential and future uses. These potential and future uses matter in the mixed pleasures of entertainment genomics, or "genotainment." Personal genomics has become a private matter of consumption at home, not inside the world of health-care professionals who can monitor or control it. There are certain paradoxes of freedom here as consumers choose their own ethnic identity, and the companies trying to reach them mobilize identities and then fold consumer choices into their technical databases. In the process they create new kinship networks, new relationships, and new identities. DNA functions with more authority than a father's personal recollections, trumps family history, and retains full potency no matter how many generations have passed. It furthermore generates entirely new and unknown relatives who fill in previously unrecognized social gaps.

The "negative branding" of DNA is in full force today and threatens all the benefits of genomic medicine that geneticists have struggled to establish since 1950. The scientific community is right to worry about these promotions, which undermine public trust and ascribe more to DNA than it can possibly provide. The stories have left the laboratory behind, and their deep appeal is persuasive regardless of the technical details. I close with a quote from what Dorothy Nelkin and I said in *The DNA Mystique:* "The danger, then, is not that inflated promises threaten to backfire on the scientific community, but that such promises will long outlive their scientific utility. Designed to appeal to popular interests, they quickly acquire a life of their own."[56]

CHAPTER FOURTEEN

Creating a "Better Baby"

The Role of Genetics in Contemporary Reproductive Practices

SHIRLEY SHALEV

SCIENTIFIC ADVANCES in assisted reproductive technology (ART) enable individuals to overcome diverse medical, social, and personal challenges and give rise to new reproductive practices and parental relations. Furthermore, many individuals and couples around the world are increasingly assuming the power of heredity and turning to diverse reproductive technologies with the hope of creating what they typically perceive to be a "better baby," that is, a child who is free of various genetic disorders, embodies their preferred sex, and potentially expresses their aesthetic criteria or other desirable traits.

This chapter offers a general overview and a critical discussion from diverse ethical and social perspectives regarding various reproductive practices that are available today for those who seek to gain more control over the genetic makeup of their prospective child, either by using their own gametes (and applying various genetic testing and preimplantation diagnosis techniques) or by purchasing donated gametes thought to have certain genetic qualities in the global free market (that is greatly facilitated by the Internet). These practices seem to reflect social perceptions that certain genetic traits are better and more valuable than others, as well as common misunderstandings often disseminated by the media regarding their power to explain human conduct. The discussion thus explores the emerging phenomenon of global gamete exchange and its profound implications on parental relations, as well as the terminology employed to describe them and negotiate their meanings. Consequently, it examines the debate about the growing interest in personal genotype and the role it plays in the establishment of our identity and overall understanding of our phenotype.

The Power of Kinship and Genetic Ties

Contemporary reproductive practices are associated with both male and female infertility, which is estimated to affect 8 to14 percent of all couples of reproductive age worldwide. It was reported that in the United States there were 7.3 million women, or 12 percent of all women of childbearing age, who availed themselves of infertility services in 2002. Over 1 million children were born worldwide following the use of reproductive technologies, while ART-born infants account for more than 1 percent of all children born in the United States and 18 percent of all multiple births. The overall cost of diagnosing and treating infertility and its outcomes is estimated to be more than 5 billion U.S. dollars per year.[1]

In pursuing options to meet diverse infertility challenges, individuals and couples are required to weigh and balance the benefits, risks, and costs of various medical procedures, such as in vitro fertilization (IVF), intracytoplasmic sperm injection (ICSI), preimplantation genetic diagnosis (PGD), and gamete cryopreservation and donation. These reproductive practices have been at the core of vigorous scholarly and public debates over the diverse political, economic, and social forces that shape their use and regulation, as well as their far-reaching implications on freedom of choice and reproductive freedom.[2]

The decision by individuals and couples to avail themselves of these reproductive options is often affected by the ways they evaluate their own congenital traits, conceptualize human life, define parental relations, and value the genetic tie between parents and offspring. The quest for a genetic tie, as well as social values attributed to kinship and genetics, seems to underlie the use of many reproductive practices that are primarily aimed at ensuring the genetic relation of at least one parent to the prospective child. Indeed, individuals are willing to subject themselves to tremendous physical and emotional risks,[3] to assume significant financial burden, and often to overcome considerable logistic constraints in order to establish, ultimately, genetic-based parental relations and assure the use of their own genetic makeup. Many patients, particularly women (who are often the ones being treated even in cases of male-factor infertility), endure painful fertility treatments and deal with considerable hardships in the hope that they will conceive a child who will carry their own, or their partner's, genetic traits.[4] In spite of alternative paths for parenthood (e.g., adoption), the growing use of ART for the purpose of establishing a genetic tie suggests that the driving force behind the use of many reproductive practices available in the fertility industry today is not necessarily the desire to have a child or to become a parent, but rather to become a genetic parent by having one's own genetic offspring.

The emphasis on achieving continuity of one's heredity can also be illustrated by historical shifts that followed the introduction of new reproductive technologies that opened up more possibilities for those who hoped for their own genetic child. For example, many infertile couples preferred undergoing IVF cycles instead of applying for adoption once IVF was introduced. Similarly, ICSI was increasingly selected over sperm donation in the hope of resolving male-factor infertility and achieving genetic fatherhood. If such medical practices are viewed in this way, it can be argued that their power goes beyond merely offering additional reproductive options, thus potentially marking new social expectations and reproductive norms that are becoming widely accepted and far more binding. Hence, ironically, the very same practices that are typically aimed at liberating human reproduction and broadening the spectrum of reproductive choices might end up in fact constraining reproductive freedom in some cases.[5]

Likewise, the complex practice of third party reproduction in general,[6] and the use of surrogacy (traditional or gestational) in countries where it has become medically and legally available, in particular,[7] can also be perceived to be another means of advancing genetic reproduction. This is especially the case for heterosexual men who seek to override the incapability of their female spouses to conceive their genetic child, but also for single heterosexual men who have no spouse or homosexual men who seek to override the lack of a female partner given their sexual preferences (also commonly known as "social infertility"). Fertility programs in the United States are encouraged to treat all requests for assisted reproduction equally, regardless of marital status or sexual orientation.[8] Thus marital status and sexual orientation of men and women are no longer considered an inevitable obstacle to reproduction. Indeed, some gay or lesbian couples decide to have one child from each partner in order to allow both partners to have a genetically related child. Other homosexual couples leave it to fate to assign their genetic tie as they perform a kind of genetic raffle, mixing samples of the semen of both partners to allow themselves equal chances to become the genetic father of the child they wish to raise together.[9]

In addition, many individuals are choosing ART in order to preserve their genetic makeup for their own future use, particularly given personal circumstances that might jeopardize their chances for genetic reproduction at a later stage, such as significant illness, chemotherapeutic treatments, military service, or exposure to hazardous chemicals. Although the option to bank sperm and embryos has been available for quite a long time, only recently have scientific advances made it possible for women to bank their unfertilized eggs. Ultimately this will allow them to override the constraints of their "biological clock" and make use of their genetic traits later on in their lives without having to choose a male partner at the time of banking. Thus the option of oocyte and ovarian tissue cryopreservation is also appealing for nonmedical, social reasons (e.g., career or delay of marriage) as growing numbers of women are deciding to secure potential future use of their genetic makeup despite the high costs and potential physical risks involved in this procedure. Moreover, this practice appears to be another step in an already rising phenomenon of postponing childbearing to a later stage in life (with own or donor egg). For example, between 1996 and 2006, the birth rate for women between the ages of 37 and 39 increased by 70 percent, and between the ages of 40 to 44 by 50 percent, in the United States.[10] Pregnancies beyond the ages of 40, 50, 60, and even 70 are increasingly being reported despite the health risks (for both mother and fetus) that are associated with advanced maternal age.[11]

Furthermore, the longing for continuation of germline legacy can be expressed not only in the desire for genetic offspring at a later stage in one's life but also by extension beyond one's lifetime, as demonstrated in the increasing demand worldwide for posthumous reproduction. The growing phenomenon of posthumous gamete retrieval in recent years is associated with diverse civilian and military contexts as partners and parents of deceased or comatose men and women are requesting the harvesting of loved ones' gametes in order to reproduce a child who will carry their genetic makeup postmortem.[12]

Finally, the quest to use one's own genetic makeup is also shared by some infertile individuals, same-sex couples, and bereaved parents who desperately hope to conceive their own genetic child one day through the futuristic option of human reproductive cloning. This controversial possibility also relates to the contentious debate about the therapeutic and reproductive potential of human embryonic stem-cell research, and the ongoing debate about related laws and regulations.[13]

In summary, contemporary reproductive practices seem both to reflect and to reinforce the common desire for genetic reproduction and the social perceptions related to reproducing offspring that carries one's own genetic makeup. However, potential parents seem to increasingly turn to those practices beyond the purpose of ensuring their genetic relation to their child with the hope to also promote health, avoid certain disabilities, and perhaps even control some of their child's other genetic traits.

On the Way to a "Better" Baby: The Elimination of Genetic Disorders and Traits

For many people, the desire for one's own genetic child seems to exist alongside the aspiration for a healthy child. About 6 percent of infants born worldwide every year (approximately 7.9 million infants) are born with birth defects, and they occur in nearly 1 in 20 pregnancies.[14] Birth defects can be caused by chromosomal abnormalities, single-gene defects or a combination of genetic mutations. However, it should also be acknowledged that not all birth defects are necessarily genetic (environmental, random or unexplained factors may be involved).

Prenatal genetic screening is commonly performed before initiation of pregnancy to help identify individuals who have increased risk for certain genetic disorders and birth defects (e.g., Tay-Sachs disease). In some communities genetic screening is even used before marriage to identify carriers of similar genetic disorders and consequently to control their match-making in order to minimize the incidence of certain genetic diseases (e.g., the organization Dor Yesharim that offers genetic screening to members of the Jewish community prior to marriage).

Additional noninvasive screenings are routinely performed during the first and second trimesters of pregnancy (including sonographic evaluations and biochemical testing of maternal blood—commonly referred as Early Risk Assessment). Further genetic tests are also offered to women of a certain age or diagnosis to determine if their fetus is affected by any chromosomal abnormality (e.g., Down syndrome), using chorionic villus sampling or amniocentesis which both carry a small risk of miscarriage. Considerable scholarly and feminist debate has pointed out the medical and social pressure that many women are facing regarding these tests, as well as their impact on the overall experience of childbearing.[15]

Based on the discovery of cell-free fetal DNA, new fetal genetic testing is now offered in the direct-to-consumer (DTC) market and allows extraction of DNA material from maternal blood to assess the fetal status in the first trimester.[16] Non-invasive prenatal diagnosis (NIPD) is predicted to become the new standard of prenatal diagnosis given the fact that it completely eliminates the risk of miscarriage and can be performed as early as 6 to 10 weeks after gestation. However, the existing debate over NIPD also relates to several concerns regarding its potential contribution to blur the fine line between medically necessary and non-medical fetal testing, undermine informed consent given the current absence of adequate regulation, and create an impression of accuracy that is often associated with submitting blood. This concern about the ways individuals interpret the results of prenatal tests has been pointed out in light of the fact that many of those tests provide "predictive" rather than "certain" results.[17]

Thus it seems that further discussion is needed about the overall characteristics of genetic testing and screening, the nature of their results, and their common perceptions. Indeed, testing results may occasionally be either false negative,[18] inconclusive or inconsistent. For example, it has been reported that inconsistencies occurred between the genetic analysis performed on the embryo and later genetic testing of the fetus or child (e.g., when children are born with genetic diseases that PGD was supposed to avoid).[19] Moreover, it has been noted that adding interpretive labels for test results of prenatal genetic screening (e.g., "negative" or "positive"), compared to numerical results, leads to changes in risk perceptions and behavioral intentions of potential parents.[20]

This highlights the importance of comprehensive discussion and public education that will help bridge possible gaps between the actual meanings of testing results and the way they are being perceived by potential parents. Such discussion may also encourage individuals at different reproductive stages to develop more realistic expectations about different genetic tests, as well as to acquire a broader understanding about various possible outcomes (e.g., some defects may be fatal while others may be treatable, and different diseases may be expressed with various levels of severity). This is of great importance especially given the fact that potential parents tend to be vulnerable and seek a conclusive determination regarding the health of their prospective child.[21]

Furthermore, such a comprehensive approach may also contribute to avoiding a potential "myth of elimination." This suggested concept relates to the power of different diagnostic practices to potentially create conceptions of health and illness that may lead to a false sense of certainty. Indeed, genetic testing does not necessarily produce definite outcomes, and although the number of genetic diseases capable of being screened is rising constantly (new tests offering identification of over 100 genetic diseases with one specimen),[22] these tests cannot account (at present) for the full scope of existing genetic disorders. Therefore, it is important to encourage prospective parents to acknowledge that the elimination of some genetic diseases is by no means an elimination of all of them, and ultimately, no test can guarantee the birth of a healthy child.

Nevertheless, the practice of PGD (which is used in conjunction with IVF) enables genetic testing of embryos before transferring them to the uterus. Undoubtedly, the ability to identify specific genetic disorders before implantation is of great value to many individuals, especially carriers of various genetic diseases, because PGD enables them to avoid the agonies of raising a sick child or of a midterm abortion. Indeed, two-thirds of all PGD cycles in 2005 were performed in order to detect chromosomal abnormalities.[23] However, this practice is at the core of a vigorous public debate from various ethical, legal, religious, and social perspectives about the status of the embryo at different developmental stages, the legitimacy of embryo screening and selection, and the bioethical challenges involved in defining what kind of genetic disorders justify refraining from implanting embryos into the womb.[24] Furthermore, these kinds of decisions could potentially add new meanings to the definitions of "health," "illness," and "normality," which are not only rooted in the medical realm but also derived from social, ethical and cultural values.[25]

Although the use of PGD for the detection of genetic disorders is widely accepted, in some cases it is also being offered in the absence of medical concerns for the sole purpose of satisfying individual preferences for certain desirable genetic traits, thereby considered highly controversial. For example, although sex selection is occasionally applied for health-related reasons (i.e., to eliminate the risk of genetic diseases that manifest themselves in only one of the sexes), the controversy aroused by this practice is rooted mainly in its potential use without any medical indication for national or individual purposes, such as family balancing or a solid parental preference for a particular sex.[26] Indeed, a survey of 415 ART clinics in the United States suggests that while sex selection through PGD is performed to avoid a genetic disease in 58 percent of PGD clinics, it is also widely provided by 42 percent of PGD clinics in the absence of any medical indication, merely to satisfy the preference of potential parents.[27]

The ethical stances towards sex selection in the absence of medical indication tend to vary between a conservative view (that takes a firm stand against it), a permissive view (that supports it, provided that IVF was initially performed for medical reasons), and middle grounds (that regard sex selection as legitimate as long as there is a prior medical indication for PGD).[28] Accordingly, the practice of sex selection is regulated differently around the globe and while it is prohibited in some countries, it is nevertheless permitted in the United States and others.

Although PGD is currently out of reach for many potential parents, concerns have been raised regarding its future potential to contribute to an already existing phenomenon that is using advances in reproductive technology to affect gender imbalances in societies biased toward a specific gender. For example, in several Asian countries, notably India and China, where reproductive technology has been employed to eliminate females via various medical methods, the desire for males has led to an unnaturally high male/female ratio in the population. Consequently, sex selection can ultimately lead to gendercide and severe demographic outcomes with intergenerational implications for family and society.

Furthermore, as diagnostic technologies are increasingly becoming more available, the selection against or abortion of female embryos is likely to be more accessible and attainable. Indeed, it is estimated that 20 million potential females have been eliminated in India, and that males under the age of 20 exceeded females by more than 32 million in 2005 in China.[29] Moreover, it has been widely reported that there are about 160 million missing girls and women in the world.[30]

Therefore, in many ways the evolution of medical technology facilitates the decline of female births today at much earlier stages of human development. Just as the increasing availability of amniocentesis and ultrasound technology was used to shift from infanticide of baby girls after birth to earlier selective abortions of unborn fetuses (female feticide), so does the introduction of PGD allow possible elimination of female embryos at an even earlier stage prior to implantation (female embryocide).[31] It was also reported that fetal sex determination can be obtained by new NIPD testing, thereby expanding the range of possible methods that could potentially be applied for the purpose of sex identification and selection.[32] Hence, although human preference for one sex over the other is not new, it appears that the methods that are being used to pursue it have dramatically changed over time, while advances in medical technology nowadays allow sex determination and intervention at an earlier stage of the process of human development (see Figure 1).

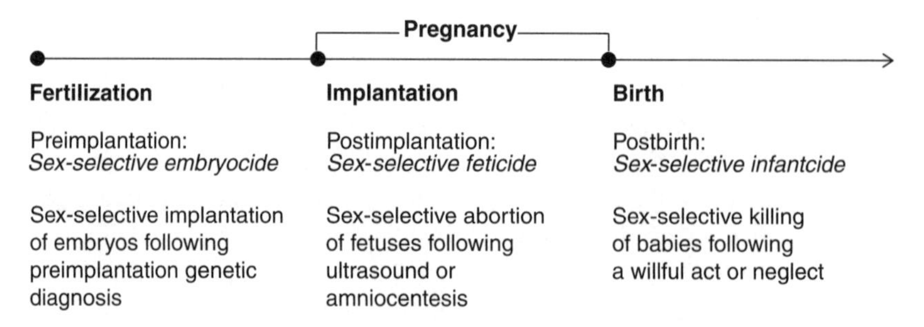

Figure 1 Sex selection along the process of human development

Finally, although sex selection does not manipulate the genome or involve any form of genetic engineering, it further demonstrates the fine line between the application of reproductive technology in order to meet the quest for a healthy child (as outlined throughout this section) and the growing medical intervention in the process of procreation for the purpose of meeting individual or national preferences (regardless of medical indications).[33] It also serves as an excellent example of the way in which contemporary uses of reproductive practices are often expressions of cultural and political contexts that extend far beyond medical considerations into the realm of social values.[34] Moreover, science and society seem to mutually affect each other because, on the one hand, social values affect the focus of and investment in scientific research, while, on the other hand, scientific advances project back on society and resonate on its practices and conceptions.

The following section further illustrates how such projections from the forefront of genetic research are increasingly making headlines and echoing a wide range of social values and conceptions regarding the role our genetic makeup plays in shaping human conduct and moral behavior. As potential parents gain more control over the specific genetic traits of their prospective offspring, such perceptions regarding the extent to which our genetic makeup could potentially affect our lives seem to have even greater impact on the way we understand and value our own genetic traits and consider them worthy of transmitting to our children.

Who Is Responsible for What? Understanding Our Phenotype through Our Genotype

Growing concerns over detection of, preference for, and selection of certain genetic characteristics are at the core of a vibrant debate. Attention to this matter may well be due to exaggerated claims made in recent scientific reports, cited and widely circulated in the media, that our genetic makeup is supposedly responsible for many human behaviors. For example, media reports suggest a genetic explanation for smoking (likelihood to smoke, difficulty in quitting, and risk for lung cancer),[35] overeating (genetic basis for obesity),[36] performing poorly on a driving test (on the basis of a particular gene variant)[37] and navigating (linkage between missing genes and orientation in space),[38] as well as bullying and aggression.[39] In addition, media reports seem to constantly associate specific genes with a seemingly interminable list of behaviors, such as taste preferences and sensitivity to bitter taste;[40] hostility, anger and pugnacity;[41] and even financial behavior of living beyond one's means and incurring credit-card debt.[42]

These accounts seem to overestimate the role of genotype as the core explanation for complex human behaviors and support the reductionist thinking that more and more areas of our lives are the outcome of lineage. Thus heritage from the past is increasingly emphasized rather than present actions. The underlying message to potential parents seems to focus at times on the genetic traits that they can potentially choose to transmit to their prospective children in order to grant them a better future, rather than on the role parental choices can play in shaping children's attitudes towards balanced nutrition, responsible driving, healthy relationships, anger management, financial management, social interaction, and other behaviors that could potentially have even greater impact on their lives.

Moreover, these claims apparently suggest that our alleged genetic shortfalls may be passed on to our children unless we take advantage of future scientific advances that may allow us someday to modify certain genetic traits.[43] For example, one day we may hypothetically be able to produce a child who will not smoke, overeat, bully, or overspend and will be blessed with wonderful taste and excellent navigation skills.

The media reports on various "breakthrough" discoveries demonstrate the power of mass communication to disseminate a conceptual framework with which to understand our genetic makeup, thereby leading to a wider ethical debate on journalistic responsibility. However, it was also noted that the responsibility for such exaggerated headlines does not lay solely on the press, but rather is the outcome of complex interrelations between scientists, industry, government and interest groups.[44]

Nevertheless, the media (print, broadcast, and virtual) have become a central source of information and a growing factor in individual decision-making processes, especially regarding health, reproduction, and other complex issues about which the majority of the general public has no personal experience nor previous knowledge.[45] Thus it appears that also in the case of genetic science the media play a major role in stimulating public awareness and disseminating information and conceptions about genetics in general, and in particular, about the rising power of isolated genes to explain a variety of human behaviors. For example, although they do acknowledge the role of cultural factors and the overall characteristics of a given relationship, media coverage has suggested that there is a genetic basis for infidelity: "Too Many One-Night Stands? Blame Your Genes";[46] "Genes May Be to Blame for Infidelity";[47] "Could Monogamy Gene Combat Infidelity?";[48] "Women's Infidelity Is All in the Genes";[49] "BU Researchers Connect Gene to Infidelity";[50] "Infidelity

Gene? Genetic Link to Relationship Difficulties Found";[51] "The Urge to Infidelity . . . It's in Her Genes";[52] "Infidelity Might Be in the Genes";[53] "The Love-Cheat Gene: One in Four Born to Be Unfaithful, Claim Scientists."[54]

Moreover, a kind of "infidelity test" was also mentioned in recent media reports as an arguable possibility that may be feasible at some future point: "If scientists identify the infidelity gene, this would raise the possibility that there might one day be a test for it. People planning to marry could then, in theory, first have their prospective partner tested to assess the risk of future betrayal."[55] This could potentially raise hopes in the power of this debatable test (or future tests) to predict our social behavior despite the realization that on the one hand, those who do not carry such an "infidelity gene" could potentially be unfaithful as well; on the other hand, those who might be identified as "carriers" of such a gene will not necessarily cheat on their partners. Therefore, even if such a test becomes available one day, its efficiency as a valuable tool for choosing a life partner remains questionable, especially given the many other psychological, cultural, and social factors that play a major role in human relationships.

Although the idea of choosing our partners on the basis of their genetic makeup is not new, such recent reports on the potential to identify particular genetic traits stretch the plausibility of this possibility. Such exaggerated claims regarding the role that certain genetic traits could potentially play in our lives are opening up new forms of negotiations as authorities debate which traits are worth investing resources in for further research. This debate is also associated with concepts of personal responsibility, because it concerns who is ultimately accountable for various social behaviors and what could potentially be done to eliminate certain human faults.

This kind of discussion is also evident in the language in which scientific findings on isolated genes are announced and reported to the general public. Media analysis of such reports suggests framing these scientific discoveries within a "discourse of accusation," that is, they tend to employ a terminology that relieves or even absolves individuals of any sense of responsibility or blame for their imperfections such as: "Can't Quit Smoking? Blame Your Genes";[56] "If You Smoke Too Much 'Blame Your Genes', Say Experts";[57] "Having Trouble Squeezing into Your Jeans? Blame It on Your Genes";[58] "Bad Driver? Blame Your Genes";[59] "Early Birds, Night Owls: Blame Your Genes";[60] "Like to Sleep Around? Blame Your Genes";[61] "Blame Your Genes for That Desperate Craving for Coffee, Research Suggests";[62] "Overeating? Blame Your Genes";[63]

"Binge Drinking? Blame It on Your Genes";[64] "Lost It All in the Stock Market? Blame Your Genes";[65] "Don't Want to Exercise? Blame Your Genes";[66] "Feeling Lonely? Genes Might Be at Fault";[67] "Cluttered Home? Blame Your Genes";[68] "Speaking in Tones? Blame It on Your Genes";[69] "Bad-Tempered Women 'Can Blame It on Genes' ";[70] "Hate Broccoli? Spinach? Blame Your Genes";[71] "If You Don't Eat Greens, Blame It on Your Genes."[72]

This discourse of accusation is being repeated all across contemporary media and suggests that the blame for the majority of human faults lies primarily in certain single genes. It further gives grounds for a misconception that detection of a particular gene in one's body serves as a sufficient indication of behavior associated with that gene, while it constitutes a new form of genetic reductionism that glorifies the power attributed to single genes as the core explanation of human conduct.

This conveys inaccurate conceptions regarding genetic causation and predictive power, while it assumes an unequivocal causal relation between genetic cause and behavioral outcome, as demonstrated by the shared headline format seen above that employs sequences of questions and answers.[73] This structure frames the fault presented (overeating, cheating, smoking, drinking, or whatever) as a question for which the identified gene is the answer. Such framing also glorifies the role of genetic science as the potential remedy for unresolved human behaviors and ultimately as our hope for redemption from ourselves.

Furthermore, the assumption that seems to underlie this framing is that since our genetic makeup has been found accountable for our misbehaviors, we may not be morally culpable. This implies that we do not control our genetic makeup (and therefore pardoned when a genetic explanation is provided to our faults), yet presumes that we do control all that is not a part of our heredity (and therefore we were supposedly held accountable to those faults prior to their genetic explanation).

This conceptualization not only provides exaggerated notions regarding the extent to which our genotype can explain our phenotype but also employs a kind of "myth of vindication," which nurtures the notion that we are clear of blame given our alleged lack of control over our misbehaviors. This suggested notion relates to a new form of genetic reductionism, which absolves us of blame and guilt over the way we look, eat, drive, and interact, thereby offering an appealing message of relief. It further implies that our presumed attempts to transform ourselves up to this point (e.g., lose weight or quit smoking) were, in fact, groundless and useless. In other words, it suggests that we cannot break free from the constraints of our genetic makeup, regardless of the strength of our will, thereby losing our moral agency. However, a counterargument to this claim could be that choice is always present, even within the constraints of a given genetic makeup, and ultimately, with choice comes responsibility. Given this view, one could argue that we must claim responsibility for our conduct and not be tempted to turn such genetic explanations into a form of genetic alibi that supposedly justifies many of our shortcomings.

Moreover, this complex accusatory discourse functions as a double-edged sword that crosses generations. One cannot be held accountable for present misconduct because it is supposedly rooted in its heredity from the past. On the other hand, one can increasingly be considered to be responsible for the genetic makeup passed on to offspring given the new possibilities offered by consumer-oriented reproductive genetics. This discourse not only suggests the potential to free our children from our congenital traits one day but also may imply that we have a responsibility to do so—among other things, to ensure that our children will not blame us for their future conduct, just as it is currently suggested that we can blame our parents.

The underlying assumption behind the discussion in this section also suggests that some genetic traits are considered "better" than others, and that potential parents should aspire to provide their prospective children with traits that are most valued. Indeed, contemporary reproductive practices enable individuals to test their genetic makeup and better control their genetic heritage, as well as to select donated gametes for their prospective offspring with their preferable genetic traits, as discussed in the following section.

"Download Baby": Creating "Better" Babies through the Global Commerce of Gametes

Individuals and couples increasingly appear to use contemporary reproductive practices in a way that allows them to obtain what they consider to be a more desirable set of genetic traits for their prospective child, within the scope of their financial capacity. As a result, the commodification of reproductive organs and tissues has flourished in recent years as sperm, ova, ovarian tissues, rented wombs, and even frozen embryos have been exchanged. The growing demand worldwide for donated gametes is typically associated with medical difficulties encountered when one seeks to use one's own gametes, the lack of an opposite-sex partner (gay or lesbian couples), or the absence of a partner altogether (single individuals).

Although a whole new set of considerations emerges once donor gametes are needed for reproduction, the fertility industry is primarily designed to meet the desire of potential parents for a healthy child. For the most part, potential sperm and egg donors are engaged in a process whereby they complete long questionnaires related to their mental, physical, and genetic history; undergo extensive physical exams, and participate in several in-person interviews and genetic counseling sessions.[74] Typically, they all go through standard testing (e.g., chromosome analysis) that also includes various infectious diseases[75] (e.g., HIV), as well as additional genetic testing based on specific ethnic background.[76] Furthermore, the question whether sperm banks can be sued under product-liability laws for failing to detect certain genetic defects in donor sperm was at the core of a lengthy litigation.[77] Similar debate regarding the view of gametes as a kind of product has also been initiated over the commercialization of certain genes along with questions regarding gene patentability, as in the case of Myriad Genetics.[78]

However, beyond concerns for the prospective child's health, potential parents also seem to be selecting their sperm or egg donors on the basis of other preferable genetic traits. Sperm banks and egg-donation agencies offer endless catalogs of donors from whom to choose according to an individual's personal, ethnic, and aesthetic preferences. Prospective parents are encouraged to narrow their search by using online menus and checklists that categorize donors by age, height, eye color, hair color and type, skin tone, blood type, body mass index, athletic skills, educational achievements, and many other parameters. In addition to medical, ethnic, and physical factors, search engines for donor gametes allow potential parents to choose different lifestyle factors (probably based on the hope that the perspective child will share some of them), such as favorite hobbies and talents (e.g., musical, athletic, artistic), favorite subjects (e.g., natural science, technology, psychology), and personal goals (e.g., traveling, improving the environment, politics).[79]

Therefore, it seems that prospective parents choose the gametes they wish to use based on their evaluations to a set of phenotypes. The growing prevalence of these practices could potentially nurture a false impression that particular gametes are causatively correlated to factors like hobbies and personal goals, when in fact, it was noted that genes are more like plants that could end up becoming different things, depending on the environment in which they are grown.[80] Hence this misleading impression might reinforce a general misunderstanding regarding the role of genetics and the extent to which it can explain our life and predict the life of our offspring. Moreover, it sows the seeds of genetic essentialism in regard to future families as it sustains potential parents' hopes, expectations, and dreams.

This growing emphasis on the genetic makeup of donors as the core criterion in the gestation of a new baby is in keeping with the general glorification of genetic makeup as a key factor that determines one's prospective quality of life, also praised and fostered by contemporary information and entertainment media.[81]

While the entire market of fertility treatments is estimated at nearly 3 billion U.S. dollars in 2004 in the United States, the industry of gamete trade alone has produced revenues of over 112 million U.S. dollars. The cost of a "regular" gamete donation typically varies but is usually around several hundred U.S. dollars for semen and around several thousand U.S. dollars for ova (sperm is usually much cheaper than eggs, given the major difference in their harvesting). However, the cost tends to be much higher for high-profile eggs and can range from 50,000 to 100,000 U.S. dollars if the donor is exceptionally good-looking, excels in sports, has unique musical skills, or is a student at an Ivy League university.[82] The commercialization of gametes has even gone one step further with the introduction of online auctions in which the sperm and eggs of highly desirable donors are supposedly sold to the highest bidder.[83]

The growing demand for gamete donations generates great financial temptation for potential donors, especially at times of financial distress. Indeed, news reports have indicated an increase in the number of men and women who apply to become sperm and egg donors.[84] This may well be due to the need for an additional source of income given the current economic crisis in the United States. Recruitment of women for egg-donation "jobs" is increasingly managed through online agencies and career websites that list them by position numbers, just like any other job in the market. This raises a vast range of ethical concerns since currently there is no national registry in the United States that keeps track of the number of donations from a single donor. One of the purposes of such a registry would potentially be to ensure that less privileged women would not be harmed by donating too many times, especially given the substantial physical risks associated with the medical procedure of egg harvesting.[85]

Indeed, financial constraints play a major role in the commercialization of gametes for both prospective donors and potential recipients. Both seem to use the global market of gametes either to offer their gametes for the highest possible price or to locate desirable donors whom they can afford. The Internet has become a central marketplace for conducting such commerce of reproductive services because it offers detailed information about the various reproductive options now available, and links different stakeholders around the world while it draws on the social and fiscal structures of capitalism, globalization, and the free market. Indeed, the Web has expanded gamete donations across countries because it facilitates overriding existing geographic, financial, and social constraints on reproductive technologies in different parts of the world (often leading to reproductive tourism). Gametes are increasingly selected online, harvested and shipped to clinics overseas. This transnational practice in the creation of babies reduces some of the costs of gamete harvesting, diagnosis, cryopreservation, storage, implantation, and pregnancy. For example, human eggs can be purchased online and fertilized in U.S. laboratories with sperm shipped in frozen containers from a prospective father overseas. The embryos produced are then frozen and sent in liquid nitrogen containers to a surrogacy agency in India to be implanted into the womb of a local gestational carrier.[86] These practices of commodification and trafficking clearly raise many ethical concerns regarding potential exploitation of unprivileged women, as well as question the very notion of their free choice.[87]

Nevertheless, the central role of the Internet in the fertility market is further illuminated by the way it facilitates selection of donated gametes based on the appearance of both donors and recipients. New search mechanisms through donor catalogs allow potential parents to match with a donor whose appearance resembles their own, their partners, or even someone famous, depending on their personal preferences and financial means. Some sperm banks and egg donation agencies apply new mathematical formulas that invite potential parents to upload their photos and to compare them with those of all available donors, free of charge. This system compares the level of similarity between donors and recipients and provides instant results that are rated on a high, medium, and low scale.[88] Some online services also claim to help potential parents predict the appearance of their prospective child by using advanced face-detection technology (an image of the prospective child is produced that is based on the pictures of its potential parents).[89] Such services demonstrate the major role that the appearance of the child seems to play in the donor-selection process.

Those who wish their child to resemble their favorite movie star or professional athlete can make use of special lists of donors who supposedly look like known celebrities (e.g., donor 11437 resembles Ben Affleck, and donor 11485 resembles Johnny Depp).[90] Sperm- and egg-donation banks address the somewhat contradictory desires of those who want their children to look like them, on the one hand, and those who wish their offspring to resemble someone else, on the other hand, presumably since they consider the other person more appealing or better looking than themselves (e.g., a beautiful model or a favorite celebrity).

It seems fair to assume that most individuals do not opt for such high-profile gametes as long as they can use their own, and they are typically not drawn into the search for donated gametes merely for the sake of "upgrading" the genetic makeup of their offspring unless they are in need of donor gametes because of their inability to use their own or those of their partner. But the growing tendency to address such genuine needs for donation by looking particularly for those gametes that supposedly offer a "better" set of genetic traits (sometimes considered even "better" than one's own) can potentially serve as evidence of contemporary social and cultural values.

Nevertheless, the very aspiration to produce better-looking children, who consequently resemble someone else rather than the potential parents themselves, seems to challenge the common desire for genetic similarity between parents and their children. As noted earlier, this desire appears to underlie the use of various ARTs, such as IVF, ICSI, and PGD, for the purpose of maximizing the probability of using one's own genome. Even those who fail to conceive and choose adoption are often guided by the child's appearance and ethnic background in comparison with their own. Therefore, one could assume that individuals in need of donor gametes will look for donors who best resemble them, which is indeed the case for some. However, others seem to choose donors whose genetic characteristics seem dramatically remote from their own presumably because they are perceived as more desirable.

Indeed, this paradox may be explained by the cultural value that frames such high-profile donor gametes not just as different from one's own but rather as better. The commerce of better genetic traits suggests the social embrace of genetic differences as long as they are perceived as equally desirable or as a kind of enhancement to one's own genetic makeup. For example, many individuals in need of surrogacy services hire gestational carriers overseas, often in third world countries, in order to reduce their overall costs. However, in many cases they refrain from using local egg or sperm donors from these countries (despite their low cost) and instead go through tremendous financial and logistic efforts to ship their preferred set of donated gametes from Western clinics, ultimately to ensure the Western appearance of their child.

Perhaps one of the most extreme expressions of the desire to create better babies with an improved set of congenital traits is the Repository for Germinal Choice (commonly known as the Nobel Prize Sperm Bank) that operated in California in the 1980s and 1990s. Founded by Robert

K. Graham, this bank sought to collect sperm from Nobel laureates in order to improve the human race and help reverse genetic decline.[91] Although the bank contributed to the birth of over 200 children, its records suggest that none were conceived using sperm of a Nobel laureate. In fact, although all sperm donations belonged to scientists, academics, athletes, artists, and businessmen, only one donor was a Nobelist. The bank's original intention was to monitor the development of these children, but only two of the families shared such information. As a result, most questions regarding this attempt to produce remarkably intelligent, better children remain unanswered.

The Nobel Prize Sperm Bank serves as an important reference point with regard to the wild motivations associated with genetic science because they are typically based on the common misconception that in choosing gametes with suposedly desirable genetic traits (in this case, intelligence) there is some kind of guarantee that they will indeed be transmitted to the next generation (e.g., that the prospective child will inherit the donor's intellectual capacities but not his less desirable traits, such as baldness). Furthermore, it glorifies the role of male sperm in determining the entire phenotype of the child because very little emphasis has been given to the genetic traits of female recipients of these high-profile donations (other than assuring that they were in good health and married to an infertile man). This appears to resonate with the cultural emphasis on the male role in human reproduction and the historic view that the embryo was derived entirely from the sperm and placed as a whole in the female womb.[92]

Such contemporary attempts to create better babies through the exchange of an improved set of gametes are also associated with profound historical and ethical challenges since they somewhat echo social Darwinism and principles deeply rooted in darker historical eras.[93] As opposed to the enforced public nature of past eugenics imposed by the agency of the state for the purpose of enhancing the nation, some contemporary practices of genetic selection function as a form of voluntary, private, commercial eugenics aimed at advancing personal preferences.[94] This approach resonates with principles of individualism, liberalism, and autonomous choice. Although these practices are mainly initiated through individual agency (of potential parents) and not by governments, some advances in genetic science and reproductive technology seem to be used to eliminate certain genetic traits because of or in the absence of medical indication, whereas other traits are singled out, favored, praised, and overpriced in the free market according to current social values and standards.

From a historical perspective, it may be claimed that what is perceived today as a potentially better baby might not have been equally appreciated in different eras. For example, the very notion of beauty (face, body shape, size) that appears to play a major role in the commerce of donated gametes has shifted dramatically over time.[95] Thus it seems that preferences for certain aesthetic traits in the fertility industry are in line with those glorified by other contemporary industries (e.g., cosmetics, fashion, diet). Furthermore, it has been argued that many of our supposedly personal desires and preferences, especially in the fields of sexuality and reproduction, are in fact realizations and projections of national values and social norms.[96]

Many contemporary reproductive practices aimed at improving the genetic makeup of a prospective child (especially in the absence of medical indication) could potentially be considered as additional forms of genetic enhancement, as well as a manifestation of the growing concern for a slippery slope on the way to designer babies. However, individual traits result from a complex interaction of genes and environment given the notable impact of environmental variation on the experessions of genes.[97] Moreover, a genetic trait that is regarded by some as undesirable and even worthy of elimination may well be appealing to others in different cultural contexts. For example, although female embryos are constantly discarded in certain third world countries, individuals in many Western countries are willing to pay considerable sums to ensure having a girl through sex selection; and although most parents would do anything to avoid a deaf child and most sperm banks reject deaf men as potential donors, one deaf couple in the United States has deliberately created a deaf child because the parents perceive deafness as a cultural characteristic rather than a disability.[98]

Thus current gamete-exchange practices have the potential to serve as new forms of discrimination as enacted through high or low demand for donor gametes of certain health-related groups (blind, deaf), ethnic groups (blacks, Jews), professional groups (athletes, musicians), or other groups that carry particular genetic traits (blue eyes, blond hair, certain height or weight). For example, it was widely published that sperm donations from redhead men are no longer accepted at the largest U.S. sperm bank due to lack of demand.[99]

Indeed, contemporary gamete-related commerce seems to be gradually changing the ways in which we think about genetics, our bodies, and the future of the human race. Furthermore, new conceptions of human diversity are emerging as these reproductive practices reflect and shape our values of what is regarded as adequate, fit, and desirable—and consequently what is considered better and worthy of reproduction. Although some claims to produce better offspring are scientifically feasible and do guide certain reproductive choices, it appears that the premise behind this practice is sometimes exaggerated or misunderstood: aiming at preferred genetic traits by opting for high-profile egg and sperm donations provides no guarantee of "better" babies and certainly no assurance of an easier or "better" life, however defined.

Last, the pursuit of better babies of the future is also associated with the quest for information regarding our genetic background of the past and with the growing demand for testing practices that allow us to explore further our genetic makeup with the hope of better understanding our present. Moreover, the ongoing request for genetic information and practices at all levels of society has been significantly stepped up by contemporary social and commercial uses of the Internet and is closely related to diverse issues of anonymity, privacy, and identity, elaborated in the following section.

The Commodification of Genetic Traits and the Establishment of Identity

Contemporary e-commerce and practices involving hidden identities greatly facilitate the search for personal genetic information, as well as the exchange of desirable gametes. Online catalogs categorize and rate the full spectrum of genetic diversity and arrange the vast array of human genetic traits so that gametes can be selected, negotiated, and paid for just like any other commodity available in the marketplace. As a result, individuals and couples around the globe shop for gametes, surrogate wombs, and even embryo adoption programs as part of their search for the ultimate genome. They create usernames and passwords to gain instant access to endless profiles of potential sperm and egg donors, login to their accounts to add donors' profiles to their online shopping carts, and provide their credit-card information to purchase additional exclusive data regarding their preferable donors, such as detailed profiles, personal interviews, and additional pictures of donors as babies and teenagers.

In a world of e-mails, tracking numbers, and online payments, individuals and couples can consume reproductive services worldwide from their workstations. Human gametes can be selected and purchased merely with a click of a mouse, and genetic parents of the same child no longer need to have sexual intercourse with one another, to meet one another, or even to know the other's name. They are connected mainly by computers without even leaving their doorsteps, mediated by websites, usernames, and numeric IDs that protect their privacy.

Since there is no national registry of gamete donations, it is difficult to estimate how many of these transactions take place annually in the United States or how many times a single donor has donated through one or several agencies. Some sperm donors have donated hundreds of times and consequently have dozens or even hundreds of offspring. Thus several concerns have been raised regarding the implications of anonymity in gamete donation, such as donors' responsibility for full disclosure of health history and future updates of health status, donors' access to information regarding the outcomes of their donation, informed consent of donors, potential for marriage between donor-conceived siblings (who are unaware of their kinship), and persons' right to know their genetic origins.[100] Indeed, many individuals who were conceived by anonymous gamete donation are left without (or very limited) information about their genetic history, while some may well not be informed at all that they were conceived through donor gametes.

As a result of these and other ethical concerns, there are now some countries that only allow non-anonymous gamete donations in order to address the essential need of donor-conceived children to track their genetic heritage, either for health purposes or merely for their sense of identity. The importance that genetic background appears to play in one's life can also be illuminated by the emergence of online voluntary registries that serve those who were conceived through anonymous gamete donation.[101]

This new phenomenon allows offspring conceived by anonymous gamete donation to look for their half siblings and genetic parents (donors). In some cases, donors are interested in finding out how many offspring they have propagated, and a few such donors have reported meeting with their children, although they are not required to do so by the anonymity protections typically provided in the contracts they signed. Thus, although the use of the Internet for contemporary commerce of gametes serves as an excellent platform for protecting the anonymity of donors (who are usually listed under donor IDs), it turns out that it also helps individuals override the constraints of anonymity and renegotiate their genetic identities.

However, the extent to which genetics is weighed into our sense of identity is also indicated by the terminology commonly used to describe our genetic origins in light of the commerce of donor sperm, eggs, and embryos. This terminology appears to challenge many of our values and beliefs, thus reflecting the power of language to legitimate social and political realities. For example, most exchanges of gametes are regarded as a form of "donation" although most donors are well paid for their sperm or eggs.[102] Despite the fact that many countries allow different forms of commercial gamete exchange, the terms commonly used to frame them remain somewhat altruistic. Such an exchange is typically referred to as taking place between "donors" and "recipients," not, for instance, between "vendors" and "vendees" or even "employees" and "employers," respectively. One can argue that these alternative terminologies could potentially better reflect the reality of gamete-related commerce in light of the growing definition of egg harvesting as a kind of part-time job (as previously noted), and given the fact that most individuals are indeed compensated for their "donation."

As in many other domains, word selection can be related to wider issues of social responsibility and may reflect potential exploitation of unprivileged individuals, as well as being a key to deeper social views regarding commodification of the human body, in general, and reproductive issues, in particular. Moreover, the power of terminology appears to shape the establishment of our genetic identity in accordance with the different terms that are being used to describe contemporary parental relations. Advanced reproductive practices enable fragmentation of parenthood among genetic, gestational, and social roles because multiple individuals may take part in the creation and upbringing of one child by contributing sperm, ovum, womb, or care—thus bringing new meanings and explanations to the definition of conception.[103] Consequently, these practices constitute new forms of parenthood that differ from each other in the type of relation that they potentially establish between children and their different kind of parents, namely, the various providers of gametes (genetic relation), pregnancy (gestational relation), and upbringing (emotional and social relation) (see Table 14.1).

Hence the terminology used in the reproductive sphere may well have a powerful impact on our understanding of genealogical trees, familial blood ties, and the notion of parentage. Moreover, it cultivates our conceptions regarding the importance of genetic ties and the role they play in making one "qualified" to be reffered to as a "parent." Our choice of words also sheds light on the extent to which a sperm or an egg donor represent any kind of parental role beyond that person's function as a provider of gametes; the social meaning of pregnancy and its capacity to constitute any kind of parental relation; the profound difference in meaning that is embedded in the term "gestational carrier" versus the term "surrogate mother" when describing a women who is carrying a child whom she is not going to raise (whether she is using her ovum or not); and the social value attributed to the continuous upbringing and care that one provides to a child, even if they are not genetically related.

Table 14.1 Typology of different forms of parental relations

Type of parental role	Genetic relation (gamete)	Gestational relation (pregnancy)	Social relation (upbringing)
Female			
"Traditional" mother	+	+	+
Designated mother through surrogacy (using own ovum)	+	–	+
Designated mother through surrogacy (using surrogate's or donor's ovum)	–	–	+
Surrogate mother (using own ovum)	+	+	–
Gestational carrier (gestation only)	–	+	–
Egg donor	+	–	–
Recipient of egg donation	–	+	+
Embryo donor	+	–	–
Recipient of embryo	–	+	+
Adopting mother	–	–	+
Male			
"Traditional" father	+	–	+
Designated father through surrogacy	+	–	+
Sperm donor	+	–	–
Recipient of sperm donation	–	–	+
Embryo donor	+	–	–
Recipient of embryo	–	–	+
Adopting father	–	–	+

Note: This typology illustrates various forms of parental relations in terms of the different roles that individuals can assume in a child's life. However, other variations and combinations may apply.

Notwithstanding the ongoing nature-nurture debate over the contributions of both genetic and social environments, it appears that referring to a sperm donor as a "father" and to an egg donor as a "mother" (and occasionally as "the" father or mother, or even as the "real" father or mother), despite their proclaimed disinterest in being involved in any way in the child's life, is in fact reinforcing another form of genetic reductionism. The entire essence of the parent-child relationship is thus being reduced to their shared genetic makeup and explained in terms that are also deeply rooted in traditional and patriarchal views.[104]

Further examination of the terminology and meaning attributed to contemporary parental relations is needed, especially given the dynamic changes of the nuclear family and in light of current living arrangements of children under the age of 18 in the United States. In 2009, 20.2 million children (27 percent) lived with only one parent, while 5.6 million children lived with at least one stepparent. Even children who lived with two parents, did not necessarily live with both biological parents but rather experienced different combinations of parental relations. For example, 5.4 million children (11 percent of all children living with two parents) lived with a biological parent and either a stepparent or adoptive parent.[105]

Although adoption cases serve as an excellent example of the merits of emotional bonding and social upbringing beyond heredity, it seems that the genetic tie between parents and their children is broadly perceived in our society as superior to other forms of relations. For example, some men (not sperm donors) are required to assume legal and financial responsibility for their genetic offspring once a genetic tie is proven. Correspondingly, DNA testing is increasingly being used when parentage has not been established or once the genetic tie between parents and their children is questioned (e.g., when an alleged parent denies paternity). Some choose to turn their backs on the child, emotionally and financially, once a DNA test indicates that the child that they are raising is most likely not genetically theirs. Thus the presence or absence of a genetic tie to the child seems to be at the heart of the social perception of parentage. Another example for the central role of genetics could be derived from the emotional management of surrogacy: many gestational carriers report that the realization that they are not genetically related to the baby that they are carrying has helped them to accept that it is not their child, as well as to maintain emotional distance throughout the pregnancy and to give the baby away after birth.[106]

The power attributed to kinship and genetic ties serves as another demonstration of the reductionist premise that places a much greater value on the genetic aspect of parenthood rather than on other parental dimensions. It thus strongly associates the role of genetics with the essence of parentage while it reinforces the power of heredity as a key factor in familial relations. This emphasis on genetics appears to further stimulate a growing interest in gaining more genetic information about kinship and congenital traits through voluntary DNA testing. Testing services are becoming increasingly accessible and affordable (DTC kits are also available online) and are being used not only to determine genetic relations (e.g., to a sibling or a parent) but also to systematically analyze one's own genetic makeup, capacities, and lineage.[107] The growing demand for such personally identified DNA analysis also raises ample concerns regarding the current absence of regulations, the security and privacy of the genetic material collected and the information derived from it, and the potential misuse of personally identifiable health information, thereby illuminating the complexity of consent, analysis, and storage of genetic information.[108]

Nevertheless, such DNA testing allows individuals to further learn about their genetic traits, discover their potential risk for various genetic diseases, determine possible relations to certain population groups, and inquire about their ancestries.[109] Hence this quest for personal genetic information may have far-reaching implications for what we think of ourselves, how we run our lives, and what we believe we are (in)capable of doing, thereby playing a growing role in the establishment of our personal identity.

Along with other components of contemporary individual identity (e.g., national, ethnic, sexual), which is constantly reconstructed at a particular point in time,[110] one might wonder whether in the near future DNA-testing results regarding our genetic makeup might end up playing a much greater role in the way we perceive ourselves and establish our subjective identity. Moreover, it may even be perceived one day as crucial for its very development and as an integral part of it (very much like obtaining genetic information regarding one's genetic parents). If so, then the hardships encountered by those conceived through anonymous gamete donation on their journey to reveal their genetic past may eventually become the reality of all members of contemporary society who also seek greater familiarity with their genetic present.

Final Note

Throughout this chapter, different forms of genetic reductionism, determinism, enhancement, and precedence have been demonstrated in relation to various reproductive practices. These include the ongoing use of reproductive technologies for the purpose of assuring kinship and genetic ties; the increasing use of genetic testing and diagnostic techniques to ensure the health of the prospective offspring and gradually to control its sex and other genetic characteristics, regardless of medical indication; the ongoing glorification of single genes and their power to explain a wide range of human conduct, as strongly suggested by the media; the global commerce of high-profile donated gametes in light of the pursuit for babies with "better" genetic traits; the attempts of individuals born through anonymous gamete donation to locate their genetic parents and reveal their genetic history; the social values attributed to the different forms of parental relations based on kinship and genetic ties; and the quest for personal genetic information through DNA testing to determine existing family relations and to gain a better understanding of one's genetic capacities, along with its overall potential implications for the establishment of identity.

The critical analysis suggested in this chapter is aimed at outlining the fundamental issues that are associated with the increasing use of these reproductive practices, however a lengthier piece would be required to discuss the entire spectrum of their potential implications to our society. Future examination is needed to further explore the complex interrelations that come into play between reproductive science, genetic research, medical ethics, bio-politics, popular media, global commerce, and social policy, in order to deepen our understanding about the role of genetics in contemporary life.

The current use of these genetic and reproductive practices, as well as the discourse and terminology involved in discussing them, seems to glorify the power of heredity as the key to understanding parental relations, personal conduct, and human potential. Thus there appear to be gradual shifts in the common desire for a healthy baby to include also the growing aspiration for a "better baby," however defined, who supposedly possesses favorable genetic traits. This accounts for many contemporary practices of testing, diagnosing, selecting, and exchanging gametes and genetic traits. Furthermore, these processes deeply affect our most fundamental beliefs about which human traits are considered valuable and worthy of passing on to our children.

As genetic capacities increasingly become the core explanation for our past, present, and future, reducing the overall human phenotype to its genotype alone not only questions our ability to release ourselves from our alleged genetic constraints (and perhaps even exceed them) but also undermines the power of other social and environmental forces to make a notable difference in our existing lives, as well as to shape future generations.

The author would like to express sincere gratitude and appreciation to Dr. Mildred Z. Solomon for her continued and valuable contribution to this chapter.

CHAPTER FIFTEEN

Forensic DNA Evidence

The Myth of Infallibility

WILLIAM C. THOMPSON

PROMOTERS OF FORENSIC DNA testing have claimed from the beginning that DNA evidence is virtually infallible.[1] In advertising materials, publications, and courtroom testimony they have claimed that DNA tests produce either the right result or no result. These claims took hold early in appellate-court opinions, which often parroted promotional hyperbole. They were bolstered by the impressive "random-match probabilities" presented in connection with DNA evidence, which suggest that the chances of a false match are vanishingly small. They were reinforced in the public imagination by news accounts of postconviction DNA exonerations. Wrongfully convicted people were shown being released from prison, while guilty people were brought to justice, all on the basis of DNA tests. With prosecutors and advocates for the wrongfully convicted both using DNA evidence successfully in court, who could doubt that it was in fact what its promoters claimed: the gold standard, a "truth machine"?[2]

The rhetoric of infallibility proved helpful in establishing the admissibility of forensic DNA tests and persuading judges and jurors of its epistemic authority.[3] It has also played an important role in the promotion of government DNA databases.[4] Innocent people have nothing to fear from being included in a database, promoters claim. Because the tests are infallible, the risk of a false incrimination must necessarily be nil. One indication of the success and influence of the rhetoric of infallibility is that until quite recently concerns about false incriminations played almost no role in policy discussions. For example, David Lazer's otherwise excellent edited volume, *DNA and the Criminal Justice System,* which offers a broad assessment of ways in which DNA evidence is transforming the justice system, says almost nothing about the potential for false incriminations.[5] The infallibility of DNA tests has, for most purposes, become an accepted fact—one of the shared assumptions underlying the policy debate.

In 2009 the National Research Council (NRC) released a scathing report on the status of forensic science. The report found serious deficiencies in the scientific foundations of many forensic science disciplines. It also found that procedures used for interpretation lack rigor, that analysts routinely take inadequate measures to avoid error and bias, and that they testify with unwarranted certainty. But the report pointedly excluded DNA testing from these criticisms. It held DNA testing up as the shining exception—an example of well-grounded forensic science that other forensic disciplines should emulate—further reinforcing DNA's status as a gold standard.[6]

In this chapter I will challenge the assumption that DNA tests are infallible. I will show that errors in DNA testing occur regularly and that DNA evidence has falsely incriminated innocent people, causing false convictions. Although I agree with the 2009 NRC report's conclusion that DNA testing rests on a stronger scientific foundation than most other forensic science disciplines, I will argue that many of the problems identified in the NRC report also apply to DNA evidence. In particular, DNA analysts take inadequate steps to avoid bias and to assess the risk of error, and they frequently overstate the statistical value of test results. Although DNA tests undoubtedly incriminate the correct person in the great majority of cases, the risk of false incrimination is high enough to deserve serious consideration in public policy debates, particularly in debates about expansion of DNA databases and debates about the need for governmental oversight of forensic laboratories. My key point is that the risk of error is higher than it needs to be. I will argue that forensic laboratories often compromise scientific rigor and quality control in order to achieve other goals, and they sometimes suppress evidence of problems in order to protect their credibility and maintain the public perception of DNA's infallibility.

Erroneous Matches

When DNA evidence was first introduced, a number of experts testified that false matches were impossible in forensic DNA testing. The claim that DNA tests are error free has been a key element of the rhetoric of infallibility surrounding DNA evidence. According to Jonathan Koehler, these experts engaged in "a sinister semantic game" in which they distinguished error by the test itself from error by the people administering and interpreting the test. They acknowledged (if pressed) that human *error* could produce false reports of DNA matches, but they emphasized that the tests themselves are error free.[7]

Sinister or not, the distinction between human error and test error is artificial and misleading, given that error-prone humans are necessarily involved in conducting and interpreting DNA tests. For those who need to assess the value of DNA evidence, such as judges, jurors, and policy makers, what matters is not whether errors arise from human or technical failings, but how often errors occur and what steps are necessary to minimize them.

The 2009 NRC report agreed that it is vital to know the error rate of forensic tests. It recognized that errors in DNA testing can occur in two ways: "The two samples might actually come from different individuals whose DNA appears to be the same within the discriminatory capability of the tests, or two different DNA profiles could be mistakenly determined to be matching." The report declared that "both sources of error need to be explored and quantified in order to arrive at reliable error rate estimates for DNA analysis."[8] This is certainly true. But I believe that the NRC report erred when it went on to assert that sufficient evidence is now available to assess the probability of a false match in DNA testing. One of the reasons that DNA testing is stronger than other forensic disciplines, according to the NRC report, is that "the probabilities of false positives have been explored and quantified in some settings (even if only approximately)."[9] But the NRC report provided no citations to support this assertion, and I know of none. The only quantification of error rates in DNA testing that I know of concerned proficiency-testing errors in the late 1980s, which have little relevance to current practices.[10] I believe that the NRC report confused rhetoric with reality when it discussed this issue.

What little we know about the potential for error in DNA testing comes almost entirely from anecdotal reports about false matches. These reports are sufficiently numerous to refute claims that errors are extremely rare or unlikely events. They are also useful for illustrating the ways in which errors can occur. But anecdotal data do not provide an adequate basis for assessing the rate of error because it is impossible to know what proportion of the errors in casework are detected—the errors we know about may be the tip of an iceberg of undetected or unreported error.

Types of Errors

One cause of false DNA matches is cross-contamination of samples. Accidental transfer of cellular material or DNA from one sample to another is a common problem in laboratories and can lead to false reports of a DNA match between samples that originated from different people. Scotland's High Court of Justiciary quashed a conviction in one case in which the convicted man (with the help of sympathetic volunteer scientists) presented persuasive evidence that the DNA match that incriminated him arose from a laboratory accident.[11] Cross-contamination is also known to have caused a number of false cold hits. For example, while the Washington State Patrol Crime Laboratory was conducting a "cold-case" investigation of a long-unsolved rape, it found a DNA match to a reference sample in an offender database, but it was a sample from a juvenile offender who would have been a toddler at the time the rape occurred. This prompted an internal investigation at the laboratory that concluded that DNA from the offender's sample, which had been used in the laboratory for training purposes, had accidentally contaminated samples from the rape case, producing a false match.[12] Similar errors leading to false database matches have been reported at forensic DNA laboratories in California, Florida, and New Jersey, as well as in New Zealand and Australia.[13] Three separate cases have come to light in which cross-contamination of samples at the Victoria Police Forensic Services Centre in Melbourne caused false cold hits. Two of those cases led to false convictions.[14]

Perhaps the most telling contamination incident occurred in Germany, where police invested countless hours searching for a mysterious woman known as the "Phantom of Heilbronn" whose DNA profile was found in evidence from a surprising variety of crimes, from murder to larceny. Her DNA was found on "guns, cigarette packs, even nibbled biscuits at crime scenes." Police sought public assistance in identifying this menace to society, and a bounty of 300,000 euros was placed on her head. It turned out that the woman in question was not a criminal at all but an employee involved in manufacturing the cotton swabs that crime laboratories use to collect DNA from crime scene samples. Accidental contamination of crime-scene samples with her DNA (which was on the swabs) caused her to be falsely implicated in dozens of crimes.[15]

A second potential cause of false DNA matches is mislabeling of samples. In 2011 the Las Vegas Metropolitan Police Department acknowledged that a mix-up of DNA samples in its forensic laboratory had caused a false conviction. The lab mistakenly switched reference samples of two men who were tested in connection with a 2001 robbery. One of the men may well have been involved in the robbery—his DNA profile matched an evidentiary sample from the crime scene. Because of the sample switch, however, this suspect was mistakenly excluded while the second man was falsely linked to the crime. Although the police now acknowledge that he was innocent, the second man was convicted and served nearly four years in prison. The error came to light when the first man's DNA profile was entered into a government offender database after he was convicted of an unrelated crime in California, and it produced a cold hit to the crime-scene sample from the 2001 Las Vegas robbery. When investigators realized that the Las Vegas lab had earlier excluded the same man as a source of that sample and had matched the sample to a different man, they realized that an error had occurred.[16]

Similar sample-labeling errors have caused false DNA incriminations in cases in California and Pennsylvania, as well as in an earlier case in Las Vegas.[17] These cases came to light during the judicial process and before conviction, but only because of fortunate happenstances. There have also been reports of systemic problems with sample labeling in Australia. A review of DNA testing by an ombudsman in New South Wales discovered that police had incorrectly transferred forensic data to the wrong criminal cases in police computer records, which on two occasions produced false DNA database matches that led to people being incorrectly charged with a crime.[18] One man was convicted before the error was discovered. Doubt was also cast on a number of convictions in Queensland when a forensic scientist who had previously worked for a state forensic laboratory publicly expressed concerns about the reliability of the lab's work. He told the *Australian* newspaper that it was not uncommon for the lab to mix up DNA samples from different cases. He said that although many such errors were caught, sample limitations made it impossible to resample or retest in some questionable cases.[19]

A sample-switch error caused a tragic delay in apprehension of a man who is believed to be the notorious Night Stalker, a serial rapist who committed over 140 sexual assaults in London. Although police became suspicious of this man relatively early during their investigation of the attacks, they did not arrest him because a DNA-testing error involving a switch of reference samples caused him falsely to be excluded as the source of biological samples found on the crime victims. The error caused a "match" to another man with the same name, but he had a solid alibi. The Night Stalker's crime spree continued for months until police eventually realized that reference samples of the two men had been switched.[20]

A third potential cause of false DNA matches is misinterpretation of test results. Laboratories sometimes mistype (i.e., assign an incorrect DNA profile to) evidentiary samples. If the incorrect evidentiary profile happens to match the profile of an innocent person, then a false incrimination may result. Mistyping is unlikely to produce a false match in cases where the evidentiary profile is compared with a single suspect, but the chance of finding a matching person is magnified (or, more accurately, multiplied) when the evidentiary profile is searched against a database.

A false cold hit of this type occurred in a Sacramento, California, rape case. A male DNA profile was developed from a swab of the victim's breast. The profile was searched against a California database. The search produced a cold hit to the profile of a man who lived in the Sacramento area, but the resulting police investigation apparently raised doubt about his guilt.[21] At that point a laboratory supervisor reviewed the work of the analyst who had typed the evidence sample. According to a report issued by the laboratory director, the supervisor determined that the analyst had "made assumptions reading and interpreting the profile of the breast swab sample that were incorrect" and "had interpreted the profile as being a mixture of DNA from a male and female, when in fact the mixture was of two males."[22]

Interpretation of DNA mixtures can be challenging under the best of circumstances, but it is particularly difficult when the quantity of DNA is limited, as was true in the Sacramento case. Under these conditions DNA tests often fail to detect all of the contributors' genetic alleles (a phenomenon known as "allelic dropout") and can sometimes detect spurious or false alleles (a phenomenon known as "allelic drop-in").[23] Determining which alleles to assign to which contributor can also be difficult, particularly when there is uncertainty about the number of contributors and whether alleles are missing. Interpretations made under these conditions are inherently subjective and hence are subject to error.[24]

A 2011 study highlighted the degree of subjectivity involved in DNA mixture interpretation and the potential it creates for false incriminations. Itiel Dror and Greg Hampikian asked seventeen qualified DNA analysts from accredited laboratories to evaluate independently the DNA evidence that had been used to prove that a Georgia man participated in a gang rape. The analysts were given the DNA profile of the Georgia man and the DNA test results obtained from a sample collected from the rape victim, but they were not told anything about the underlying facts of the case (other than scientific details needed to interpret the test results). The analysts were asked to judge, on the basis of the scientific results alone, whether the Georgia man should be included or excluded as a possible contributor to the mixed DNA sample from the victim. Twelve of the analysts said that the Georgia man should be excluded, four judged the evidence to be inconclusive, and only one agreed with the interpretation that had caused the Georgia man to be convicted and sent to prison—that is, that he was included as a possible contributor to the DNA mixture. The authors found it "interesting that even using the 'gold standard' DNA, different examiners reach conflicting conclusions based on identical evidentiary data." Noting that the analyst who testified in the Georgia case had been exposed to investigative facts suggesting that the Georgia man was guilty, they suggested that this "domain irrelevant information may have biased" the analyst's conclusions.[25] The potential for bias in DNA testing and ways to deal with it are discussed further in "Gross Negligence, Scientific Misconduct, and Fraud" below.

How Errors Are Detected

Proving that an error has occurred in DNA testing is not always easy. DNA evidence has such authority that doubts often arise about other evidence that contradicts it. Consider, for example, the case of Timothy Durham, who was accused of raping a young girl in Oklahoma City. At his trial Durham produced eleven alibi witnesses, including his parents, who all testified that he was with them attending a skeet-shooting competition in Dallas at the time at which the rape occurred. Durham also produced credit-card receipts for purchases he made in Dallas on that day. But the prosecution had something stronger: the young victim's identification and DNA evidence. Durham was convicted and sentenced to 3,000 years in prison.[26]

How can we know whether a DNA test is wrong? One way is to do additional DNA testing. Luckily for Durham, a portion of the incriminating evidence was available, and his family could afford to have it retested. The new DNA test not only excluded him as the source of the semen found on the victim but also showed that the previous DNA test had been misinterpreted. Durham is one of three men in the United States who have been convicted and sent to prison on the basis of erroneous DNA matches but later exonerated by additional DNA testing. (The other two are Josiah Sutton, who was falsely incriminated because of an error in interpretation, and Gilbert Alejandro, who was falsely incriminated because of fraud by a DNA analyst.)[27] It is important to understand, however, that retesting cannot catch every error. Some errors arising from cross-contamination of evidence, mislabeling of samples, and coincidental matches are undetectable during retesting because the new tests simply replicate the erroneous result of the first. In some cases the initial tests exhaust the available evidentiary samples and leave nothing to retest. And many defendants who are incriminated by DNA evidence find it difficult to obtain a retest even when evidence is available.[28]

A second way DNA-testing errors come to light is when laboratories make an admission of error, typically by withdrawing an erroneous laboratory report and issuing a revised report with different results. An interesting example occurred in Philadelphia in 2000. The city's crime laboratory tested samples from a rape victim and from a suspect named Joseph McNeil. The lab's initial report stated that DNA profiles matching McNeil's were found in three evidentiary samples: a vaginal swab, a cervical swab, and a seminal stain on the victim's underwear. McNeil was charged with rape and taken into custody. Although McNeil adamantly denied the crime and rejected a favorable plea bargain, his lawyer could not conceive that a DNA tests could be wrong about three different samples. After an independent expert noted some discrepancies in the lab report, however, he sought access to his client's DNA for an independent test. At that point the police lab realized that an error had been made and issued a new report exonerating McNeil. In its initial test the lab had mixed up the reference samples from McNeil and the victim. What the lab had mistakenly reported as McNeil's profile in samples found on the victim was in fact the victim's profile.[29]

A third way laboratory errors come to light is through proficiency testing. In accredited DNA laboratories analysts must take two proficiency tests per year. Generally this involves comparison of samples from known sources. The analysts typically know that they are being tested but are not told the correct results until after they have reported their conclusions. These tests have been criticized as too easy to detect problems that might arise in actual casework. Nevertheless, errors occasionally occur, generally arising from cross-contamination or sample-labeling problems, sometimes from misinterpretation of partial or degraded DNA profiles.[30] Many laboratories treat proficiency-test results as confidential records, which makes details about the frequency and nature of errors difficult to obtain.

Perhaps the best source of information on the nature and frequency of laboratory foul-ups is "contamination logs" and "corrective action files" that are maintained by some DNA laboratories. Guidelines issued by the FBI's DNA Advisory Board in 1998 recommend that forensic DNA laboratories "follow procedures for corrective action whenever proficiency-testing discrepancies and/or casework errors are detected" and "maintain documentation for the corrective action."[31] Although many laboratories have ignored these guidelines, some laboratories (probably the better ones) have kept records of instances in which, for example, samples are mixed up or DNA from one sample is accidentally transferred to another sample, causing a false match. These records are generally treated as confidential but occasionally become public when they are released under court order as part of the discovery process in criminal cases, or when news organizations file public records act requests for them.[32]

Some labs have voluminous corrective action files that show that errors occur regularly. Files from Orchid-Cellmark's Germantown, Maryland, facility, for example, showed dozens of instances in which samples were contaminated with foreign DNA or DNA was somehow transferred from one sample to another during testing. Files from the District Attorney's Crime Laboratory in Kern County, California, a relatively small lab that processes a low volume of samples (probably fewer than 1,000 per year), showed an array of errors during an eighteen-month period, including multiple instances in which (blank) control samples were positive for DNA, an instance in which a mother's reference sample was contaminated with DNA from her child, several instances in which samples were accidentally switched or mislabeled, an instance in which an analyst's DNA contaminated samples, an instance in which DNA extracted from two different samples was accidentally combined in the same tube, falsely creating a mixed sample, and an instance in which a suspect tested on two different occasions did not match himself (probably because of another sample-labeling error).[33]

In 2008 the *Los Angeles Times* obtained corrective action files from several California labs and found many instances of cross-contamination, sample mislabeling, and other problems. For example,

> Between 2003 and 2007, the Santa Clara County [California] district attorney's crime laboratory caught 14 instances in which evidence samples were contaminated with staff members' DNA, three in which samples were contaminated by an unknown person and six in which DNA from one case contaminated samples from another. The records also revealed three instances in which DNA samples were accidentally switched, one in which analysts reported incorrect results and three mistakes in computing the statistics used in court to describe the rarity of a DNA profile.[34]

The errors documented in these files have typically been detected by laboratory staff—often, but not always, before a mistaken report was issued. Consequently, forensic scientists sometimes argue that the problems recorded in corrective action files are "not really errors" because they were caught by the lab. They argue, with some justification, that these files are evidence that the laboratory's quality-control system is working to detect and correct errors when they occur. The problem with this analysis is that errors often are caught because of circumstances that are not always present when such errors occur. It will not always be the case, for example, that mistaken cold hits will implicate offenders who were too young to have committed the crime, nor will it always be the case that cross-contamination of DNA samples will produce unexpected results that flag the error. If DNA from a suspect is accidentally transferred into a "blank" control sample, it is obvious that something is wrong; if the suspect's DNA is accidentally transferred into an evidentiary sample, the error will not necessarily be obvious because there is another explanation—that the suspect contributed DNA to the evidentiary sample. The same processes that cause detectable errors in some cases can cause undetectable errors in others. Although laboratories should be encouraged to keep careful records of "unexpected events" and should be commended for doing so, the extensive catalogs of error recorded in existing files can hardly be taken as reassuring evidence that laboratory quality-control systems are working. They are a warning signal that we need to worry about similar errors that labs do not catch, although the frequency of such errors is obviously difficult to estimate.

Moreover, there is great variation among labs in the size and scope of their corrective action files. Some labs (again, probably the better ones) have extensive files documenting numerous problematic incidents and steps taken to deal with them, but other labs either fail to maintain such files or claim that their files are empty because they have never, ever had a problem requiring corrective action. Given the high frequency of incidents warranting corrective action in some very reputable DNA laboratories, it strains credibility to believe that such incidents never occur in other laboratories. A more likely explanation is that these labs choose not to document errors in order to maintain a pretense of infallibility.

An embarrassing episode at the San Francisco police crime laboratory, which came to light in 2010, supports this interpretation. An anonymous person sent letters to the San Francisco Public Defender's Office and to the American Society of Crime Laboratory Directors' Laboratory Accreditation Board (ASCLD-LAB), which had issued a "certificate of accreditation" to the San Francisco laboratory. The letters alleged that laboratory managers had inappropriately covered up a sample-switch error that had occurred when the lab was processing DNA evidence in a homicide case. In response to an inquiry from the ASCLD-LAB, the laboratory managers wrote a letter denying that any such error had occurred. During a subsequent inspection of the laboratory, however, representatives of the ASCLD-LAB found evidence that the error had indeed occurred and that the laboratory staff had falsified laboratory records to cover it up. It is not clear that the error materially affected the test results in the homicide case. Nevertheless, the lab managers seemed intent on preventing defense lawyers from discovering that a problem had occurred. In order to avoid disclosing a seemingly minor problem, the lab managers contravened an important quality-control procedure recommended by the FBI's DNA Advisory Board and lied to their accrediting agency.[35] This is a clear instance of a laboratory putting the appearance of infallibility ahead of good laboratory practice (and basic honesty). Similar incidents in which laboratory managers suppressed evidence of DNA-testing errors have been reported at the Maine State Police Crime laboratory, the U.S. Army Criminal Investigation Laboratory, the North Carolina State Bureau of Investigation, and the Houston Police Department Crime Laboratory.[36]

Gross Negligence, Scientific Misconduct, and Fraud

Since the mid-1990s news reports have offered a continuing stream of stories about gross negligence, scientific misconduct, and fraud in forensic laboratories.[37] A number of these problems have affected DNA testing, including the following:

- The Houston Police Department shut down the DNA and serology section of its crime laboratory in 2003 after a television exposé revealed serious deficiencies in the lab's procedures that were confirmed by an outside audit. Two men who were falsely convicted on the basis of botched lab work were released from prison after subsequent DNA testing proved their innocence. In dozens of cases DNA retests by independent laboratories failed to confirm the conclusions of the Houston lab. An independent investigation found that the laboratory had failed for years to employ proper scientific controls, had routinely misrepresented the statistical significance of DNA matches, and in some cases had suppressed exculpatory test results.[38] The unit reopened under new management in 2006. In 2008, however, the head of the DNA unit was forced to resign in the face of allegations that she had helped DNA analysts in the unit cheat on proficiency tests.[39]
- In Virginia postconviction DNA testing in the high-profile case of Earl Washington Jr. (who was falsely convicted of capital murder and came within hours of execution) contradicted DNA tests on the same samples performed earlier by the State Division of Forensic Sciences. An outside investigation concluded that the state lab had botched the analysis of the case, had failed to follow proper procedures, and had misinterpreted its own test results.[40] In a second

capital case postconviction reviews found that the state lab had overstated the value of the DNA evidence that incriminated the defendant and had improperly dismissed as inconclusive results that were strongly exculpatory.[41] In a third capital case the state analyst grossly overstated the statistical significance of a DNA match.[42]

- In North Carolina the *Winston-Salem Journal* published a series of articles in 2005 documenting numerous DNA-testing errors by the North Carolina State Bureau of Investigation.[43] In 2010 an independent audit of this lab by two FBI laboratory supervisors found that lab analysts had withheld or misrepresented the results of tests for the presence of blood in more than 200 cases.[44]
- A multi-year investigation by the McClatchy news organization, beginning in 2005, revealed that an analyst at the U.S. Army Criminal Investigation Laboratory had a history of cross-contaminating samples, had violated laboratory protocols, and had falsified test results. An independent investigation found significant problems in one-quarter of all the cases this analyst had handled. Laboratory managers failed to disclose these problems to lawyers involved in the relevant cases and took other steps to cover up these problems.[45]
- DNA analysts at a number of other laboratories have been fired for falsification of test results, including labs operated by the FBI, Orchid-Cellmark, the Office of the Chief Medical Examiner in New York City, and Bexar County, Texas. Fraud allegations were also leveled against an analyst at the Chicago Police Department Crime Laboratory.[46]

The most common form of misconduct in DNA testing is shading of scientific findings to make them more coherent or more consistent with what the analyst believes to be true. For example, the analyst may fail to report minor (or seemingly minor) discrepancies between profiles, problems with controls, or other inconsistencies among findings and may justify this as an effort to avoid confusing lawyers and jurors with "irrelevant" information. The problem with this practice is that the analyst's conception of what is true (and therefore what is "relevant" and "irrelevant") is often colored by investigative facts communicated by police officers and prosecutors. When I asked one analyst to explain why she had decided to disregard a discrepancy between two "matching" DNA profiles in a rape/robbery case, she responded: "I know it's a good match—they found the victim's purse in [the defendant's] apartment."

DNA analysts are often well informed about the underlying facts of the cases they process—as those facts are reported by the police. In cases

I have reviewed, there often are comments in the case file such as the following (from a rape case in Virginia): "This [man] is suspected in other rapes but they can't find the [victims]. Need this case to put [him] away." Or this, from an aggravated assault case in California: "Suspect—known crip gang member—keeps 'skating' on charges—never serves time. This robbery he gets hit in head with bar stool—left blood trail. Miller [the deputy district attorney who was prosecuting the case] wants to connect this guy to scene w/DNA."

Information of this type may well influence analysts' interpretations of test results, particularly in cases where the results are somewhat ambiguous or otherwise problematic. Because interpretive bias of this kind can operate unconsciously, I hesitate to label it scientific misconduct, although the failure of forensic scientists to implement rigorous procedures to guard against such bias is surely bad scientific practice. The 2009 NRC report recognized that interpretive bias is a significant problem in forensic science as a whole but did not acknowledge that it is also a problem for forensic DNA testing.

Procedures have been proposed for reducing bias by temporarily "blinding" analysts to unnecessary information when they are analyzing and interpreting DNA tests, but the forensic science community has yet to accept that such procedures are necessary or even desirable. Part of the problem is confusion over the forensic scientist's role in the judicial process. Some believe that it is appropriate to consider a broad range of investigative facts (such as the purse in the apartment) in drawing conclusions about forensic evidence. As one put it, "if this 'bias' leads to the truth, is it really a bias?" Others believe (implausibly) that they can control any bias by act of will.[47]

Bias shades into intentional scientific misconduct when analysts begin suppressing or misrepresenting their findings. An independent investigation of the Houston Police Department Crime Laboratory found many instances of this type.[48] Some of the problems in Virginia and North Carolina and at the U.S. Army DNA laboratory also fall into this category. The guilty analysts appear to have been motivated partly by a desire to help police and prosecutors convict the "right" people and partly by a desire to cover up shortcomings in their own scientific work.

Production pressures are also an important factor. Several of the analysts who were fired for fraud were caught falsifying laboratory records in order to cover up the failure of scientific controls in their assays, and particularly to hide the presence of positive results in blank samples that are included in the assays to detect contamination. As discussed earlier, cross-contamination of samples is a common event in forensic laboratories, but it can be embarrassing for analysts, particularly if it happens too often, because it raises questions about their technical competence and care in handling samples. Furthermore, because the entire assay must be redone if a control sample signals the presence of contaminating DNA (even if the contamination appears to have affected only the control), these incidents are important setbacks for an analyst trying to keep up with a demanding workload.[49] There is evidence that the army analyst who falsified results was striving to maintain his reputation as the most productive analyst in the laboratory.[50] A colleague of the Orchid-Cellmark analyst who was fired for falsifying results told me that the analyst in question strove always to be at or near the top of a chart posted on the laboratory wall that tracked analysts' productivity (measured by samples and cases successfully completed).

Analysts working in a pressured environment may be tempted to cut corners in order to keep on schedule and thereby make themselves look good, particularly if they know (or think they know) on the basis of other investigative facts that they have reached the correct conclusion about which samples match. Analysts in the Houston Police Department Crime Laboratory simply dispensed with running blank control samples in their assays, which is clearly a dangerous and unacceptable scientific practice but undoubtedly sped up their work: there is no need to redo assays when controls fail if one has no controls. From misconduct of this sort it is perhaps not a very big step to the more blatant falsifications of analysts like Fred Zain, Joyce Gilchrist, and Pamela Fish, who are alleged to have faked or misrepresented results of entire tests in order to incriminate people they thought were guilty.[51]

For a number of years I have urged forensic scientists to adopt blind procedures for interpreting DNA test results. My main concern is with unconscious bias in interpretation, which I believe is a widespread problem. But I believe that blind procedures would also reduce the temptation to falsify DNA data, misrepresent findings in laboratory reports, and ignore evidence of problems in assays. Analysts who do not know whether their tests point to the "right" person will (I believe) be more cautious and rigorous in their interpretations and more honest in acknowledging uncertainty and limitations in their findings.

Coincidental Matches

The impressive numbers that accompany DNA evidence contribute greatly to its persuasive power. Often called random-match probabilities (RMPs), the numbers represent the frequency of a particular DNA profile in a reference population. Statistician Bruce Weir has estimated that the average probability that two unrelated people will have the same thirteen-locus DNA profile is between 1 in 200 trillion and 1 in 2 quadrillion, depending on the degree of genetic structure in the human population.[52] Numbers this small make it seem that the chances the wrong person will "match" are unworthy of consideration. But this impression is incorrect for several reasons.

First, RMPs describe only the chances of a random unrelated person having a particular DNA profile; they have nothing to do with the likelihood of the wrong person being reported to match for other reasons, such as cross-contamination of samples, mislabeling of samples, or error in interpreting or recording test results. RMPs quantify the likelihood of one possible source of error (coincidental matches) while ignoring other events that can also cause false incriminations and often are more likely to do so.

Second, extremely low RMPs, like those computed by Weir, apply only in the ideal case in which the lab finds a match between two complete single-source DNA profiles. The evidence in actual cases is often less than ideal. Evidentiary samples from crime scenes frequently produce incomplete or partial DNA profiles that contain fewer genetic alleles (characteristics) than complete profiles and are therefore more likely to match someone by chance. A further complication is that evidentiary samples are often mixtures. Because it can be difficult to tell which alleles are associated with which contributor in a mixed sample, there often are many different profiles (not just one) that could be consistent with a mixed sample, and hence the chances of a coincidental match can be much higher.

To illustrate these points, consider the DNA profiles shown in Table 15.1. Forensic laboratories typically "type" samples using commercial test kits that can detect genetic characteristics (called alleles) at various loci (locations) on the human genome. The most commonly used forensic DNA tests examine loci that contain short tandem repeats (STRs), which are sections of the human genome where a short sequence of genetic code is repeated a number of times. (They are called short *tandem* repeats because these short repeating units occur on both sides of the DNA double helix). Although everyone has STRs, people tend to vary in the number of times the genetic code at each STR repeats itself, and each possible variant is called an allele. Generally there are between six and eighteen possible alleles at each locus. Each person inherits two of these alleles, one from each parent, and the pair of alleles at a particular locus constitutes a genotype. The complete set of alleles detected at all loci for a given sample is called a DNA profile.[53]

Profile A in Table 15.1 is a complete thirteen-locus DNA profile, while profiles B and C are partial profiles of the type often found when a limited quantity of DNA, degradation of the sample, or the presence of inhibitors (contaminants) makes it impossible to determine the genotype at every locus. Because partial profiles contain fewer genetic markers (alleles) than complete profiles, they are more likely to match someone by chance.[54] The chance that a randomly chosen U.S. Caucasian would match the profiles shown in Table 15.1 is 1 in 250 billion for profile A, 1 in 2.2 million for profile B, and 1 in 16,000 for profile C.[55]

Because profiles D and E contain more than two alleles at some loci, they are obviously mixtures of DNA from at least two people. Profile A is consistent with profile D (i.e., every allele in profile A is included in profile D), which means that the donor of profile A could be a contributor to the mixture. But many other profiles would also be consistent. At locus D3S1358, for example, a contributor to the mixture might have any of the following genotypes: 15,16; 15,17; 16,17; 15,15; 16,16; 17,17. Because so many different profiles may be consistent with a mixture, the probability that a noncontributor might by coincidence be included as a possible contributor to the mixture is far higher in a mixture case than in a case with a single-source evidentiary sample. Among U.S. Caucasians approximately 1 person in 790,000 has a DNA profile consistent with the mixture shown in profile D. Thus the RMP for mixed profile D is higher than the RMP for single-source profile A by five to six orders of magnitude. When partial profiles like profiles B and C are also mixtures, the RMPs can be high enough to include thousands, if not millions, of people as possible donors. RMPs greater than 1 in 100 are sometimes reported in such cases.

A third important caveat about extremely low RMPs like those reported by Weir is that they are estimates of the probability of a coincidental match among random individuals who are unrelated to the donor of the sample in question. In actual cases the pool of possible suspects is likely to contain individuals who are related to one another. For example, a man might falsely be accused of a crime that was actually committed by a brother, uncle, or cousin. In such cases the probability of a false incrimination due to a coincidental match is much higher than the RMP might suggest. Consider again profile A in Table 15.1. Although this profile would be found in only 1 in 250 billion unrelated individuals, the probability of finding this profile in a relative of the donor is far higher: 1 in 14 billion for a first cousin; 1 in 1.4 billion for a nephew, niece, aunt, or uncle; 1 in 38 million for a parent or child; and 1 in 81,000 for a sibling. In cases involving partial and mixed profiles, the chances of a coincidental

Table 15.1 Matching DNA profiles

	STR locus												
Profile	D3S1358	vWA	FGA	D8S1179	D21S11	D18S51	D5S818	D13S317	D7S820	CSF1PO	TPOX	THO1	D16S539
A	15,16	17,18	21,22	13,14	29,30	14,17	11,12	11,12	8,10	11,12	8,11	6,9	11,12
B	15,16	17,18		13,14	29,30		11,12	11,12		11,12	8,11		
C	15,16	17		13,14	30		11,12	11	8,10				
D	15,16,17	17,18	21,22	13,14,15	29,30	12,13 14,17	11,12	11,12	8,9, 10,12	11,12	8,9,11	6,7,9	11,12,13
E	15,16,17	17,18	21,23	13,14,15	29,30	12,17	11,12	11,12	8,9,10	11,12	8,9		

match to a relative of the donor can, commensurately, be higher by orders of magnitude than for a complete single-source profile like profile A.

A fourth important caveat about the impressive RMPs that often accompany forensic DNA evidence is that the risk of obtaining a match by coincidence is far higher when authorities search through millions of profiles in a DNA database looking for a match than when they compare the evidentiary profile to the profile of a single individual who has been identified as a suspect for other reasons. As an illustration, suppose that a partial DNA profile from a crime scene occurs with a frequency of 1 in 10 million in the general population. If this profile is compared with that of a single innocent suspect who is unrelated to the true donor, the probability that it will match is only 1 in 10 million. Consequently, if one finds such a match when one tests an individual who is already suspected for other reasons, it seems safe to assume that the match was no coincidence. By contrast, in searches through a database as large as the FBI's National DNA Index System (NDIS), which reportedly contains over 8 million profiles, there are literally millions of opportunities to find a match by coincidence. Even if everyone in the database is innocent, there is a substantial probability that one (or more) will match the profile with a general-population frequency of 1 in 10 million. Hence a match obtained in such a database search may well be coincidental, particularly if there is little or no other evidence against a matching individual.[56]

When the estimated frequency of the DNA profile is 1 in *n,* where *n* is a number larger than the earth's population, some people assume that the profile must be unique, an error that statistician David Balding has called the "uniqueness fallacy."[57] In such cases the expected frequency of duplicate profiles is less than one, but it never falls to zero no matter how rare the profile is. If the frequency of a profile is 1 in 10 billion, for example, then the expected likelihood of finding a duplication in a population of 250 million unrelated individuals is about 1 in 40. This may sound like a low risk, but in a system in which thousands of evidentiary profiles with frequencies on the order of 1 in 10 billion are searched each year against millions of database profiles, coincidental matches will inevitably be found.[58]

Indeed, a large number of coincidental DNA matches have already been found in database searches. The British Home Office has reported that between 2001 and 2006, 27.6 percent of the matches reported from searches of the United Kingdom's National DNA Database were to more than one person in the database. According to the report, the multiple-match cases arose "largely due to the significant proportion of crime scene sample profiles that are partial."[59] In other words, officials were frequently searching for profiles like profiles B and C in Table 15.1 that would be expected to match more than one person in a database of millions. But the frequent occurrence of DNA matches to multiple people surely makes the point that a DNA match by itself is not always definitive proof of identity.

False incriminations arising from such coincidental matches have occurred in both the United Kingdom and the United States. In 1999 the DNA profile of a sample from a burglary in Bolton, England, was matched in a database search to the profile of a man from Swindon, England. The frequency of the six-locus profile was reported to be 1 in 37 million. Although the Swindon man was arrested, doubts arose about the identification because he was disabled and apparently lacked the physical ability to have committed the Bolton crime. Testing of additional genetic loci excluded him as the source of the sample and proved that the initial 1-in-37-million match was simply a coincidence. As David Balding points out, this kind of coincidence is not particularly surprising because "the match probability implies that we expect about two matches in the United Kingdom (population ≈ 60 million), and there could easily be three or four."[60]

In 2004 a Chicago woman was incriminated in a burglary by what turned out to be a coincidental cold hit. The woman's lawyer told the *Chicago Sun-Times* that it was only her strong alibi that saved the woman from prosecution: "But for the fact that this woman was in prison [for another offense at the time the crime occurred] . . . I absolutely believe she'd still be in custody."[61]

A similar error came to light in 2010 in an Ohio burglary prosecution. The homeowner had confronted the burglar, whom he described as short, stout, and balding, and had yanked some hair from his scalp. DNA typing of tissues attached to the hair produced a six-locus partial DNA profile with an RMP of 1 in 1.6 million. Ten years later a database search matched this profile to one Steven Myers, who was described as a tall, skinny 25-year-old and who had no known connection to the town where the burglary had occurred. Despite the mismatch between the homeowner's description of the burglar and Myers, who would have been only 15 at the time of the crime, Myers was indicted and spent seven months in jail awaiting trial. Luckily for him, the hair samples were still available. Retesting produced results at additional loci that excluded him as the donor, and he was released.[62]

Misleading Statistics

DNA analysts sometimes present misleading statistics that overstate the value of the DNA evidence. For example, in cases where a suspect's profile is being compared with a mixture, analysts sometimes present the frequency of the suspect's profile rather than the frequency of profiles that would be included as possible contributors to the mixture. This practice is misleading because the relevant issue in such a case is the probability of a random match to the mixture, not the probability of a random match to the suspect. In a case where a suspect with profile A was matched to a mixture like profile D, the relevant statistic is 1 in 790,000, not 1 in 250 billion.

Before the scandal broke in 2003, the Houston Police Department Crime Laboratory routinely presented the wrong statistic in mixture cases. In the case of Josiah Sutton, for example, the laboratory reported an RMP of 1 in 690,000 (the frequency of Sutton's profile) when the probability of a random match to the mixed evidentiary sample was approximately 1 in 15. (Also, because Sutton was one of two men who were falsely accused of the crime, the chance the lab would find a coincidental match to at least one of them was approximately 1 in 8.)[63]

Although the proper way to compute statistics in mixture cases has been widely known since at least 1992, when it was discussed in a report by the National Research Council, the practice of presenting the suspect's profile frequency in mixture cases has been surprisingly persistent. I have seen instances of it in many cases, including a capital case in South Carolina that I reviewed in 2010.

A more subtle problem arises when a suspect's profile (such as profile A) is compared with a partial profile in which some of the suspect's alleles are missing (such as profile E). Any true discrepancy between profiles means that they could not have come from the same person, but an analyst may well attribute discrepancies like those between profiles A and E to technical problems in the assay or to degradation of sample E and therefore declare A to be a possible contributor to mixture E despite the discrepancies. The problem then becomes how to assign statistical meaning to such a partial match.

At present there is no generally accepted method. The approach laboratories typically use is to compute the frequency of genotypes at loci where the two profiles match and simply ignore loci where they do not. This approach has been strongly criticized for understating the likelihood of a coincidental match (and thereby overstating the value of the DNA evidence), but it remains the most common approach in cases of this type and is currently used throughout the United States.[64]

Fallacious Statistical Conclusions

Another persistent problem has been fallacious testimony about the meaning of a DNA match. Analysts sometimes give testimony consistent with a logical error called the "prosecutor's fallacy" (or, alternatively, the "fallacy of the transposed conditional") that confuses the RMP with a different statistic known as the source probability. The RMP is the probability that a random unrelated person would match an evidentiary sample. The source probability is the probability that a person with a matching DNA profile is the source of the evidentiary sample. The RMP can be estimated by the DNA analyst using purely scientific criteria; the source probability can be assessed only on the basis of all the evidence in the case, including nonscientific evidence. Hence, although forensic scientists can properly present RMPs (if they compute them correctly), it is improper for them to testify about source probabilities. But sometimes they do so anyway.[65]

For example, when a defendant named Troy Brown was prosecuted for rape in Nevada, the analyst testified that his DNA profile matched the DNA profile of semen found on the victim, and that the RMP was 1 in 3 million. Prompted by the prosecutor, she went on to testify that this meant that there was a 99.999967 percent chance that Brown was the source of the semen, and only a .000033 percent chance that he was not. On the basis of this testimony, the prosecutor argued that the DNA evidence by itself proved Brown's guilt beyond a reasonable doubt. When Brown's case was accepted for review by the U.S. Supreme Court in 2009, a group of twenty "forensic evidence scholars" filed an amici curiae brief discussing problems with the DNA analyst's testimony. The Supreme Court described those problems correctly in its resulting opinion, although it dispensed with the case on procedural grounds without considering whether fallacious testimony of this type violates a defendant's constitutional rights.[66]

Statistical Accuracy: Independence Assumptions

Thus far I have been assuming that the statistical estimates computed by forensic laboratories are accurate, but there is still some uncertainty about that due largely to the refusal of the FBI to allow independent scientists to perform statistical analyses of the DNA profiles in the National DNA Index System (NDIS). Forensic laboratories typically base their frequency estimates not on NDIS or any other large database containing millions of profiles but on published statistical databases that contain a few hundred profiles from "convenience samples" of members of each major racial or ethnic group. To generate a number like 1 in 2 quadrillion from a statistical database that consists of a few hundred profiles requires an extrapolation based on strong assumptions about the statistical independence of various markers.[67]

When DNA evidence was first introduced in the late 1980s and early 1990s, a heated debate arose about the independence assumptions. Although many forensic and academic scientists were comfortable with these assumptions, some prominent critics expressed concern that the independence of the markers might be undermined by population structure—the tendency of people to mate with those who are genetically similar to themselves within population subgroups. By 1992 the dispute about statistical independence had led several appellate courts to rule DNA evidence inadmissible under the *Frye* standard, which requires that scientific evidence be generally accepted in the scientific community as a condition for its admissibility in jury trials.[68]

Although the exclusion of DNA evidence affected relatively few cases, it created a sense of crisis in the forensic science community and led to a flurry of research designed to test the extent of population structure and, by extension, the independence of the markers. By the mid-1990s new data had assuaged the worst fears about the extent of population structure, and criticism began to fade. The 1996 NRC report on DNA evidence recognized the potential importance of population structure, but it concluded on the basis of the data available at the time that the effect was likely to be modest and could be addressed by using a small correction factor, called theta, in computing match probabilities. Since that time statistical estimates based on assumptions of independence have routinely been admissible (with or without the theta correction).[69]

But troubling questions about statistical independence linger for several reasons. First, the growing use of large government databases for identification of unknown profiles has made it more important than it was in the past to know precisely how rare matching profiles are. When the scientific community reached closure on the issue in the 1990s, DNA testing was used primarily for confirming or disconfirming the guilt of individuals who were already suspects. In cases where DNA of a person who is already a suspect is found to match the DNA of the perpetrator, it probably does not matter very much whether the frequency of the matching profile is really 1 in 10 trillion, say, rather than 1 in 10 billion or 1 in 10 million. Any of these probabilities is low enough to effectively rule out the theory of a coincidental match and therefore justify a conviction. When a suspect is identified in a search of a large database, however, the precise rarity of the matching profile is much more important. In such cases the DNA evidence that identifies the suspect may constitute the only evidence against that person. Hence it is crucial to know whether the suspect is the only person with the matching profile. If the frequency is really 1 in 10 trillion, then the likelihood that any other human will have the profile is extremely low, but the likelihood is not nearly as low if the frequency is 1 in 10 billion; and if the frequency is 1 in 10 million, then the suspect is certainly not the only person with the matching profile. Hence whether a conviction is justified may well depend on the precise rarity of the profile.

The relatively small size of available statistical databases makes it impossible to perform sensitive tests of the statistical independence of markers across multiple loci. Such tests could be conducted if population geneticists were given access to the DNA profiles (with identifying information removed) in the large offender databases used for criminal identification. For example, Bruce Weir published an analysis of a government database from the state of Victoria, Australia, that contained 15,000 profiles.[70] He found no evidence inconsistent with the standard assumptions on which statistical calculations are based, but according to one critic, even that database was too small to do "rigorous statistical analysis" of independence across six or more loci. Weir and other experts have suggested that the DNA profiles in FBI's CODIS system be made available (in anonymized form) for scientific study. Weir told the *Los Angeles Times* that the independence assumptions relied on for computing profile frequencies should be tested empirically using the national database system: "Instead of saying we predict there will be a match, let's open it up and look."[71]

The 1994 DNA Identification Act, which gave the FBI authority to establish a national DNA index, specifies that the profiles in the databases may be disclosed "if personally identifiable information is removed, for population statistics databases, for identification research, or for quality control purposes."[72] Requests for access to anonymized (deidentified) profiles in state databases for purposes of statistical study by independent experts have been made by defense lawyers in a number of criminal cases but so far have been vigorously and successfully resisted. According to the *Los Angeles Times*, the FBI has engaged in "an aggressive behind-the-scenes campaign" to block efforts to obtain access to database profiles or information about the number of matching profiles in databases.[73]

In December 2009 a group of thirty-nine academics (including the author of this chapter and one of the editors of this volume) signed an open letter published in *Science* calling for the FBI to "release anonymized NDIS profiles to academic scientists for research that will benefit criminal justice." The letter argued that disclosure of the profiles would "allow independent scientists to evaluate some of the population genetic assumptions underlying DNA testing using a database large enough to allow . . . powerful tests of independence within and between loci, as well as assessment of the efficacy of the theta factor used to compensate for population structure." The letter also pointed to a number of other scientific questions that could be answered through analysis of the NDIS data, including questions about how match probabilities are affected by the number of relatives in the database and questions about the degree to which DNA profiles cluster because of identity by descent. Furthermore, analysis could provide insight into the frequency and circumstances in which certain kinds of typing errors occur. To date the FBI has published no scientific findings derived from the NDIS data and has yet to release the data to any independent scientists for review.[74]

The continuing uncertainty about the accuracy of statistical estimates is not a neutral factor in weighing the chances of a false incrimination due to coincidence. Some people mistakenly assume that statistical uncertainty "cancels out"—that is, that the estimates may be too low but also may be too high, so our ignorance of the truth is unlikely to harm criminal defendants. Statistician David Balding has demonstrated mathematically that this position is fallacious. The extreme estimates produced by forensic laboratories depend on the assumption of perfect knowledge about the frequency of DNA profiles, and to the extent that our knowledge is uncertain, the estimates should be considerably less extreme. Hence Balding declares that "ignoring this uncertainty is always unfavourable to defendants."[75]

Intentional Planting of DNA

The ability of criminals to neutralize or evade crime-control technologies has been a persistent theme in the history of crime. Each new method for stopping crime or catching criminals is followed by the development of countermeasures designed to thwart it. For example, the development of ignition locks did not solve the problem of car theft because criminals quickly learned to defeat the locks by hot-wiring cars, stealing keys, and other tactics that led to the development of additional protective devices (steering-wheel bars, locator beacons), which eventually proved vulnerable to further criminal countermeasures. The history of safecracking has been a virtual arms race between safe manufacturers looking to build ever-safer boxes and criminals finding more advanced ways to break in. It would hardly be surprising, therefore, if criminals sought ways to avoid being identified by DNA tests.[76]

Police officials have expressed concern about that very issue. Between 1995 and 2006, a period when DNA testing was becoming more common, the clearance rate for rape cases reportedly declined by 10 percent. Asked to explain this trend, a number of police officials suggested that criminals have become more sophisticated about evading detection. Police officials have also suggested that television shows like *CSI* can serve as tutorials on getting away with crime, although there is no good empirical evidence to prove this claim.[77]

There are anecdotal reports of criminals trying to throw investigators off the track by planting biological evidence. An accused serial rapist in Milwaukee reportedly attempted to convince authorities that another man with the same DNA profile was responsible for his crimes by smuggling his semen out of the jail and having accomplices plant it on a woman who then falsely claimed to have been raped. It occurred to me, and must have occurred to some criminals, that the rapist would have been more successful had he planted another man's semen on his actual victims. Semen samples are not difficult to obtain. In a park on the campus where I teach, semen samples in discarded condoms can be found regularly (particularly in springtime). Perhaps I have been studying DNA testing too long, but I cannot pass that area without wondering whether the young men who leave those biological specimens could be putting their futures at risk. And there are other items besides semen that might be used to plant an innocent person's DNA at a crime scene. Clothing the person wore, a cigarette the person smoked, or a glass from which the person drank could all, if placed at a crime scene, create a false DNA link between an innocent person and a crime. When such planting occurs, will the police be able to figure it out? Will a jury believe that the defendant could be innocent once a damning DNA match is found? I have strong doubts on both counts and, consequently, believe that intentional planting of DNA evidence may create a significant risk of false incriminations.

As with the other risks, this one is magnified by the growing use of DNA databases. If someone plants your DNA at a crime scene, it might throw police off the trail of the true perpetrator, but it is unlikely to incriminate you unless your profile is in the database. The authorities are likely to search the profile of the crime-scene sample against a database, but if your profile is not in the database, they will find no match and will be left with just another unknown sample. Suppose, however, that you are unlucky enough to have your profile in the database. In that case the police will likely find it, at which point they will have something far better than an unknown sample—they will have a suspect. Given the racial and ethnic disparities that exist in databases, that suspect is disproportionately likely to be a minority-group member.[78]

The expansion of databases increases the number of people who risk being falsely incriminated in this manner. The seriousness of this risk is obviously difficult to assess. It depends on how frequently criminals engage in evidence planting, whose DNA they plant, how often the planted DNA is detected, and how often its detection leads to criminal charges and conviction, among other factors. One can only guess how often these events occur, but it would be foolish to assume that these events will not occur or have not occurred already. Consequently, this risk is one that must be weighed against the benefits of database expansion.

In the future, more sophisticated criminal countermeasures could compromise the effectiveness of DNA testing as a crime-fighting tool. A researcher at the University of Western Australia has studied the effects of contaminating simulated crime scenes with a concentrated solution of amplicons (short fragments of DNA copied from the DNA in a biological sample). She used a standard test kit of the type employed by forensic DNA laboratories and a procedure known as the polymerase chain reaction (PCR) to create highly concentrated solutions of DNA fragments from the core CODIS loci. She then tested the effects of spraying this solution about a room using a small atomizer. She found, not surprisingly, that the concentrated solution of amplicons was detected by standard STR tests and produced profiles that could easily be mistaken for the profiles of typical forensic samples. What is more interesting (and disturbing) is that the DNA profile of the amplicons was, under some conditions, detected preferentially over the DNA profile of actual biological samples in the room. For example, when amplicons from person A were spritzed with the atomizer over a bloodstain from person B, and a sample from the bloodstain was typed using standard STR procedures, the result sometimes appeared to be a mixture of DNA from person A and person B, but sometimes it appeared to consist entirely of DNA from person A—in other words, the contaminating DNA from the atomizer was the only profile that was detected. This prompted a warning that criminals could use this technique to commit "DNA forgery" and to fraudulently plant DNA with the intention of implicating an innocent person.[79]

Kary Mullis, who invented the PCR, anticipated this potential misuse of the technique. In a conversation I had with him in 1995, Mullis jokingly discussed creating a company called "DN-Anonymous" that would sell highly amplified solutions of DNA from celebrities, or from large groups of people, that criminals could use to cover their tracks. Although

Mullis was not serious about doing this himself, he predicted that someone would do so within the next ten years. As far as I know, Mullis's prediction has yet to come true, but it may be only a matter of time before materials designed to stymie DNA tests (by planting other people's DNA at crime scenes) become available for sale on the Internet along with kits designed to thwart drug tests.

Improving DNA Evidence

Do innocent people really have nothing to fear from DNA evidence? It should now be clear to readers that this claim is overstated. Cross-contamination of samples, mislabeling, and misinterpretation of test results have caused (and will continue to cause) false DNA matches. Coincidental matches and intentional planting of evidence create added risks of false incrimination. These risks are magnified for people whose profiles are included in government DNA databases. We know less than we should about the nature and scope of these risks, and we have done far less than we should to minimize and control these risks.

The 2009 NRC report identified significant problems with the "culture" of forensic science. It found that the field is too strongly influenced by law enforcement and insufficiently connected to academic science. It recommended that crime laboratories be separated from law-enforcement control and that a new federal agency called the National Institute of Forensic Science (NIFS) be established. The NIFS would oversee the field, fund research designed to improve the validity and reliability of forensic methods, establish best-practice standards, and investigate problems. Although the NRC report pointedly excluded DNA testing from its criticism of other forensic science techniques, I believe that this chapter makes it clear that many of the "culture" problems in other domains of forensic science are also problems for forensic DNA testing. An agency like the NIFS is needed as much to improve DNA testing as it is to address deficiencies in other forensic science disciplines.[80]

The great advantage that DNA testing has over other disciplines is the ability to estimate RMPs. Forensic scientists cannot at present estimate the chances of a coincidental match in latent print analysis, tool-mark analysis, or trace-evidence comparison (or any other forensic discipline) the way they can with DNA evidence. As the NRC report recognized, however, RMPs are only one factor affecting the value of DNA evidence. Even that factor is shadowed by lingering uncertainty, although the uncertainty could be resolved if the FBI were willing to give independent scientists access to NDIS profiles.

For DNA evidence to achieve the gold-standard status it purports to have, several steps are necessary. Forensic laboratories need to be more open and transparent about their operations. Independent scientists should be given access to all databases for purposes of scientific study. Laboratories should be required to keep careful records of errors, problems, and other unexpected events, and those events should be investigated carefully. Just as crashes and near misses in aviation are examined carefully (by a government agency) to determine what can be learned from them and how such episodes can be avoided, false incriminations and near false incriminations like the many discussed here should be examined and evaluated.

Greater efforts to assess the frequency and source of errors are also needed. There is no good reason (other than lack of resources) that laboratories are not subjected to realistic external, blind proficiency tests in which analysts must type samples that appear to be part of routine casework without knowing that they are being tested. There should also be a public program of research that monitors the operation of government databases in order to assess the frequency and causes of false cold hits. There is no good reason not to record and disclose information about how many searches are conducted, how discriminating the searches are, and how many produce cold hits, as well as the number of cold hits that are confirmed or disconfirmed by subsequent evidence.

More rigorous standards for interpretation and reporting of test results are also needed, along with a mechanism to enforce them. The failure of forensic scientists to adopt blind procedures for interpretation is a particularly important problem. We also need better mechanisms for monitoring and evaluating expert testimony.

Finally, we need better institutional mechanisms for investigating allegations of serious negligence and misconduct. Inadequate investigative efforts are part of the reason that the scandalous scientific misconduct in the Houston Police Department Crime Laboratory went on for more than a decade without correction. At present, investigations of alleged misconduct are typically conducted by entities that not only lack scientific expertise but also have serious conflicts of interest.[81] The district attorney's office that relied on the evidence to convict defendants is often called on to investigate allegations that the evidence was fraudulent or mistaken or overstated. It would be far better if the investigation could be conducted by an independent state or federal agency with appropriate scientific expertise.

Whether there is political support for the creation of the NIFS or something like it remains to be seen. An agency of this type would clearly be helpful in achieving the goals just outlined. In the meantime, it is important for the academic community to adopt a more realistic view of DNA evidence. Those who continue to promote the myth of its infallibility may well be undermining efforts to make it better.

CHAPTER SIXTEEN

Nurturing Nature

How Parental Care Changes Genes

MAE-WAN HO

Ruth Hubbard against Genetic Determinism

It has been thirty years since I first met Ruth Hubbard and her husband George Wald at the conference "Towards a Liberatory Biology" in Bressanone in the Italian Alps.[1] From a broad sociopolitical perspective, Hubbard was already a leading light in the radical critique of genetic determinism—the idea that organisms are hardwired in their genetic makeup. As a research scientist who had worked on visual pigments for many years, she was by no means unaware of the hormones and enzymes encoded by genes that enable an organism to transform energy, grow, and develop in a certain way, but she insisted that there are social determinants for what people are, or are perceived to be, that are much more powerful than biology and genes.

I suspect that she was getting rather impatient with the anodyne and frequently opaque rhetoric of sociologists who fail to come to grips with the real issues, not to mention the obfuscation by some "bioethicists" who were a contradiction in terms. The unsuspecting public was left at the mercy of slick propaganda from vested interests intent on profiting by blaming people's ills on their genes and selling them both the diagnosis and appropriate remedies: abortion for the unborn, gene drugs and gene therapies for adults scared witless after having tested positive for genes that would allegedly give them incurable diseases. Ruth Hubbard's book, coauthored with Elijah Wald, *Exploding the Gene Myth: How Genetic Information Is Produced and Manipulated by Scientists, Physicians, Employers, Insurance Companies, Educators and Law Enforcers,* is admirable for delivering its important message clearly, succinctly, and with punch and panache, true to how she is in real life.[2]

How Scientific and Social Critiques Converge

My critique of genetic determinism is based more on science than on politics, which I take broadly to be reliable knowledge of nature that enables us to live sustainably with it.[3] That is certainly not to understate the large influences that society and politics have on science and, more to the point, what passes as science, which can be very much mistaken and unreliable, as is the case of genetic determinism. Science is what we live by and hence has large implications for how we live and choose to live.

My critique converges with Hubbard's because social and environmental influences are indeed powerful determinants on how we grow and develop, precisely as Hubbard has been saying, so much so that they can mark and change our genes for life. That is what the Human Genome Project to sequence the entire human and other genomes has ended up telling us, in spite of the fact that it was inspired and promoted by genetic determinism.

The new genetics of the "fluid genome" had already emerged by the early 1980s, long before the human genome project was conceived.[4] It belongs in the organic paradigm of spontaneity and freedom ("quantum jazz biology")"[5] that defies any kind of determinism, whether biological or environmental.

A Decade of the Human Genome Yields Next to Nothing

More than ten years ago, President Bill Clinton announced the first draft of the human genome sequence and said that it would "revolutionize the diagnosis, prevention and treatment of most, if not all human diseases." Francis Collins, then director of the genome agency at the U.S. National Institutes of Health, said that genetic diagnosis of diseases would be accomplished in ten years and that treatments would enter the market perhaps five years after that.[6]

The anticlimax came just eight months later when the complete map was announced.[7] Chief gene sequencer Craig Venter admitted, "We simply do not have enough genes for this idea of biological determinism to be right." The environment, he said, is critical.

More than ten years later, genomics research has yielded no cures, and the hope of identifying genes for common diseases is fast receding. Nina Paynter and her research team at Brigham and Women's Hospital in Boston looked at 101 genetic variants (single-nucleotide polymorphisms) from whole genome scans that had been linked to heart disease. Together, these turned out to be of no value in predicting the disease among 19,000 women in a study that tracked their health for twelve years.[8] In contrast, family history was the most significant predictor, as it had been before genomics. As Harold Varmus, now director of the National Cancer Institute, said: "Genomics is a way to do science, not medicine."[9]

Demise of Genetic/Biological Determinism

The assumption that genes are stable and insulated from environmental influences is pivotal in neo-Darwinian theory, the root and stem of genetic/biological determinism.[10] It was inspired by Weismann's theory of the germplasm, which was flawed from the start. Plants do not have separate germ cells at all; every somatic cell is potentially capable of becoming a germ cell, which is why plants can be propagated from cuttings. Most animals also do not have germ cells that separate from the rest of the body early in development. Furthermore, there is no evidence that genes in germ cells are stable or immune from environmental influences. We now know that the environment can directly affect the germ cells in the developing fetus, giving rise to the grandparent effects. Toxic environmental substances such as bisphenol A and other endocrine disrupters specifically affect germ cells in the developing fetus.[11] Even more surprisingly, sperm cells are efficient vehicles for carrying foreign (altered) genes into egg cells at fertilization,[12] and these foreign genes can be expressed in embryos developed from the fertilized eggs.[13]

Evidence that genes are neither stable nor immune from direct environmental influence has been accumulating almost since the inception of genetic engineering—applied to unraveling the detailed molecular machinery of genetics—began in the mid-1970s. To their astonishment, molecular geneticists soon witnessed classical genetics being turned upside down on their lab benches. They found exceptions and violations to every tenet of classical genetics. In direct contradiction to the concept of a relatively static genome with linear causal chains emanating from genes to the organism and the environment, they discovered constant crosstalk between genome and environment. Feedback from the environment not only determines which genes are turned on where, when, by how much, and for how long, but also marks, moves, and changes the genes themselves. By the early 1980s molecular geneticists had already coined the term "the fluid genome"[14] to capture what I later described as a molecular "dance of life" necessary for survival.[15]

Just when we finally got used to thinking that a gene in molecular genetics was a coding sequence (eventually "read out" as an amino-acid sequence forming proteins) equipped with various control regions for start and stop that would determine how actively the gene is expressed, when, where, and for how long, we had to think again. New research is revealing how such "genes" are in bits dispersed throughout the genome, interweaving with bits of other genes.[16] As genes are intertwined, so are their functions. Multiple DNA sequences may serve the same function, and conversely, the same DNA sequence can have different functions. It is futile to try to define a gene or a separable function for any piece of DNA. This is ultimately why genes for common diseases can never be found. Incidentally, this is also why genetic modification is both dangerous and futile: human genetic engineers do not know the steps of this incredibly complex molecular dance of life.[17] All they can do, even now, is to follow and marvel at some of the footprints of this dance, marks left on the DNA and the histone proteins bound to the DNA, a script that an individual will pass on to the next generation.

Perpetuation of the Myth of Genetic Determinism in Academia

Mainstream genetics research during the decades since the discovery of DNA's double helix in 1953 has focused on identifying "genes" or a "genetic predisposition" for every "trait," real or imaginary.[18] Imaginary traits are rife in the hybrid discipline of evolutionary psychology, long dedicated to inventing stories on "selective advantage" for each of the "traits" so that the corresponding gene could become "fixed" in the population by neo-Darwinian natural selection.

Another hybrid discipline, behavioral genetics, formerly dedicated to studies based on identical twins, began identifying DNA (gene) markers for behavior and indeed claimed to have found one for an increased tendency toward violent behavior in boys who experienced maltreatment in childhood.[19] The gene encoding the enzyme monoamine oxidase A (MAOA)—involved in the metabolism of neurotransmitters—exists in two variants: one expressing high activity, the other, low activity. Although all boys in the study showed increased "disposition towards violence" if they were maltreated as children, those with low enzyme activity were purported to show an increase in violence. The researchers claimed a weak residual effect due to the low-activity MAOA while conceding the large effect of the environment. But even this alleged genetic predisposition soon faded away as more data became available.[20]

Behavioral geneticists are not the only scientists wasting time and resources chasing will-o'-the-wisp gene markers. The project to map genetic predisposition to diseases was the main rationale for the $3 billion Human Genome Project that, decades later, has delivered next to nothing, basically because it is not genomic DNA but epigenetic environmental influences that overwhelmingly affect our health and well-being.[21]

Epigenetic Inheritance

The term "epigenetic" came from epigenesis, the process whereby an organism with differentiated organs, tissues, and cells develops from a relatively featureless egg. Developmental geneticist and evolutionist Conrad Waddington invented the concept of the "epigenetic landscape" to represent the dynamic structure of the developmental system that defines the range of nonrandom changes for evolution.[22] This was the sense in which we Ho and Saunders used "epigenetic" in 1979.[23] Nowadays "epigenetic" usually refers to a heritable change that does not involve DNA sequence alteration,[24] but that use of the term is rapidly becoming obsolete because of epigenetic mechanisms that actually change DNA sequences directly or via an RNA intermediate that undergoes editing, alternative splicing, and other processes coupled with reverse transcription.[25]

Epigenetic inheritance is effectively the inheritance of acquired traits, usually attributed to Jean-Baptiste de Lamarck (1744–1829),[26] and it has come into its own in maternal effects.

New research has abundantly confirmed the overriding importance of environmental influences across disciplines from nutrition to toxicology, and most dramatically in brain development.[27]

Neither Genetic nor Environmental Determinism Rules

For as long as anyone can remember, people have been debating whether it is our genetic makeup or the environment that determines who we are. New research findings on how maternal care has a lasting influence on her offspring's behavior that persists for generations are telling us that this is definitely not the right question to ask. The epigenetic interplay between genes and the environment puts the ball right back into our court. The question we should be asking is perhaps this: how can we give everyone the best opportunity in life?

Maternal effects on development are well known and demonstrated across many species. In mammals the long period of gestation and postnatal mother-child relationship provides maternal influences that extend well into the adult life of the offspring.

Prenatal stress and malnutrition experienced by the mother affects her neuroendocrine system and, in turn, the development of the nervous system in the fetus.[28] The care received (usually from the mother, but possibly by surrogates) during early infancy can produce changes in the development of the nervous system that regulates its response to novelty and social behavior.[29] Thus the maternal environment experienced by a developing organism can play a critical role in shaping its adult behavior.

Infant rhesus macaques socially isolated for periods of three to twelve months play much less, are highly aggressive with peers, perform poorly in learning and cognitive discrimination tasks, and are inhibited and fearful of novelty.[30] These behavioral patterns continue into adulthood and affect reproductive success, particularly in artificially reared females, who display high rates of infant abuse, neglect, and infanticide. Maternally deprived macaques also have an elevated hypothalamic-pituitary-adrenal (HPA) response to stress, impairments in learning and social behavior, and altered serotonergic systems (which regulate anxiety), suggesting that it is the disruption of the mother-infant relationship rather than the general consequence of social isolation that contributes to these effects.

In humans, environmental adversity occurring early in life is associated with an increased risk of both physical and psychiatric disorders in adulthood. The experience of childhood abuse and neglect has been demonstrated to increase the rates of diabetes and cardiovascular disease, as well as susceptibility to drug abuse, depression, schizophrenia, and anxiety-related disorders.[31]

There is substantial evidence that lack of parental care or childhood abuse can contribute to subsequent criminal behavior.[32] A study sponsored by the U.S. National Institute of Justice showed that a child who experienced neglect or physical abuse was 53 percent more likely to be arrested as a juvenile and 38 percent more as an adult compared with a child who was not neglected or abused. Another study found that 68.4 percent of male inmates from a New York State correctional institution reported childhood abuse or neglect: 71.2 percent for violent offenders and 61.8 percent for nonviolent offenders.

It has been estimated that up to 70 percent of abusive parents were themselves abused,[33] and 20 to 30 percent of abused infants are likely to become abusers. These findings in humans have been replicated in experiments on primates.[34]

Clearly the environment plays a large role, but it does not determine whether children will grow up to be criminals, any more than their genetic makeup determines what they will become. More important, changing the environment can often undo the harm that individuals or their parents have experienced in early life.

Epigenetics of Maternal Behavior

Researchers at McGill University in Montreal, Canada, and Columbia University in New York have been studying maternal behavior in rats for many years. They have found that mother rats that care adequately for their pups and others that do not do so shape their offspring's response to stress accordingly for the rest of their lives, correlated with different states of expression in relevant genes.[35]

The mother rat licks and grooms (LG) her pups in the nest and while nursing them also arches her back. Some (high-LG) mothers do that more often than others (low-LG). The offspring of high-LG mothers grow up less fearful and more able to cope with stress than those of low-LG mothers, and involves the HPA pathway of response to stress. The magnitude of the HPA stress response is a function of the corticotrophin-releasing factor (CRF) secreted by the hypothalamus, which activates the pituitary-adrenal system. The pituitary-adrenal system is in turn modulated by glucocorticoid secreted in the hypothalamus, which feeds back to inhibit CRF synthesis and secretion, thus dampening the HPA response and restoring homeostasis.

The adult offspring of high-LG mothers show increased glucocorticoid expression in the hippocampus and enhanced sensitivity to glucocorticoid feedback. This enhanced sensitivity is due to the increased expression of glucocorticoid receptors (GRs), boosted in turn by the increased expression of the transcription factor NGF-1-A that binds to the promoter of the GR gene. These differences in gene expression are accompanied by significant differences in DNA methylation (addition of methyl groups) of the GR promoter, with low methylation from offspring of high-LG mothers correlating with high expression, and high methylation from offspring of low-LG mothers correlating with low expression. The researchers also found significantly higher acetylation of histone in chromatin protein around the GR gene (consistent with active gene expression) in the offspring of high-LG mothers than in the offspring of low-LG mothers.

Interestingly, cross-fostering the offspring of low-LG to high-LG mothers and vice versa at day one after birth induced changes in the offspring in line with the foster mother, with correlated changes in the gene expression states. Foster parents can influence their children biologically.

It turns out that the different gene expression states are acquired during the first week of life and persist into adulthood. Pups of both high-LG and low-LG mothers start out practically the same. Just before birth the entire region of the GR promoter is unmethylated in both groups because most gene marks are erased in the germ cells. Changes develop according to the behavior of the mother within the critical period of the first week of life and remain stable thereafter.

Nevertheless, these changes in DNA methylation and histone acetylation can be reversed, even in adults, as demonstrated by the rather drastic method of infusing chemical activators or inhibitors into the brain, with concomitant changes in the adult's response to stress.[36] Thus infusing the histone deacetylase inhibitor trichostatin A (TSA) into the brains of offspring from low-LG mothers increased histone acetylation and decreased methylation of the GR promoter, thus boosting GR expression to levels indistinguishable from those in the brains of offspring from high-performing mothers. When these offspring were tested for anxiety levels, they performed like offspring from high-LG mothers.

On the other hand, injecting methionine, the precursor of S-adenosyl methionine (SAM), the cofactor of DNA methylase, into the brains of offspring from high-LG mothers increased methylation of the GR promoter to levels the same as those of offspring from low-performing mothers, thereby decreasing GR expression and causing them to switch their behavior accordingly to resemble that of offspring from low-LG mothers. Thus epigenetic states are stable but dynamic. They are truly plastic and give no support to any kind of determinism, genetic or environmental.

Maternal Care and Sex Hormones

What predisposes mothers to be caring or otherwise? Apparently the female offspring inherit the characteristics of their mothers with regard to maternal care, not genetically but epigenetically. The hippocampus is the "emotion center" of the brain. It is vulnerable to stress and richly supplied with receptors for the sex and reproductive hormones, and maternal care is regulated by those hormones.

In rats, the researchers found oxytocin receptors linked to the expression of maternal behavior.[37] Oxytocin (OT) is a hormone secreted by the posterior pituitary gland and stimulates the contraction of the uterus and ejection of milk. Variations in OT receptor levels in critical brain regions, such as the medial preoptic area (MPOA) of the hypothalamus, are associated with differences in maternal care. OT receptor binding in the MPOA is increased in high-LG compared with low-LG mothers. Furthermore, differences in OT receptor binding in the MPOA between high-LG and low-LG females are dependent on estrogen, which is eliminated by ovariectomy and reinstated with estrogen replacement. However, whereas ovariectomized high-LG females respond to estrogen with an increase in OT receptor binding, low-LG females show no such effect. Studies with mice suggest that estrogen regulation of OT receptor binding in the MPOA requires the α-subtype of the estrogen receptor (ERα). ERα is a transcription factor that regulates gene transcription on binding estrogen. The cellular response to estrogen depends on the amount of ERα present.

The researchers found that by day six after birth, ERα expression in the MPOA of female offspring from high-LG mothers is significantly increased compared with that of female offspring from low-LG mothers, and this state continues into adulthood and is correlated with the female offspring of high-LG and low-LG mothers becoming high-LG and low-LG mothers accordingly. This epigenetic state perpetuates itself via the female line until and unless it is disrupted by environmental intervention.

Cross-Fostering Reverses the Damage

One effective environmental intervention is cross-fostering. The biological offspring of high- and low-LG mothers were reciprocally exchanged within twelve hours of birth and reared to adulthood. When these offspring were examined, ERα expression in the MPOA of the adult females born to low-LG mothers but cross-fostered to high-LG mothers became indistinguishable from that of the normal biological offspring of high-LG mothers; conversely, ERα expression in the MPOA of adult females born to high-LG mothers but reared by low-LG mothers resembled that of normal biological offspring of low-LG mothers. Cross-fostering in itself had no effect; exchanging offspring between two low-LG mothers or two high-LG mothers did not alter the expression of ERα in the MPOA of the offspring.

Correlated with the high and low ERα expression in the MPOA were significant differences in the methylation of cytosine-guanine (CpG) sites across the entire ERα promoter. Overall, significantly elevated levels of methylation were found in the promoter of offspring with low ERα expression in the MPOA compared with high ERα expression in the MPOA.

Maternal Care Influences Brain Development and Many Gene Functions

Obviously, maternal care influences more than a few genes. Prior to the findings detailed in the preceding sections, the McGill University team found that in rats, increased anxiety in response to stress in the offspring of low-LG mothers is associated with decreased neuronal development and density of synapses in the hippocampus. The offspring of high-LG mothers, on the other hand, show increased survival of neurons and synapses in the hippocampus and improved cognitive performance under stressful conditions.[38] Researchers at the University of Amsterdam and Leiden University in the Netherlands have also found that the pyramidal neurons in layers 2 and 3 of the brain cortex from high- and low-LG adult rats have different morphologies.[39] The high-LG rat neurons have more slender "dendritic trees"—the branching processes receiving inputs from other neurons—with fewer branches than those from low-LG rats. The density of dendritic spines (small projections from the surfaces of the dendritic trees) is also significantly lower in high-LG rats. These observations suggest a rather extensive influence of maternal care on brain development and gene expression.

In order to examine the effect of high- and low-LG mothers and TSA or methionine infusion on gene expression, the four different treatment groups were compared with their respective control groups using microarrays to monitor changes in 31,099 unique messenger RNA transcripts.[40] A total of 303 transcripts (0.97 percent) were altered in the offspring of high-LG mothers compared with offspring of low-LG mothers: 253 transcripts (0.81 percent) upregulated and 50 transcripts (0.15 percent) downregulated. TSA treatment of offspring of low-LG mothers altered 543 transcripts (1.75 percent): 501 transcripts (1.61 percent) upregulated and the remaining 42 transcripts (0.14 percent) downregulated. Methionine treatment of offspring of high-LG mothers changed 337 transcripts (1.08 percent), with 120 (0.39 percent) upregulated and 217 (0.7 percent) downregulated.

The results suggest that maternal care during the first week of life determines the expression of hundreds of genes in the adult offspring, but changes in gene expression are nevertheless reversible even into adulthood. Caring mothers tend to activate more genes in their offspring than mothers that do not provide adequate care. TSA treatment results predominantly in gene activation, and methionine treatment results predominantly in silencing genes.

Epigenetic Effects of Enriched Environment

Researchers at Tufts University School of Medicine, Boston, Massachusetts, and Rush University Medical Center, Chicago, Illinois, have demonstrated that exposure of 15-day-old mice to two weeks of an enriched environment that includes novel objects, increased social interactions, and voluntary exercise enhances long-term potentiation not just in the mice but also in the future offspring of female mice through early adolescence, even if the offspring never experienced the enriched environment.[41] Long-term potentiation (LTP) is a persistent increase in strength of synapses between neurons following high-frequency stimulation and is a form of synaptic plasticity known to be important for learning and memory. The effect of the enriched environment lasts for about two months and is not canceled by cross-fostering, indicating that the transgenerational effect occurs before birth, during embryogenesis. The effect is age dependent because it cannot be induced in adult mice. In both generations of mice, LTP induction is accompanied by the new appearance of a whole new signalling pathway, the cAMP/p38 MAP (mitogen activated protein) kinase-dependent signalling cascade. If the effect occurs in humans, it means that an adolescent's memory can be influenced by environmental stimulation experienced by the mother when she was young.

Epigenetic Footprints of Childhood Trauma in Humans

Are the epigenetic footprints of maternal care identified in detailed animal studies relevant to the human species? Michael Meaney and his colleagues at McGill University have extended their findings in rats to humans. They examined epigenetic differences in a neuron-specific promoter of the glucocorticoid receptor in postmortem hippocampi (twelve in each group) obtained from suicide victims with a history of childhood abuse, suicide victims with no history of childhood abuse, and nonsuicide victims who died from other causes, none of whom had a history of childhood abuse.[42]

They found decreased expression of glucocorticoid receptor and increased DNA methylation of the specific promoter that binds the transcription factor NGF1-A in suicide victims with a history of childhood abuse compared with suicide victims without childhood abuse, who were indistinguishable from controls. These are the same epigenetic footprints that the team had previously discovered in rodents that did not provide adequate maternal care.

Psychiatric disorders such as major depression and posttraumatic stress disorder are commonly connected with disorders of the cardiovascular, metabolic, and immune systems. Recent studies suggest that accelerated aging of cells may be an explanation. Telomeres are DNA repeats that cap the ends of chromosomes and make them more stable, and they shorten with each cell division, making them a marker for biological age. Physiological stress such as radiation and toxins, oxidative stress, and cigarette smoke can shorten telomeres.

The body responds to stress by the coordinated activities of several systems, including the HPA axis, the sympathetic nervous system, and the immune system. They mobilize energy and prepare the individual to cope with the stress. Chronic stress, however, can damage the endocrine, immune, and metabolic systems and may result in shortening the telomeres.

Individuals giving care to patients with Alzheimer's disease experience chronic stress, and when their white blood cells were examined, the telomeres were found to be shortened. The same telomere shortening has been linked to pessimism in healthy postmenopausal women and in patients with unipolar and bipolar mood disorders.

Researchers at Butler Hospital and Brown University Medical School, Providence, Rhode Island, have now found that stress in childhood due to maltreatment also leads to telomere shortening.[43] Telomere shortening is a major risk factor for a range of adverse conditions, including major depression, anxiety disorders, and substance abuse.

The researchers tested thirty-one adults (twenty-two women and nine men) aged 18 to 64 years, recruited via advertisement in the community for a larger study of stress reactivity and psychiatric symptoms. Of these, twenty-one reported no history of childhood maltreatment, and ten reported a history of moderate or severe childhood maltreatment. None had acute or unstable medical illness, endocrine diseases, or ongoing treatment with drugs that might influence HPA axis functions.

The maltreatment group did not differ significantly from the control group with respect to age, sex, smoking status, body mass index (a measure of obesity), hormonal contraception use in female subjects, race, education, socioeconomic status, or perceived stress. The maltreatment group had significantly shorter telomeres than the control group and was associated with both physical neglect and emotional neglect. The sample size was small, so there was no association of telomere length with age in the sample, which made the association with childhood abuse or neglect all the more significant.

Implications for Health

Although the epigenetic effects of maternal (parental) care have been worked out in the most detail in rodents, there is a potential for similar effects in other species, including primates and humans, as recent evidence indicates.

In humans, a lack of parental care or childhood abuse can contribute to subsequent criminal behavior. Furthermore, lack of parental care and parental overprotection ("affectionless control") are also risk factors for depression, adult antisocial personality traits, anxiety disorders, drug use, obsessive-compulsive disorders, and attention-deficit disorders.[44] Conversely, people who reported high levels of maternal care were found to have high self-esteem, low anxiety, and less salivary cortisol in response to stress. Longitudinal studies demonstrated that mother-child attachment is crucial in shaping the cognitive, emotional, and social development of the child. Throughout childhood and adolescence, secure children are more self-reliant and self-confident and have more self-esteem. Secure infants also have better emotional regulation, express more positive emotion, and respond better to stress. Infant disorganized attachment has been associated with the highest risk of developing later psychopathology, including dissociative disorders, aggressive behavior, conduct disorder, and self-abuse.

Nutrition, Environmental Enrichment, and Mental Health

The dramatic effects of TSA and methionine infusion in altering gene expression patterns in rats also have obvious implications for drug intervention or, better yet, intervention and prevention through adequate nutrition.[45] Epigenetic drugs such as inhibitors of DNA methylation or histone deacetylation lack specificity[46] and may have unintended and untoward side effects.

In rats, dietary L-methionine has been shown to be crucial for normal brain development, and its deficiency has been implicated in brain aging and neurodegenerative disorders. Synthesis of SAM (a cofactor for DNA methyl transferase) is dependent on the availability of dietary folates, vitamin B_{12}, methionine, betaine, and choline. Developmental choline deficiency alters SAM levels and global and gene-specific methylation, and prenatal choline availability has been shown to affect neural cell proliferation and learning and memory in adulthood.[47] Several studies have shown that additional dietary factors, including zinc and alcohol, can affect the availability of methyl groups for SAM formation and thereby influence CpG methylation. Maternal methyl supplements positively affect the health and longevity of the offspring.

Other studies have shown that certain dietary components may act as histone deacetylase inhibitors (HDACis), including diallyl disulfide, sulforaphane, and butyrate. For example, broccoli, which contains high levels of sulforaphane, has been associated with H3 and H4 acetylation in peripheral blood mononuclear cells in mice three to six hours after consumption.

HDACis are an active area of research as anti-inflammatory and neuroprotective agents in autoimmune diseases, such as lupus and multiple sclerosis. Sodium butyrate has been shown to have antidepressant effects in mice.

The new findings on environmental intervention, such as environmental enrichment to reverse the damages of social isolation and fostering to reverse the harm of parental neglect, are indicative of the huge potential for saving our children with the appropriate social policies. All in all, these remarkable findings on the epigenetic effects of maternal care show how important it is for societies to look after the welfare of children and mothers to be in order to ensure both the mental and the physical health of future generations.

Conclusion

The Unfulfilled Promise of Genomics

JEREMY GRUBER

> This landmark achievement will lead to a new era of molecular medicine, an era that will bring new ways to prevent, diagnose and treat disease.
>
> —White House Press Statement, June 25, 2000

> It is my hope and expectation that over the next one or two decades—or however long it takes—genomic discoveries will lead to an increasingly long list of health benefits for all the world's peoples.
>
> —Francis Collins, National Institutes of Health Director, June 25, 2010

These are exciting times for genomics research. New tools and techniques are increasingly allowing researchers to dig ever deeper into the workings of biological processes. Scientists are discovering new disease pathways, identifying genes linked to Mendelian disorders and allowing for early intervention, making some advances in the use of biomarkers for pharmacogenomics and therapeutic interventions with cancer patients, and offering parents a growing list of pre- and postnatal screenings for inherited abnormalities.

Perhaps the greatest advances have been made in sequencing technologies, where both size and price have dropped dramatically, although added costs from preparing and isolating DNA and assembling the raw data once it is sequenced are rarely mentioned. Even so, considering that not so long ago the Human Genome Project cost $3 billion and took ten years to complete, those advances in sequencing technologies are clearly an impressive technological feat.

As the science progresses, however, it has become clear that the major challenge for the future will not be sequencing technologies and broad public access to them but rather the cost and difficulty of interpreting and applying the huge amounts of data they generate and their inherent limitations. There continues to be a large gap between basic research and clinical applications, and that gap has become filled with exaggeration, hyperbole, and outright fraud. Just as eugenicists in the twentieth century became entranced with the work of Gregor Mendel and sought to apply principles of genetics to social theory, so too have molecular biologists and the academic, commercial, and policy communities in which they operate become ensconced in a worldview that sees the field of genomics as the most fundamental mechanism for improving the human condition.

These advocates offer a view of scientific progress that is often completely at odds with how science naturally progresses: incrementally in stops and starts. Scientists make observations, mold those observations into a hypothesis, and test that hypothesis, gather data, and continue to refine hypotheses to reflect their findings. However, the current era of biotechnology has followed a somewhat different path full of hubris and bordering on faith. Despite James Watson's famous declaration in his selling of the Human Genome Project that "we used to think our fate was written in the stars. Now we know it is written in our genes"[1] and Francis Collins's declaration on the ten-year anniversary of the project's completion that a long list of health benefits for "all the world's peoples" was sure to come, the promise of a "revolution" in clinical medicine has not yet materialized. The Human Genome Project was sold with brazen predictions about its future contribution to improve health care in America. Ten years later the financial and policy mechanisms that enabled it have become entrenched, and this worldview continues to emanate from every level of the scientific and policy establishment, seeking to capture the public's imagination while fertilizing a zeitgeist that declares this the "age of biology."

The Role of the Media and Public Understanding

It is not hard to see why. We are regularly inundated with media reports that link gene variants to everything from behavioral traits such as social networking to rather common diseases or conditions such as Alzheimer's disease and obesity.[2] Media accounts of criminal forensics often read like an episode of the television series *CSI: Crime Scene Investigation.* For many in the media establishment, genetic causation is conceptually simple to understand and explain to an uninformed public, even if such reporting puts them completely at odds with the complexity of causality (see Chapter 9) and the incremental nature of scientific advancement. But journalists, who usually have no scientific training or experience, often get their stories and information from press releases published by industry, government, or interest groups rather than from the scientists themselves. Biotechnology company marketing departments, well trained in media relations, mold the media and by extension the public's view of the state of the science with "breakthrough" announcements and individual stories of how new research is "saving lives." The spin, which usually starts long before press reports begin, is designed to exploit a natural fear of disease and offer the public a narrative full of hope and belief in the power of science (see Chapter 13).

News reports find a sympathetic audience because the public's basic knowledge regarding genetics is extremely poor. Most American adults came of age before genetics was part of the standard education curriculum. Even today, there is a "wide scale deficiency in genetics knowledge of high school students."[3] A 2011 comprehensive study of high-school-level coursework related to genetics in all fifty states found a whopping 85 percent of states' standards receiving scores of "inadequate."[4] Even when genetics is taught in schools, it is generally limited to basic principles of Mendelian inheritance and genetic transmission—exploring binary traits like eye color, for example—that leave students with a rather all-or-nothing view of genetics. Even where reliable associations are found, they do not correspond to the kind of major influences implied in high-school biology courses. Even the high-school curriculum recommended by the National Institutes of Health (NIH) furthers this "sell" by including activities such as "Molecular Medicine Comes of Age . . . Role-play pharmaceutical company employees to develop new drugs—exemplifies benefits of molecular genetics."[5] As a result, even in our ordinary discourse genetic identity has become a substitute for race, and we find ourselves using the word "genetic" to describe just about anything allegedly immutable.[6]

Far less often do the media take a critical look at findings that leap from correlation to causation. A case in point is the controversy over the supposed discovery of the "longevity" gene by a group of American researchers. Research over the past decade had seemed to show a causal link between proteins called sirtuins and lifespan. Experiments on earthworms and fruit flies—commonly used as models to examine the biology of human aging—suggested that an extra dose of the naturally occurring enzymes could prolong life by up to 50 percent. Despite reflecting a severe misunderstanding of evolution by so directly tying a mutation in such organisms to humans, these early results unleashed a flood of new research, much of which backed up the original findings. They also spawned a major media storm and a flourishing market in dubious health products claiming to boost sirtuins and thus slow down one's biological clock. Then researchers in the United Kingdom found a basic flaw in the designs of these studies: they failed to account for all the possible differences between genetically manipulated organisms and the "natural" ones against which they were compared. Once this discrepancy was accounted for, further studies found no added affect of sirtuins on longevity.[7]

The State of the Science

We are still only at the dawn of the development of scientific knowledge about the relationships among genes, human health, and the environment. Although research has identified a few single dominant genes that are strongly associated with disease risk, such as the *BRCA1* and *BRCA2* genes associated with breast and ovarian cancer, even the vast majority of cases of breast cancer cannot be attributed to genetic causes. Finding new alleles that predict susceptibility to most common diseases remains elusive. Our knowledge about most identified disease-related genes is far from sufficient to make reliable predictions about relative risk. The more we discover about human genetics, the more we learn about the small effect of most individual genes on human traits. Indeed, even when genetic variants that mediate disease are combined, they typically confer low relative risks, with the exception of extremely rare single-gene disorders. They become even less significant when they are balanced against the fact that everyone is at absolute risk for common multifactorial diseases. Genomics research may be offering us more things to measure, but only a very small fraction of these are medically meaningful and can be turned into new diagnostics.

Scientists are gaining increasing knowledge that environmental influences are far more important than specific alterations in the genetic codes of common diseases, despite the fact that our tools for accurately measuring environmental exposure continue to underestimate its influence (see Chapter 16). This underestimation of environmental influences is magnified by the fact that many of the most important environmental risk factors act by multiple mechanisms and affect many diseases. These discoveries have spawned an entirely new field of research, epigenetics, that studies the influence of environmental exposures on transcriptional regulation, but the field is in such an infant stage that its parameters are not yet set; epigenetic phenomena are probably whole different suites of things. If you consider, however, that U.S.-born Japanese men have a rate of colon cancer twice that of native-born Japanese men, and U.S.-born Japanese women have a colon cancer rate 40 percent higher than that of their counterparts born in Japan,[8] you will have a fairly good idea of how dramatic the relative contributions of the environment can be. As a result, family medical history remains the simplest applied "genomic" tool in practice today because it reflects not only shared genetic variation but also shared exposures, shared responses to environmental factors, and shared behaviors.

The scientific establishment has offered many reasons for the failure to identify susceptibility genes, including too-small study samples, inappropriate control groups, and biases in the studies, but the more obvious answer may simply be a reflection of the complexity of common diseases with risks that depend on gene-gene and gene-environment interactions. In the past decade large pharmaceutical companies rushed to take part in the "biotechnology revolution" by buying biotechnology companies and absorbing their research and development departments or simply refocusing their own R&D on similar pursuits. But as pharmaceutical and biotechnology companies have increasingly focused their research and development investments on genomics, there has been a corresponding and precipitous drop in productivity. They have been unable to sustain sufficient innovation to replace the loss of revenues due to patent expirations for successful products. Criticisms of this unsustainable downward trend have largely focused on a mixture of excessive regulation, rising costs, shorter product life cycles, and internal inefficiencies. Even if these factors are accepted as correct, however, they simply cannot explain why between 1998 and 2008 the output of new molecular entities (NMEs) dropped by almost 50 percent, and the success of late-stage clinical trials dropped equally dramatically.[9] Biotechnology's role in this downward spiral has not gone unnoticed: "The larger presence of newly entered biotechnology firms in the US, largely geared towards exploratory research rather than exploitation of known compounds and mechanisms of action, makes the US system more oriented towards riskier research and markets."[10]

A clearer understanding of what genomics can achieve will become possible only when more research is completed. The truth is that we just do not know enough yet. That is not to say that genomics research is not valuable or should not continue; it is and it should, but the results to date do suggest that there is a substantial discrepancy between the grandiose perceptions of the state of the science and the reality and that what we are witnessing is far from a "genomics revolution" in medicine. In short, biotechnology has overpromised and underdelivered.

The biotechnology industry is simply asking for more time and equally laying blame on a medical establishment that is slow to embrace new technologies and treatments (they curiously fail to lay blame on the health-insurance industry, which has shown an equal amount of caution). It is largely true that most physicians will declare the value of the information being generated by new research while simultaneously lacking the knowledge or training to incorporate it into practice. But it is equally true that physicians are regularly inundated with new clinical applications, which they must necessarily prioritize on the basis of likely efficacy. It is not hard to understand why the medical establishment has not rushed to embrace genomics when there remains a scarcity of current data showing that genomic approaches to medicine will actually protect or improve patient health today (see Chapter 11). The Evaluation of Genomic Applications in Practice and Prevention, a Centers for Disease Control and Prevention working group, has evaluated over 200 new genomic applications in medicine and found that only one delivered a quantifiable health benefit.[11] Even this application, a test for an inherited type of colorectal cancer, was favored not for its value to the patient but for its value to family members who might be at higher risk if they had the tested mutation.

The Commercialization of Biotechnology

Why the imbalance? The elephant in the room is that biotechnology development first and foremost is being driven by a large amount of speculative investments and political support. Any explanation of why the natural euphoria of scientists working in the field of genomics has so utterly transformed the focus of medical research in their direction must acknowledge the salient fact that often the most vocal supporters of genomics research are those who stand to benefit financially from it. Since the early 1980s the biotechnology industry has undergone extremely rapid growth, with its current collective market cap hovering at just under $400 billion.[12] The industry's primary cheerleader, the Biotechnology Industry Organization (BIO), has done a remarkably effective job of selling its product to investors and policy makers alike. Beginning in the early 1990s, industry supporters sought substantial increases in NIH funding to be directed toward biomedical research as they began linking such research to job creation and economic growth generally. By the late 1990s Congress had set forth a five-year plan to double the budget of the NIH.

That plan saw its greatest advocate in President Clinton, who dedicated five paragraphs of his 1998 State of the Union address to biomedical research. He declared:

> As part of our gift to the millennium, I propose a 21st Century Research Fund for path-breaking scientific inquiry, the largest funding increase in history for the National Institutes of Health, the National Science Foundation, and the National Cancer Institute.
>
> We have already discovered genes for breast cancer and diabetes. I ask you to support this initiative so ours will be the generation that finally wins the war against cancer and begins a revolution in our fight against all deadly diseases.[13]

In subsequent years the NIH's budget doubled, with annual increases averaging $2.9 billion and creating the richest research organization in the world. The United States now spends more than any other country on genomics research. The NIH spent $4.9 billion on genomics research in 2008 alone, and private investment was twice that amount.[14] And it is certainly no coincidence that the overarching focus on genomic-centered research has followed the career trajectory of the current director of the NIH, Francis Collins, who had so successfully sold the "benefits" of the Human Genome Project to Congress and two administrations as the director of the National Human Genome Research Institute.

As venture-capital investment has begun to wane and as the biotechnology industry has failed to live up to its own hype, supporters have gone back to Congress and state governments for even more funding through a long-term strategic plan of reforms, declaring that

> what's needed is a policy environment that incentivizes the magnitude of investment necessary to translate the scientific potential that resides in the thousands of American biotech companies into the breakthrough cures, treatments, enhanced agricultural products, vaccines to defend against bioterrorism and revolutionary biofuels that can transform society. Only by transforming the policy environment can we create a robust innovation economy that helps America compete globally by maintaining our position as world leader in biotechnology research and development. And only by investing in biotech today can we discover the new treatments and cures that will not only save lives, but reduce long-term health care costs by keeping people healthier and reducing chronic disease.[15]

They have found many willing listeners in all levels of government. Absent an appreciation of the complex relationship between basic science research and clinical applications, governments have embraced the life sciences as an engine of economic growth. Many states are going to great lengths to attract biotechnology companies by issuing bonds and providing tax breaks, loan guarantees, and other incentives. BIO was able to get $1 billion in tax credits for biotech companies embedded in the 2011 federal health-care legislation and has been the primary force behind new bills introduced in Congress that would increase funding and lower regulatory hurdles, such as the personalized-medicine bill sponsored by Congresswoman Anna Eshoo (who not coincidentally represents the 14th Congressional District in California, where large segments of the industry are located). That may prove to be only the tip of the iceberg. The Obama administration's Office of Science and Technology Policy declared in late 2011 its intent to move forward with a "National Bioeconomy Blueprint," asking the biotechnology industry and biomedical researchers for ideas on how to "harness biological research innovations to address national challenges in health, food, energy, and the environment" and to make "changes in regulatory policy to move life sciences breakthroughs from the lab to the market."[16] Needless to say, the industry immediately heralded the decision.[17]

The promotion of genomic-based medical research has by no means been limited to private industry, though. Universities should be places where healthy skepticism of claims about science and its applications are pursued. But more than almost any other high-technology business, the biotechnology industry maintains extremely close ties with leading academic institutions where much of the basic research in genomics occurs. Academic researchers form much of the initial staff of biotech start-up companies, and as these companies grow, they form lucrative partnerships with universities that include everything from collaborative research to consulting relationships.[18] Many such researchers sit on the advisory boards of biotechnology companies. As a result, academic researchers are often the biggest industry boosters. In 2003, when BIO testified before the House Subcommittee on Health to encourage, among other things, greater investment in biomedical research and transfer of federally funded research to universities and the private sector, it was represented not by a company CEO or a trade-organization advocate but by Phyllis Gardner, senior associate dean for education and student affairs at Stanford University.[19] In her testimony she declared:

> At Stanford University alone, over 1,200 "spin-off" companies have been established by current or former students and faculty. Recognized early on by then University President Fred Terman as an important strategy for seed funding of translational research and innovation, the vast majority of these companies were founded with technologies initially developed under

> government funding. . . . This thriving business ecosystem, in turn enables further R&D initiatives and two-way technology flow between academia and industry. Stanford's Office of Technology Licensing has a robust record of licensing university patents, with royalty income that flows back to the university and the individual inventor.[20]

Perhaps nowhere has the intersection of all these factors been better illustrated than with the rise of the direct-to-consumer (DTC) genetic-testing industry, which offers genetic testing directly to the public (see Chapter 13). The industry has become a flashpoint in debates over policy because it lies at the confluence of basic research in genomics, public health, and the market economy. These companies offer individuals the opportunity to discover whether their genomes possess single-nucleotide polymorphisms and, in some cases, known Mendelian variants associated with disease and cancer risk, nutrient metabolism, drug response and metabolism, and recessive carrier states, among others. They further offer risk-assessment services, which look at several genes simultaneously to give probabilities of disease development over one's lifetime, and offer diet and lifestyle recommendations on the basis of these genetic test results. Some of these companies are outright fraudulent,[21] but even reputable DTC companies are in a difficult position: they attempt to respond to and market their services to the growing social interest over personal information while at the same time communicating the current limitations of what we can learn from genetic information. This may be one of the reasons they regularly caution that while what they are offering is often the best of the current state of the science, it isn't medical advice. But it is easy to overstate the significance of genetic test results, particularly when these companies know that fewer consumers will purchase their products unless they can directly see the benefit of that information. So these companies regularly make and market suggestive statements to the public. The website of 23andMe states, "Take a more active role in managing your health."[22] Navigenic's website offers "A New Look for a Healthier Future."[23] The website of deCODE Genetics promises to "deCODE your health."[24] And Pathway Genomics claims, "It's Now Possible to Know How Genes May Affect Your Health."[25] Every member of the industry makes both explicit and implicit claims that knowing your genetic information will demonstrably improve your health.

By offering its products directly to the public in an environment without regulatory oversight, this very small segment of the biotechnology industry has become the unwilling face of an industry that prefers to conceal its scientific limitations. Additional questions of relaxed privacy standards and inadequate consent procedures, particularly for the "bad actors" of the industry, began to leave a mark beyond just the DTC industry. Indeed, a high-profile investigation by the U.S. General Accounting Office in 2010 produced a damning report whose title says it all: "Direct-to-Consumer Genetic Tests: Misleading Test Results Are Further Complicated by Deceptive Marketing and Other Questionable Practices." The firestorm that accompanied the report included a congressional hearing that lambasted the industry, with one member going so far as to call DTC "snake oil" that could cause individuals to make radical and unfounded medical decisions.[26] Many longtime congressional observers were struck by the vitriol from both sides of the political aisle. Although that characterization was certainly inaccurate, since reputable DTC companies often do not only represent the current state of the science, but do so with some of the top experts and state-of-the-art facilities in the industry, the implication was valid. Even if the information provided by DTC companies might be educational in nature, and even if there was no direct harm from such companies, as medical decisions based on reported results from such companies would be unlikely without the balancing intervention of a physician, the fact remained that there was no proven clinical application for almost all the basic information being given to the consumer. Ironically, some of the most well respected of these companies have responded to such criticisms by conducting their own basic research.

Indeed, the marketing of such testing goes far beyond the companies themselves and is reflective of the relationship between academic institutions and commercial entities and the marketing of a genome-centric view of health to the public. In May 2010 the University of California at Berkeley announced that it would be sending incoming freshman a cotton swab with which to send in a DNA sample to be tested for three gene variants as part of a program titled "Bring Your Genes to Cal" that would focus on genetics and personalized medicine. Optional lectures on genetics were to follow only after a decision to participate was made. Originally the program planned to engage the services of a DTC company to conduct the testing and included prizes of more comprehensive testing services from such companies until mounting criticism forced the university to abandon these elements. However, it was soon discovered that the professor who had created and would be running the program was the founder of several commercial biotech companies (and had served as a consultant for others), including a genetic-testing company founded less than a year earlier.[27] He had also been on a tenure committee that had denied tenure to a scientist who had been critical of a deal several years earlier between Berkeley and Novartis, and he was heavily criticized for having a conflict of interest because of his biotech ties.[28]

Finding the Right Balance

The story of the "Bring Your Genes to Cal" program is emblematic of a culture that is devoted to fostering a version of science that is often at odds with the current state of development. All of this points to the resounding conclusion that much of the discipline is far from mature enough to move beyond the research laboratory into the doctor's office. In the interim, we must take a more critical look at disease causation and refocus our priorities. Given the many complex interactions that underlie almost all human diseases, even improving existing approaches to identifying and modifying genetic risk factors will often have significantly less value than modifying nongenetic risk factors.

Thus even with new insights to come, the idea that we should be prioritizing identification of susceptibility to the debilitating effects of smoking, for example, rather than working to lower rates of smoking overall in the population is flawed, and the same goes for other aspects of diet and lifestyle. The social distribution of many of these conditions heavily weighs toward identification of societal rather than genetic origins. The fact is that the vast majority of morbidity and premature mortality in the developed world comes from just three distinct modifiable behaviors, disproportionate food and alcohol consumption, smoking status, and sedentary behavior, and one highly modifiable environmental factor, pollution.[29]

In the balance there would be a significantly higher public health benefit from disease prevention by prioritizing research and policy in addressing these factors. Indeed, as Figure 2 clearly demonstrates, common multifactorial diseases such as colon cancer, stroke, coronary heart disease, and type 2 diabetes are highly preventable by lifestyle modifications.

The state of public health in the global South is even more obviously determined by economic and social conditions. According to the World Health Organization,

> Although some of these major risk factors (e.g. tobacco use or overweight and obesity) are usually associated with high-income countries, in fact, more than three quarters of the total global burden of diseases they cause already occurs in low- and middle-income countries. The poorest countries face a double burden as they still face a high and concentrated burden from poverty, undernutrition, unsafe sex, unsafe water and sanitation, iron deficiency and indoor smoke from solid fuels.[30]

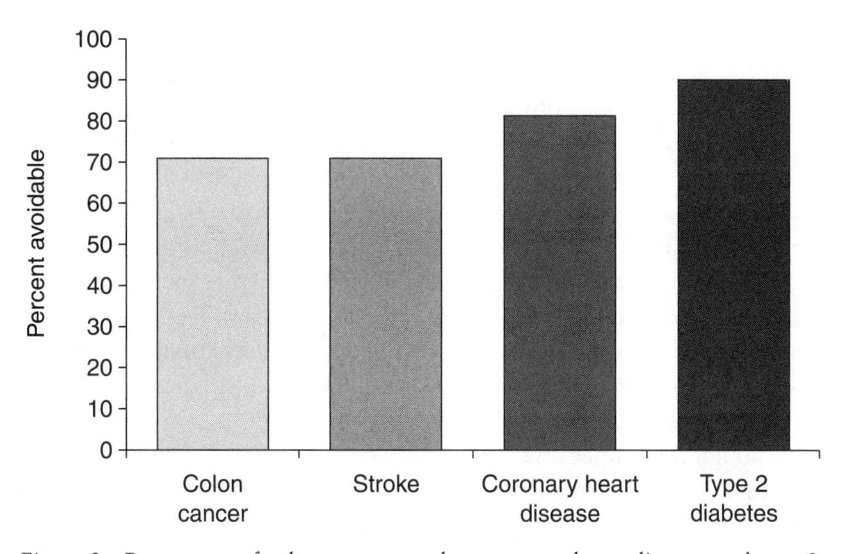

Figure 2 Percentage of colon cancer, stroke, coronary heart disease, and type 2 diabetes potentially preventable by lifestyle modifications

Yet policy and investment fails to reflect these hard realities. In 2011 the NIH devoted $1.17 billion to the basic behavioral and social sciences, about 25 percent of what was spent on genomics research.[31] Private investment was certainly much less as well. And no single public health project has ever come anywhere close to mirroring the $3 billion spent on the Human Genome Project. A decade after the mapping of the human genome, genomics research continues to top the field. Seven of the top thirteen researchers in all fields in 2010 worked in genomics, as identified by the frequency with which published papers are cited.[32] Ironically, three of those seven researchers came from Iceland's deCODE Genetics, which filed for bankruptcy in 2009 before reconstituting itself the next year as a private company.

By creating unreasonable expectations, genomics research and its supporters have monopolized a considerable amount of public and private resources, both financial and political, and its overemphasis has limited the ability of researchers to focus on hundreds of common diseases, including asthma, lung and kidney disease, and many forms of cancer, not to mention bacterial infections, such as tuberculosis, which killed 2.9 million people worldwide in 2009 alone.[33]

The promise of genomics may have provided policy makers with a simple narrative of basic health research investment, but it has led to poor decision making on their part and has proved to be an insufficient standard bearer in the fight to improve the human condition. Even if individuals often lead undisciplined lives and behaviors can be difficult to modify, given that even a small improvement in our ability to alter behavior, for example, could yield significant benefits, a reappraisal of funding priorities would benefit the promotion of human health. Public health policy and funding must find a more reasonable balance between the real promise of long-term basic genomics research and more immediate evidence-based projects and policies. Genomic research that does show more immediate, large-scale, and practical applications, such as the real promise of public health genomics to address malaria vulnerability[34] for the betterment of developing countries, should certainly be included in such calculations.

Although those operating out of pure economic interest share much of the blame for the current exaggerated position genomics hold in overall research focus, it is ultimately the scientists and researchers themselves who bear much of the responsibility. The current system for assessing research productivity, combined with the demands to publish and attract both private and governmental research funding, puts enormous pressure on researchers to make, publicize, and defend "breakthrough" discoveries. This is compounded by the added pressure of journals to publish "impact" articles. As a result, few genomics researchers publicly speak out, and the resulting void has been filled with a distortion of science without parallel in any other discipline.

So far, research in molecular genetics has unlocked many mysteries of basic human biology, but it has proved to be just one of many possible high-tech tools that show substantial benefit to those unfortunate few who have rare genetic disorders but less direct near-term benefit to the rest of us. Perhaps that will change with new discoveries to come. Our collective challenge is to ensure that as biotechnology development does continue, as it inevitably will in directions we cannot even begin to anticipate, the public needs informed and unbiased coverage of both its successes and failures. Only by so doing can we steer biotechnology development toward the advancement of public health in the United States and around the world.

NOTES

SELECTED READINGS

ACKNOWLEDGMENTS

CONTRIBUTORS

INDEX

Notes

Introduction

1. Norriss S. Hetherington, "Isaac Newton's Influence on Adam Smith's Natural Laws in Economics," *Journal of the History of Ideas* 44 (1983): 497–505, at 498.
2. Philosopher Karl Popper famously took issue with this characterization of the origin of scientific discovery. See Popper, *The Logic of Scientific Discovery* (New York: Harper and Row, 1968).
3. J. A. G. Rogers, "Locke's *Essay* and Newton's *Principia,*" *Journal of the History of Ideas* 39 (1978): 217–232, at 226.
4. P. Casini, "Newton's *Principia* and the Philosophers of the Enlightenment," *Notes and Records of the Royal Society of London* 42 (1988): 35–52, at 47–48.
5. Richard Hofstadter, *Social Darwinism in American Thought,* rev. ed. (Boston: Beacon Press, 1955), 3.
6. J. D. Watson and F. H. C. Crick, "A Structure for Deoxyribose Nucleic Acid," *Nature* 171 (1953): 737–738.
7. Francis Crick, "On Protein Synthesis" (paper delivered at a meeting of the Society of Experimental Biology, 1957), http://www.genomenewsnetwork.org/resources/timeline/1957_Crick.php; Crick, "On Protein Synthesis," *Symposia of the Society for Experimental Biology* 12 (1958): 138–163. See note 14.
8. F. Jacob and J. Monod, "Genetic Regulatory Mechanisms in the Synthesis of Proteins," *Journal of Molecular Biology* 3 (1961): 318–356.
9. F. H. C. Crick, Leslie Barnett, S. Brenner, et al., "General Nature of the Genetic Code for Proteins," *Nature* 192 (1961): 1227–1232.
10. Francis Crick, "Centeral Dogma of Molecular Biology," *Nature* 227 (1970): 561–563.

11. S. N. Cohen, A. C. Y. Chang, H. W. Boyer, et al., "Conbstruction of Biologically Functional Bacterial Plasmids *in vitro,*" *Proceedings of the National Academy of Sciences of the United States of America* 70 (1973): 3240–3244.
12. Craig Venter, Mark D. Adams, Eugene W. Myers, et al., "The Sequence of the Human Genome," *Science* 291 (2001): 304–351.
13. Matt Ridley, *Francis Crick: Discoverer of the Genetic* Code (New York: HarperCollins, 2006), 70.
14. F. H. C. Crick, "The Biological Replication of Macromolecules," in *Symposium for the Society for Experimental Biology: The Biological Replication of Macromolecules,* XII, ed. F. K. Sanders (New York: Academic Press, 1958), 138–163.
15. Proteins operate in many ways in the process of turning information from DNA into the synthesis of a polypeptide molecule, including their role in the ribosome—the staging area for the assembly of amino acids into proteins.
16. Martin Richards, "How Distinctive Is Genetic Information?," *Studies in History and Philosophy of Science Part C: Studies in History and Philosophy of Biological and Biomedical Sciences* 32, no. 4 (December 2001: 663–687, at 668.
17. Barry Commoner, "Unraveling the DNA Myth," *Harpers Magazine* 304 (2002): 30–47, at 44.
18. Chih-Yen King and Ruben Diaz-Avalos, "Protein-Only Transmission of Three Yeast Prion Strains," *Nature* [Letters] 428 (2004): 319–322.
19. John Dupré, "What Genes Are and Why There Are Not Genes for Race," in *Revisiting Race in a Genomic Age,* ed. Barbara A. Koenig, Sandra Soo-Jin Lee, and Sara S. Richardson (New Brunswick, NJ: Rutgers University Press, 2008), 39–55, at 41.
20. Sheldon Krimsky, *Genetic Alchemy: The Social History of the Recombinant DNA Controversy* (Cambridge, MA: MIT Press, 1982).
21. Steven Lindow developed a genetically mutant strain of the bacterium *Pseudomonas syringae* with a gene excised. The selected gene coded for a protein that served as a nucleating locus for ice crystals. Without the ice-nucleating protein in microorganisms spread out over an agricultural field, ice formation would be delayed until the temperature reached about 5–7°F below freezing, sufficient to protect crops from frost damage. See "The Release of Genetically Modified Organisms into the Environment: The Case of Ice Minus," in *Environmental Hazards: Communicating Risks as a Social Process,* by Sheldon Krimsky and Alonzo Plough (Dover, MA: Auburn House, 1988), 75–129.
22. Alternative splicing occurs when RNA makes different messenger RNAs from an identical DNA sequence; a single gene can code for multiple proteins.
23. Food and Drug Administration, "Premarket Notice Concerning Bioengineered Foods: Proposed Rule," *Federal Register* 66 (January 18, 2001): 4706–4738, at 4728, http://www.fda.gov/Food/GuidanceComplianceRegulatoryInformation/GuidanceDocuments/Biotechnology/ucm096149.htm?utm.
24. Mingyao Li, Isabel X. Wang, Yun Li, et al., "Widespread RNA and DNA Sequence Differences in the Human Transcriptome," *Science* 333 (2011): 53–58.

25. G. R. Gaffney, S. F. Lurie, and F. S. Berlin, "Is There Familial Transmission of Pedophilia?," *Journal of Nervous Mental Disorders* 172 (1984): 546–548.
26. Benjamin Weiser, "Court Rejects Judge's Assertion of a Child Pornography Gene," *New York Times,* January 28, 2011.
27. M. Walter, J. Witzel J, C. Wiebking, et al., "Pedophilia Is Linked to Reduced Activation in Hypothalamus and Lateral Prefrontal Cortex during Visual Erotic Stimulation," *Biological Psychiatry* 62 (2007): 698–701.
28. Richards, "How Distinctive Is Genetic Information?," 673.
29. James H. Fowler, Laura A. Baker, and Christopher T. Downs, "Genetic Variation in Political Participation," *American Political Science Review* 102 (2008): 233–248, at 233.
30. Jay Joseph, "The Genetics of Political Attitudes and Behavior: Claims and Refutations," *Ethical Human Psychology and Psychiatry* 12 (2010): 200–217.
31. G. Gottlieb, "Normally Occurring Environmental and Behavioral Influences on Gene Activity: From Central Dogma to Probabilistic Epigenesis," *Psychological Review* 105 (1998): 792–802.
32. James H. Fowler and Christopher T. Dawes, "Two Genes Predict Voter Turnout," *Journal of Politics* 30 (2008): 579–594, at 580.
33. Joseph, "The Genetics of Political Attitudes and Behavior: Claims and Refutations," at 211.
34. Evelyn Fox Keller, *The Century of the Gene* (Cambridge, MA: Harvard University Press, 2000), 62.
35. Ruth Hubbard, "The Theory and Practice of Genetic Reductionism—From Mendel's Laws to Genetic Engineering," in *Towards a Liberatory Biology,* ed. Steven Rose (New York: Allison and Busby, 1982).
36. Ibid., 73.
37. Ruth Hubbard and Elijah Wald, *Exploding the Gene Myth: How Genetic Information Is Produced and Manipulated by Scientists, Physicians, Employers, Insurance Companies, Educators, and Law Enforcers* (Boston: Beacon Press, 1993), 2.
38. Hubbard, "Theory and Practice of Genetic Reductionism," 65.

1. The Mismeasure of the Gene

Epigraph source: S. J. Gould, *The Mismeasure of Man,* rev. ed. (New York: Norton, 1996), 86.

1. J. Loeb, *The Organism as a Whole* (New York: G. P. Putnam's Sons, 1916); W. B. Cannon, *The Wisdom of the Body* (New York: W. W. Norton, 1932); E. Baldwin, *Dynamic Aspects of Biochemistry* (New York: Macmillan, 1948).
2. J. D. Watson and F. H. C. Crick, "A Structure for Deoxyribonucleic Acid," *Nature* 171 (1953): 737–738.
3. J. D. Watson, *The Double Helix* (New York: Athenaeum, 1968).
4. B. Maddox, *Rosalind Franklin* (New York: HarperCollins, 2002).
5. Watson, *Double Helix,* 115.
6. E. Schrödinger, *What Is Life?* (New York: Macmillan, 1945), 20–22.

2. Evolution Is Not Mainly a Matter of Genes

1. Daniel C. Dennett, *Darwin's Dangerous Idea: Evolution and the Meanings of Life* (New York: Simon and Schuster, 1995), 21.
2. According to a 1992 article in *Creation,* a magazine published by Creation Ministries International, "Thirteen species of finches live on the Galápagos, the famous island group visited by Charles Darwin in the 1830s. The finches have a variety of bill shapes and sizes, all suited to their varying diets and lifestyles. The explanation given by Darwin was that they are all the offspring of an original pair of finches, and that natural selection is responsible for the differences. . . . Surprisingly to some, this is the explanation now held by most modern creationists. It would not need to be an 'evolutionary' change at all, in the sense of giving any evidence for amoeba-to-man transformation." See Carl Wieland, "Darwin's Finches: Evidence Supporting Rapid Post-Flood Adaptation," *Creation,* June 1992, 22–23.
3. Isaac Salazar-Ciudad, "On the Origins of Morphological Disparity and Its Diverse Developmental Bases," *BioEssays* 28 (2006): 1112–1122.
4. Stuart A. Newman and Wayne D. Comper, "'Generic' Physical Mechanisms of Morphogenesis and Pattern Formation," *Development* 110 (1990): 1–18; G. Forgacs and S. A. Newman, *Biological Physics of the Developing Embryo* (Cambridge: Cambridge University Press, 2005).
5. Olivia Judson, "Let's Get Rid of Darwinism," *New York Times,* July 15, 2008, http://judson.blogs.nytimes.com/2008/07/15/lets-get-rid-of-darwinism/index.html.
6. William Paley, *Natural Theology* (London: Printed for R. Faulder by Wilks and Taylor, 1802), 1.
7. Immanuel Kant, *Critique of Judgement* [1790], trans. J. H. Bernard (New York: Hafner, 1951), 248.
8. Jean-Baptiste Lamarck, *Zoological Philosophy: An Exposition with Regard to the Natural History of Animals* [1809], trans. Hugh Elliot (Chicago: University of Chicago Press, 1984); discussed in Stuart A. Newman and Ramray Bhat, "Lamarck's Dangerous Idea," in *Transformations of Lamarckism: From Subtle Fluids to Molecular Biology,* ed. Snait Gissis and Eva Jablonka (Cambridge, MA: MIT Press, 2011), 157–169.
9. Gerry Webster and Brian C. Goodwin. *Form and Transformation: Generative and Relational Principles in Biology* (Cambridge: Cambridge University Press, 1996), 8.
10. Robert Shapiro, "Astrobiology: Life's Beginnings," review of *First Life: Discovering the Connections between Stars, Cells, and How Life Began,* by David W. Deamer (Berkeley: University of California Press, 2011), *Nature* 476 (2011): 30–31.
11. Stuart A. Newman, "Evolution: The Public's Problem and the Scientists'," *Capitalism Nature Socialism* 19 (2008): 98–106.
12. Darwin's paternal grandfather, the famous pottery manufacturer and industrial magnate Josiah Wedgwood, tried out as many as 5,000 ceramic formulations before hitting on the iconic Jasper ware, which eventually made his

fortune. See the website of the Wedgwood Museum, Stoke-on-Trent, United Kingdom, http://www.wedgwoodmuseum.org.uk/learning/discovery_packs/2179/pack/2184/chapter/2344.

13. Stuart A. Newman, Gabor Forgacs, and Gerd B. Müller, "Before Programs: The Physical Origination of Multicellular Forms," *International Journal of Developmental Biology* 50 (2006): 289–299.
14. Pierre-Gilles de Gennes, "Soft Matter," *Science* 256 (1992): 495–497.
15. These advances are described throughout Forgacs and Newman, *Biological Physics of the Developing Embryo.*
16. Antonís Rokas, Dirk Krüger, and Sean B. Carroll, "Animal Evolution and the Molecular Signature of Radiations Compressed in Time," *Science* 310 (2005): 1933–1938; S. A. Newman, "The Developmental-Genetic Toolkit and the Molecular Homology-Analogy Paradox," *Biological Theory* 1 (2006): 12–16.
17. Malcolm S. Steinberg, "Cell Adhesive Interactions and Tissue Self-Organization," in *Origination of Organismal Form: Beyond the Gene in Developmental and Evolutionary Biology,* ed. G. B. Müller and S. A. Newman (Cambridge, MA: MIT Press, 2003), 137–163.
18. Michael Krieg, Yohanna Arboleda-Estudillo, Pierre-Henri Puech, et al., "Tensile Forces Govern Germ-Layer Organization in Zebrafish," *Nature Cell Biology* 10 (2008): 429–436.
19. Mary-Lee Dequéant and Olivier Pourquié, "Segmental Patterning of the Vertebrate Embryonic Axis," *Nature Reviews Genetics* 9 (2008): 370–382.
20. The theory of the dynamics of chemical and biochemical oscillations is one of the areas of nonlinear physics, mentioned earlier, that did not exist in Darwin's time or, for that matter, until the mid-twentieth century.
21. Stuart A. Newman and Ramray Bhat, "Dynamical Patterning Modules: A 'Pattern Language' for Development and Evolution of Multicellular Form," *International Journal of Developmental Biology* 53 (2009): 693–705.
22. Ramray Bhat and Stuart A. Newman, "Snakes and Ladders: The Ups and Downs of Animal Segmentation," *Journal of Biosciences* 34, no. 2 (2009): 163–166.
23. Céline Gomez, Ertuğrul M. Özbudak, Joshua Wunderlich, et al., "Control of Segment Number in Vertebrate Embryos," *Nature* 454 (2008): 335–339.
24. James A. Fowler, "Control of Vertebral Number in Teleosts—An Embryological Problem," *Quarterly Review of Biology* 45 (1970): 148–164; David W. Osgood, "Effects of Temperature on the Development of Meristic Characters in *Natrix Fasciata*," *Copeia* 1 (1978): 33–47.
25. See Forgacs and Newman, *Biological Physics of the Developing Embryo,* for additional examples of developmental processes dependent on nonlinear and other self-organizing physical processes, and Newman and Bhat, "Dynamical Patterning Modules," for the role of physicomolecular "dynamical patterning modules" in the early evolution of the animals.
26. See Eva Jablonka and Marion J. Lamb, *Evolution in Four Dimensions: Genetic, Epigenetic, Behavioral, and Symbolic Variation in the History of Life* (Cambridge, MA: MIT Press, 2005), for a discussion of some determinants

that act beyond the level of DNA to influence phenotypes and their evolution, and R. Gorelick and M. D. Laubichler, "Genetic = Heritable (Genetic ≠ DNA)," *Biological Theory* 3 (2008): 79–84, for an attempt to generalize the concept of the gene to include factors other than DNA.

3. Genes as Difference Makers

1. Kim Sterelny and Paul Griffiths, *Sex and Death: An Introduction to Philosophy of Biology* (Chicago: University of Chicago Press, 1999); Lenny Moss, "One, Two (Too?), Many Genes?," *Quarterly Review of Biology* 78 (2003): 57–67.
2. Horace Freeland Judson, "Talking about the Genome," *Nature* 409 (2001): 769. Judson proposes as a solution to this problem that we "revive and put into public use the term 'allele.' Thus, 'the gene for breast cancer' is rather the allele, the gene defect—one of several—that increases the odds that a woman will get breast cancer" (ibid.). It seems to me, however, that the equation of allele with gene defect risks perpetuating precisely the same confusion: it is not the allele itself that is responsible for the phenotypic difference, but the difference between alleles.
3. Moss, "One, Two (Too?), Many Genes?," 60.
4. Quoted in ibid., 62; Wilhelm Johannsen, "Some Remarks about Units of Heredity," *Hereditas* 4 (1923): 140.
5. John Dupré, "Deconstructing the Gene," *Critical Quarterly* 48 (2006): 118.
6. Of course, the concept "normal" is itself fraught with difficulty and is subject to ambiguities that primarily have to do with persistent confusion between properties of individuals and those of populations. But most commonly, it too is understood as a relational property, pertaining to comparison not between individuals but with the statistical norm of a population. Émile Durkheim, e.g., wrote, "L'état de santé, tel qu'elle le peut définir, ne saurait convenir exactement à aucun sujet individuel, puisqu'il ne peut être établi que par rapport aux circonstances les plus communes" (The state of health, insofar as it can be defined, never conforms exactly to that of an individual subject, but can only be established in relation to the most common circumstances). Émile Durkheim, *Les règles de la méthode sociologique* [1894], 16th ed. (Paris: Presses Universitaires de France, 1967), 62. See also Ian Hacking, *The Taming of Chance* (Cambridge: Cambridge University Press, 1990), 160–164.
7. The full quotation in the original reads as follows: "On voit qu'un fait ne peut être qualifié de pathologique que par rapport à une espèce donnée. Les conditions de la santé et de la maladie ne peuvent être définies in abstracto et d'une manière absolue. La règle n'est pas contestée en biologie; il n'est jamais venu à l'esprit de personne que ce qui est normal pour un mollusque le soit aussi pour un vertébré. Chaque espèce a sa santé, parce qu'elle a son type moyen qui lui est propre, et la santé des espèces les plus basses n'est pas moindre que celle des plus élevées. . . . Le type de la santé se confond avec celui de l'espèce. On ne peut même pas, sans contradiction, concevoir une espèce qui, par elle-même et en vertu de sa constitution fondamentale, serait irrémédiablement malade.

Elle est la norme par excellence et, par suite, ne saurait rien contenir d'anormal." Durkheim, *Les règles de la méthode sociologique,* 4.

8. E.g., Canguilhem writes, "We think with Goldstein that the norm concerning pathology is above all an individual norm." Georges Canguilhem, *Essai sur quelques problèmes concernant le normal et le pathologique,* Publications de la Faculté des Lettres de l'Université de Strasbourg (Clermont-Ferrand: Imprimerie "La Montagne," 1943), 72. See also his essay "Le concept et la vie" [1966], in *Etudes d'histoire et de philosophie des sciences,* 3rd ed. (Paris: Vrin. 1975), 364.
9. For an extensive discussion of the relation between the language of medical genetics and that of classical genetics, see Barton Childs, *Genetic Medicine: A Logic of Disease* (Baltimore: Johns Hopkins University Press, 1999).
10. This is not to suggest that maintaining such a diet is an easy task, or that the almost inevitable relapses are not without dire risks. Probably the best discussions of the history and politics of PKU, as well as of risks associated with its treatment, are to be found in Diane B. Paul, "The History of Newborn Phenylketonuria Screening in the U.S.," in *Promoting Safe and Effective Genetic Testing in the United States: Final Report of the Task Force on Genetic Testing,* ed. N. A. Holtzman and M. S. Watson (Baltimore: Johns Hopkins University Press, 1998), 137–160; Diane B. Paul, "A Double-Edged Sword," *Nature* 405 (2000): 515; and Diane B. Paul and Paul J. Edelson, "The Struggle over Metabolic Screening," in *Molecularizing Biology and Medicine: New Practices and Alliances, 1910s–1970s,* ed. S. de Chadarevian and H. Kamminga (Amsterdam: Harwood Academic Publishers, 1998), 203–220.
11. G. A. Jarvis, "Phenylpyruvic Oligophrenia: Introductory Study of 50 Cases of Mental Deficiency Associated with the Excretion of Phenylpyruvic Acid," *Archives of Neurologic Psychiatry* 38 (1937): 944–963; H. Bikel, J. Gerrard, and E.M. Hickmans, "Influence of Phenylalanine Intake on Phenylketonuria," *Lancet* 265 (1953): 812–813.

4. Big B, Little b

1. E. J. Jablonka and M. J. Lamb, *Evolution in Four Dimensions: Genetic, Epigenetic, Behavioral, and Symbolic Variation in the History of Life* (Cambridge, MA: MIT Press, 2005); R. C. Lewontin, *The Triple Helix: Gene, Organism, and Environment* (Cambridge, MA: Harvard University Press, 2000); D. S. Moore, *The Dependent Gene: The Fallacy of "Nature vs. Nurture"* (New York: Times Books, 2002); L. Moss, *What Genes Can't Do* (Cambridge, MA: MIT Press, 2003); D. Noble, *The Music of Life: Biology Beyond Genes* (New York: Oxford University Press, 2006); J. S. Robert, *Embryology, Epigenesis, and Evolution: Taking Development Seriously* (Cambridge: Cambridge University Press, 2004).
2. E. F. Keller, *The Century of the Gene* (Cambridge, MA: Harvard University Press, 2000); E. Pennisi, "DNA Study Forces Rethink of What It Means to Be a Gene," *Science* 316 (2007): 1556–1557.

3. For an example of the belief in blending, see Fleeming Jenkin, review of "The Origin of Species," *North British Review* 46 (1867): 277–318.
4. G. Mendel, "Versuche über Pflanzenhybriden," *Verhandlungen des naturforschenden Vereines in Brünn, Bd. IV für das Jahr 1865,* Abhandlungen, 3–47, 1866.
5. A. H. Sturtevant, "The Behavior of the Chromosomes as Studied through Linkage," *Zeitschrift für induktive Abstammungs- und Vererbungslehre* 13 (1915): 234–287; cited in E. A. Carlson, *The Gene: A Critical History* (Philadelphia: W. B. Saunders Co., 1966), 69.
6. R. A. Sturm and T. N. Frudakis, "Eye Colour: Portals into Pigmentation Genes and Ancestry," *Trends in Genetics* 20 (2004): 327–332, at 327.
7. Ibid.
8. R. Lickliter and H. Honeycutt, "Developmental Dynamics: Toward a Biologically Plausible Evolutionary Psychology," *Psychological Bulletin* 129 (2003): 819–835; M. J. Meaney, "Epigenetics and the Biological Definition of Gene × Environment Interactions," *Child Development* 81 (2010): 41–79.
9. G. Gottlieb, D. Wahlsten, and R. Lickliter, "The Significance of Biology for Human Development: A Developmental Psychobiological Systems View," in *Handbook of Child Psychology,* 5th ed., ed. W. Damon, vol. 1, *Theoretical Models of Human Development,* ed. R. M. Lerner (New York: Wiley, 1998), 233–234; R. Lickliter and H. Honeycutt, "Rethinking Epigenesis and Evolution in Light of Developmental Science," in *Oxford Handbook of Developmental Behavioral Neuroscience,* ed. M. S. Blumberg, J. H. Freeman, and S. R. Robinson (New York: Oxford University Press, 2010), 30–47; Meaney, "Epigenetics and the Biological Definition of Gene × Environment Interactions"; D. S. Moore, "Espousing Interactions and Fielding Reactions: Addressing Laypeople's Beliefs about Genetic Determinism," *Philosophical Psychology* 21 (2008): 331–348; K. Stotz, "With 'Genes' like That, Who Needs an Environment? Postgenomics's Argument for the 'Ontogeny of Information,'" *Philosophy of Science* 73 (2006): 905–917.
10. For an extended argument in support of this claim, see Moore, *Dependent Gene.*
11. Keller, *Century of the Gene;* E. M. Neumann-Held, "The Gene Is Dead—Long Live the Gene: Conceptualizing Genes the Constructionist Way," in *Sociobiology and Bioeconomics: The Theory of Evolution in Biological and Economic Theory,* ed. P. Koslowski (Berlin: Springer-Verlag, 1998), 105–137; Stotz, "With 'Genes' like That."
12. F. S. Collins, "Medical and Societal Consequences of the Human Genome Project" (Shattuck Lecture), *New England Journal of Medicine* 341 (1999): 28–37.
13. For an example of this argument, see S. Oyama, *The Ontogeny of Information: Developmental Systems and Evolution,* 2nd ed. (Durham, NC: Duke University Press, 2000).
14. Noble, *Music of Life,* 7–8.

15. For more information on this phenomenon, see D. L. Black, "Mechanisms of Alternative Pre–Messenger RNA Splicing," *Annual Review of Biochemistry* 72 (2003): 291–336.
16. R. Gray, "Death of the Gene: Developmental Systems Strike Back," in *Trees of Life: Essays in Philosophy of Biology,* ed. P. Griffiths (Dordrecht: Kluwer Academic Publishers, 1992); Noble, *Music of Life.*
17. E. T. Wang, R. Sandberg, S. Luo, et al., "Alternative Isoform Regulation in Human Tissue Transcriptomes," *Nature* 456 (2008): 470–476, at 470.
18. J. M. Johnson, J. Castle, P. Garrett-Engele, et al., "Genome-Wide Survey of Human Alternative Pre-mRNA Splicing with Exon Junction Microarrays," *Science* 302 (2003): 141–144; Moore, "Dependent Gene," 80.
19. Q. Pan, O. Shai, L. J. Lee, et al., "Deep Surveying of Alternative Splicing Complexity in the Human Transcriptome by High-Throughput Sequencing," *Nature Genetics* 40 (2008): 1413–1415; Wang et al., "Alternative Isoform Regulation."
20. M. E. Dinger, K. C. Pang, T. R. Mercer, et al., "Differentiating Protein-Coding and Noncoding RNA: Challenges and Ambiguities," *PLoS Computational Biology* 4 (2008): 1–5, e1000176. doi:10.1371/journal.pcbi.1000176.
21. Keller, *Century of the Gene,* 69; Pennisi, "DNA Study Forces Rethink."
22. Jablonka and Lamb, *Evolution in Four Dimensions.*
23. E. F. Keller, *The Mirage of a Space between Nature and Nurture* (Durham, NC: Duke University Press, 2010).
24. X. Estivill, "Complexity in a Monogenic Disease," *Nature Genetics* 12 (1996): 348–350; C. R. Scriver and P. J. Waters, "Monogenic Traits Are Not Simple: Lessons from Phenylketonuria," *Trends in Genetics* 15 (1999): 267–272.
25. Moss, *What Genes Can't Do,* 44–45.
26. P. E. Griffiths and K. Stotz, "Genes in the Postgenomic Era," *Theoretical Medicine and Bioethics* 27 (2006): 499–521; H. Rolston, "What Is a Gene? From Molecules to Metaphysics," *Theoretical Medicine and Bioethics* 27 (2006): 499–521.
27. S. F. Gilbert and D. Epel, *Ecological Developmental Biology* (Sunderland, MA: Sinauer Associates, 2009); F. Ramus, "Genes, Brain, and Cognition: A Roadmap for the Cognitive Scientist," *Cognition* 101 (2006): 247–269.

5. The Myth of the Machine-Organism

1. Jacques E. Dumont, Fréderic Pécasse, and Carine Maenhaut, "Crosstalk and Specificity in Signalling: Are We Crosstalking Ourselves into General Confusion?," *Cellular Signalling* 13 (2001): 457–463, at 457; Emmanuel D. Levy, Christian R. Landry, and Stephen W. Michnick, "Signaling through Cooperation," *Science* 328 (2010): 983–984, at 983.
2. Bruce J. Mayer, Michael L. Blinov, and Leslie M. Loew, "Molecular Machines or Pleiomorphic Ensembles: Signaling Complexes Revisited," *Journal of Biology* 8 (2009): 81.1–81.8, at 81.2.

3. Richard Dawkins, *The Blind Watchmaker: Why the Evidence of Evolution Reveals a Universe without Design* (New York: W. W. Norton, 1996), 171.
4. Stephen Rothman, *Lessons from the Living Cell: The Limits of Reductionism* (New York: McGraw-Hill, 2002), 265.
5. Barry J. Grant, Alemayehu A. Gorfe, and J. Andrew McCammon, "Large Conformational Changes in Proteins: Signaling and Other Functions," *Current Opinion in Structural Biology* 20 (2010): 142–147, at 143; Vladimir N. Uversky, "The Mysterious Unfoldome: Structureless, Underappreciated, yet Vital Part of Any Given Proteome," *Journal of Biomedicine and Biotechnology,* article ID 568068 (2010), doi:10.1155/2010/568068, at 2; Jörg Gsponer and M. Madan Babu, "The Rules of Disorder, or, Why Disorder Rules," *Progress in Biophysics and Molecular Biology* 99 (2009): 94–103.
6. A. Keith Dunker, Christopher J. Oldfield, Jingwei Meng, et al., "The Unfoldomics Decade: An Update on Intrinsically Disordered Proteins," *BMC Genomics* 9, suppl. 2 (2008): 1–26, at 1–2.
7. Fedor Kouzine and David Levens, "Supercoil-Driven DNA Structures Regulate Genetic Transactions," *Frontiers in Bioscience* 12 (2007): 4409–4423, at 4409.
8. Peter Fraser and James Douglas Engel, "Constricting Restricted Transcription: The (Actively?) Shrinking Web," *Genes and Development* 20 (2006): 1379–1383, at 1379.
9. Jonathan B. Chaires, "Allostery: DNA Does It Too," *ACS Chemical Biology* 3, no. 4 (2008): 207–209, at 207; Christophe Lavelle, "Forces and Torques in the Nucleus: Chromatin under Mechanical Constraints," *Biochemistry and Cell Biology* 87 (2009): 307–322, at 308.
10. Tom Tullius, "DNA Binding Shapes Up," *Nature* 461 (2009): 1225–1226, at 1225.
11. For more on genes, DNA, and chromosomes, with a special emphasis on the epigenetic revolution, see my essays at "What Do Organisms Mean?," http://natureinstitute.org/txt/st/mqual.
12. Urs Albrecht and Jürgen A. Ripperger, "Clock Genes," http://unifr.ch/biochem/assets/files/albrecht/publications/AlbrechtRipperger.pdf.
13. Daniel R. Hyduke and Bernhard Ø. Palsson, "Towards Genome-Scale Signalling-Network Reconstructions," *Nature Reviews Genetics* 11 (2010): 297–307 at 297, doi:10.1038/nrg2750, emphasis added; Barbara McClintock, "The Significance of Responses of the Genome to Challenge," Nobel lecture (December 8, 1983), 193, http://www.nobelprize.org/nobel_prizes/medicine/laureates/1983/mcclintock-lecture.pdf, emphasis added.
14. The distinction between the genes in these two stories is not neat or clean. For example, the idea of information is routinely incorporated into the causal story, if not always coherently. In any case, the distinction I am drawing has at least a little in common with Lenny Moss's distinction between "Genes-D" and "Genes-P." See Lenny Moss, *What Genes Can't Do* (Cambridge, MA: MIT Press, 2003).
15. Timothy Lenoir, *The Strategy of Life: Teleology and Mechanics in Nineteenth-Century German Biology* (Chicago: University of Chicago Press, 1982); E. S.

Russell, *The Interpretation of Development and Heredity: A Study in Biological Method* (1930; reprint, Freeport, NY: Books for Libraries Press, 1972).

16. Paul Weiss, "The Cell as Unit," *Journal of Theoretical Biology* 5 (1963): 389–397, at 395.
17. Paul Weiss, "From Cell to Molecule," lecture (1960), reprinted in *The Molecular Control of Cellular Activity,* ed. John M. Allen (New York: McGraw-Hill, 1962), 1–72, at 1.
18. Russell, *Interpretation of Development and Heredity,* 287.
19. Ibid., 267–268; M. Delbrück, "Aristotle-totle-totle," in *Of Microbes and Life,* ed. Jacques Monod and Ernest Borek (New York: Columbia University Press, 1971), 50–55, at 55.
20. Weiss, "From Cell to Molecule," 3.
21. Weiss, "Cell as Unit," 396.
22. Weiss, "From Cell to Molecule," 6.
23. Weiss, "Cell as Unit," 395.
24. Richard P. Feynman, Robert B. Leighton, and Matthew Sands, *The Feynman Lectures on Physics,* vol. 1 (Reading, MA: Addison-Wesley, 1962), 4-2.
25. Sean B. Carroll, *Endless Forms Most Beautiful: The New Science of Evo Devo and the Making of the Animal Kingdom* (New York: W. W. Norton), 2005.
26. Steve Talbott, "Can the New Science of Evo-Devo Explain the Form of Organisms?," *NetFuture,* no. 171 (December 13, 2007); latest version available at http://natureinstitute.org/txt/st/mqual.
27. The patterned coordination of which I speak can be thought of, in an older terminology, as the "formal cause" of the organism. The modern objection to formal causes seems to amount to little more than the complaint that they are not efficient causes—and it is the attempt to picture them as efficient causes that makes them seem mystical.
28. As an example of how an organism can be studied as a unity, see Craig Holdrege, "What Does It Mean to Be a Sloth?," http://natureinstitute.org/nature/sloth.htm. Other examples are available at http://natureinstitute.org/nature.

6. Some Problems with Genetic Horoscopes

1. Richard Dawkins, *The Selfish Gene* (Oxford: Oxford University Press, 1976).
2. Dorothy Nelkin and M. Susan Lindee, *The DNA Mystique: The Gene as a Cultural Icon* (New York: W. H. Freeman, 1995).
3. See the interview in L. Jaroff, "The Gene Hunt," *Time,* March 20, 1989, 62–67.
4. GenePlanet, http://www.geneplanet.com/.
5. Nic Fleming, "Rival Genetic Tests Leave Buyers Confused," *Sunday Times,* September 7, 2008, http://www.timesonline.co.uk/tol/news/uk/science/article4692891.ece.
6. Eva Jablonka and Marion Lamb, *Evolution in Four Dimensions: Genetic, Epigenetic, Behavioral, and Symbolic Variation in the History of Life* (Cambridge, MA: MIT Press, 2005), chap. 2.

7. D. B. Goldstein, "Common Genetic Variation and Human Traits," *New England Journal of Medicine* 360 (2009): 1696–1698.
8. Ibid., 1696.
9. Based on M. J. West-Eberhard, *Developmental Plasticity and Evolution* (New York: Oxford University Press, 2003).
10. Jablonka and Lamb, *Evolution in Four Dimensions;* S. Gilbert and D. Epel, *Ecological Developmental Biology and Epigenesis: An Integrated Approach to Embryology, Evolution, and Medicine.* (Sunderland, MA: Sinauer, 2009).
11. For discussion see Jablonka and Lamb, *Evolution in Four Dimensions.*
12. W. K. Smits, O. P. Kuipers, and J. W. Veening, "Phenotypic Variation in Bacteria: The Role of Feedback Regulation," *Nature Reviews Microbiology* 4 (2006): 259–271.
13. R. B. Wickner, H. K. Edskes, E. D. Ross, et al., "Prion Genetics: New Rules for a New Kind of Gene," *Annual Review of Genetics* 38 (2004): 681–707.
14. G. Grimes and K. Aufderheide, "Cellular Aspects of Pattern Formation: The Problem of Assembly," *Monographs in Developmental Biology* 22 (1991): 1–94.
15. T. Cavalier-Smith, "The Membranome and Membrane Heredity in Development and Evolution," in *Organelles, Genomes, and Eukaryote Phylogeny: An Evolutionary Synthesis in the Age of Genomics,* ed. R. P. Hirt and D. S. Horner (Boca Raton, FL: CRC Press, 2004), 335–351.
16. S. Henikoff and M. Smith, "Histone Variants and Epigenetics," in *Epigenetics,* ed. D. C. Allis, T. Jenuwein, D. Reinberg, et al. (Cold Spring Harbor, NY: Cold Spring Harbor Laboratory Press, 2007), 249–264.
17. M. F. Fraga, E. Ballestar, M. F. Paz, et al., "Epigenetic Differences Arise during the Lifetime of Monozygotic Twins," *Proceedings of the National Academy of Sciences of the United States of America* 102 (2005): 10604–10609.
18. M. D. Anway, A. S. Cupp, M. Uzumcu, et al., "Epigenetic Transgenerational Actions of Endocrine Disruptors and Male Fertility," *Science* 308 (2005): 1466–1469; D. Crews, personal communication.
19. B. R. Carone, L. Fauquier, N. Habib, et al., "Paternally Induced Transgenerational Environmental Reprogramming of Metabolic Gene Expression in Mammals," *Cell* 143 (2010): 1084–1096.
20. E. Bernstein and C. D. Allis, "RNA Meets Chromatin," *Genes and Development* 19 (2005): 1635–1655.
21. K. D. Wagner, N. Wagner, H. Ghanbarian, et al., "RNA Induction and Inheritance of Epigenetic Cardiac Hypertrophy in the Mouse," *Developmental Cell* 14 (2008): 962–969.
22. E. Jablonka and G. Raz, "Transgenerational Epigenetic Inheritance: Prevalence, Mechanisms, and Implications for the Study of Heredity and Evolution," *Quarterly Review of Biology* 84 (2009): 131–176.
23. J. Reinders, B. B. H. Wulff, M. Mirouze, et al., "Compromised Stability of DNA Methylation and Transposon Immobilization in Mosaic Arabidopsis Epigenomes," *Genes and Development* 23 (2009): 939–950; F. K. Teixeira,

F. Heredia, A. Sarazin, et al., "A Role for RNAi in the Selective Correction of DNA Methylation Defects," *Science* 323 (2009): 1600–1604.

24. E. Avital and E. Jablonka, *Animal Traditions: Behavioural Inheritance in Evolution* (Cambridge: Cambridge University Press, 2000).
25. Gilbert and Epel, *Ecological Developmental Biology and Epigenesis.*
26. Avital and Jablonka, *Animal Traditions.*
27. I. Zilber-Rosenberg and E. Rosenberg, "Role of Microorganisms in the Evolution of Animals and Plants: The Hologenome Theory of Evolution," *FEMS Microbiology Reviews* 32 (2008): 723–735.
28. S. Stern, T. Dror, E. Stolovicki, et al., "Transcriptional Plasticity Underlies Cellular Adaptation to Novel Challenge," *Molecular Systems Biology* 3 (2007): 455–469.
29. Goldstein, "Common Genetic Variation and Human Traits."

7. Cancer Genes

Epigraph sources: Peyton Rous, "Surmise and Fact on the Nature of Cancer," *Nature* 183 (1959): 1357–1361, at 1361; Upton Sinclair, *I, Candidate for Governor: And How I Got Licked* [1935] (Berkeley: University of California Press, 1994), 109.

1. Ana M. Soto and Carlos Sonnenschein, "The Somatic Mutation Theory of Cancer: Growing Problems with the Paradigm?," *BioEssays* 26 (2004): 1097–1105.
2. Erwin Schrödinger, *What Is Life? The Physical Aspect of the Living Cell* (New York: Cambridge University Press, 1945).
3. Jacques Monod, *Chance and Necessity: An Essay on the Natural Philosophy of Modern Biology* (New York: Knopf, 1971).
4. D. Hull, *The Philosophy of Biological Science* (Englewood Cliffs, NJ: Prentice-Hall, 1974).
5. See, e.g., Anne V. Buchanan, Samuel Sholtis, Joan Richtsmeier, et al., "What Are Genes 'For' or Where Are Traits 'From'"? What Is the Question?," *BioEssays* 31 (2009): 198–208; Graziano Pesole, "What Is a Gene? An Updated Operational Definition," *Gene* 417 (2008): 1–4; Leslie A. Pray, "What Is a Gene? Colinearity and Transcription Units," *Nature Education* 1 (2008), http://www.nature.com/scitable/topicpage/What-is-a-Gene-Colinearity-and-Transcription-430; Jean Gayon, "The Concept of the Gene in Contemporary Biology: Continuity or Dissolution?," in *The Influence of Genetics on Contemporary Thinking,* ed. Anne Fagot-Largeault, Shahid Rahman, and Juan Manuel Torres (Dordrecht: Springer, 2007), 81–95; section three of Gayon's chapter is titled "What Is a Gene in Contemporary Biology? A Comprehensive Picture"; Mark B. Gerstein, Can Bruce, Joel S. Rozowsky, et al., "What Is a Gene, Post-ENCODE? History and Updated Definition," *Genome Research* 17 (2007): 669–681; Paul E. Griffiths and Karola Stotz, "Genes in the Postgenomic Era," *Theoretical Medicine and Bioethics* 27 (2006): 499–521; Helen Pearson, "What Is a Gene?," *Nature* 441 (2006): 398–401, at 398; Rolston Holmes, "What Is a Gene? From Molecules to

Metaphysics," *Theoretical Medicine and Bioethics* 27 (2006): 471–497; Subhash C. Lakhotia, "What Is a Gene?," *Resonance* 2 (1997): 44–53.

6. Scott F. Gilbert and S. Sarkar, "Embracing Complexity: Organicism for the 21st Century," *Developmental Dynamics* 219 (2000): 1–9; Ana M. Soto and Carlos Sonnenschein, "Emergentism as a Default: Cancer as a Problem of Tissue Organization," *Journal of Biosciences* 30 (2006): 103–118; B. Rosslenbroich, "Outline of a Concept for Organismic Systems Biology," *Seminars in Cancer Biology* 21 (2011): 156–164.
7. Scott F. Gilbert, "Mechanisms for the Environmental Regulation of Gene Expression: Ecological Aspects of Animal Development," *Journal of Biosciences* 30 (2005): 65–74.
8. D. J. Barker, "The Origins of the Developmental Origins Theory," *Journal of Internal Medicine* 261 (2007): 412–417.
9. C. H. Waddington, "The Epigenotype," *International Journal of Epidemiology* 41 (2012): 10–13.
10. W. Bechtel, "The Evolution of Our Understanding of the Cell: A Study in the Dynamics of Scientific Progress," *Studies in History and Philosophy in Science* A 15 (1984): 309–356.
11. Ana M. Soto, Carlos Sonnenschein, and Paul-Antoine Miquel, "On Physicalism and Downward Causation in Developmental and Cancer Biology," *Acta Biotheoretica* 56 (2002): 257–274.
12. Robert C. King, William D. Stansfield, Pamela Khipple Mulligan, et al., *A Dictionary of Genetics* (New York: Oxford University Press, 2006), 285.
13. Carlos Sonnenschein and Ana M. Soto, "Theories of Carcinogenesis: An Emerging Perspective," *Seminars in Cancer Biology* 18 (2008): 372–373.
14. See Percival Pott, *Chirurgical Observations Relative to the Cataract, the Polypus of the Nose, the Cancer of the Scrotum, the Different Kinds of Ruptures, and the Mortification of the Toes and Feet* (London, 1775), 63–68, at 67. A reprint and discussion of this chapter ("Cancer Scrota") appears in M. Potter, "Percival Pott's Contribution to Cancer Research," *National Cancer Institute Monographs* 10 (1963): 1–13.
15. Henry C. Pitot, *Fundamentals of Oncology,* 4th ed. (New York: M. Dekker, 2002), 41.
16. See Theodor Boveri, *Zur Frage der Entstehung maligner Tumoren* [1914]; English translation, *On the Question of the Origin of Malignant Tumors,* trans. and annotated by Henry Harris (Cold Spring Harbor, NY: Cold Spring Harbor Laboratory Press, 2008); also found at the website of the *Journal of Cell Science,* http://jcs.biologists.org/content/121/Supplement_1/1.full; see also Michel Morange, *A History of Molecular Biology,* trans. Matthew Cobb (Cambridge, MA: Harvard University Press, 1998); and Joan H. Fujimura, *Crafting Science: A Sociohistory of the Quest for the Genetics of Cancer* (Cambridge, MA: Harvard University Press, 1996).
17. See Carlos Sonnenschein and Ana M. Soto, *The Society of Cells: Cancer and Control of Cell Proliferation* (New York: Bios Scientific Publishers, 1999).
18. Melinda Cooper, "Regenerative Pathologies: Stem Cells, Teratomas and Theories of Cancer," *Medicine Studies* 1 (2009): 55–66, at 63.

19. Bruce Alberts, Alexander Johnson, Julian Lewis et al., eds., *Molecular Biology of the Cell,* 5th ed. (New York: Garland, 2008), 1103, 1244; Robert A. Weinberg, "Positive and Negative Controls on Cell Growth," *Biochemistry* 28 (1989): 8263–8269.
20. Sonnenschein and Soto, "Theories of Carcinogenesis," 372–373.
21. See ibid.; Ana M. Soto and Carlos Sonnenschein, "The Tissue Organization Field Theory of Cancer: A Testable Replacement for the Somatic Mutation Theory," *BioEssays* 33 (2011): 332–340.
22. C. Shih, B. Z. Shilo, M. P. Goldfarb, et al., "Passage of Phenotypes of Chemically Transformed Cells via Transfection of DNA and Chromatin," *Proceedings of the National Academy of Sciences of the United States of America* 76 (1979): 5714–5718.
23. Christopher Greenman, Philip Stephens, Raffaella Smith, et al., "Patterns of Somatic Mutation in Human Cancer Genomes," *Nature* 446 (2007): 153–158.
24. Ibid., 153.
25. Marco Gerlinger, Andrew J. Rowan, Stuart Horswell, et al., "Intratumor Heterogeneity and Branched Evolution Revealed by Multiregion Sequencing," *New England Journal of Medicine* 366 (2012): 883–892, at 883.
26. Sonnenschein and Soto, "Theories of Carcinogenesis," 373.
27. Carlos Sonnenschein and Ana M. Soto, "The Death of the Cancer Cell," *Cancer Research* 71 (2011): 4334–4337; Soto and Sonnenschein, "Tissue Organization Field Theory of Cancer."
28. Theodosius Dobzhansky, 1973. "Nothing in Biology Makes Sense Except in the Light of Evolution," *American Biology Teacher* 35 (1973): 125–129.
29. Maricel V. Maffini, Ana M. Soto, Janine M. Calabro, et al., "The Stroma as a Crucial Target in Rat Mammary Gland Carcinogenesis," *Journal of Cell Science* 117 (2004): 1495–1502.
30. Maricel V. Maffini, Janine M. Calabro, Ana M. Soto, et al., "Stromal Regulation of Neoplastic Development: Age-Dependent Normalization of Neoplastic Mammary Cells by Mammary Stroma," *American Journal of Pathology* 167 (2005): 1405–1410.
31. B. W. Booth, C. A. Boulanger, L. H. Anderson, et al., "The Normal Mammary Microenvironment Suppresses the Tumorigenic Phenotype of Mouse Mammary Tumor Virus-neu-Transformed Mammary Tumor Cells," *Oncogene* 30 (2011): 679–689.
32. Gina Kolata, "In Long Drive to Cure Cancer, Advances Have Been Elusive," *New York Times,* April 24, 2009.
33. Carlos Sonnenschein and Ana M. Soto, "Why Systems Biology and Cancer?," *Seminars in Cancer Biology* 21 (2011): 147–149.
34. M. Drack and O. Wolkenhauer, "System Approaches of Weiss and Bertalanffy and Their Relevance for Systems Biology Today," *Seminars in Cancer Biology* 21 (2011): 150–155.
35. M. A. O'Malley and J. Dupré, "Fundamental Issues in Systems Biology," *BioEssays* 27 (2005): 1270–1276.

36. Kurt Saetzler, Carlos Sonnenschein, and Ana M. Soto, "Systems Biology beyond Networks: Generating Order from Disorder through Self-Organization," *Seminars in Cancer Biology* 21 (2011): 165–174.
37. Donald E. Ingber and Michael Levin, "What Lies at the Interface of Regenerative Medicine and Developmental Biology?," *Development* 134 (2009): 2541–2547.
38. P. E. Griffiths, "Genetic Information: A Metaphor in Search of a Theory," *Philosophy of Science* 68 (2001): 394–412; G. Longo, "Critique of Computational Reason in the Natural Sciences," in *Fundamental Concepts in Computer Science,* ed. E. Gelenbe and J.-P. Kahane (Singapore: Imperial College Press/World Scientific, 2009), 43–70.
39. Rosslenbroich, "Outline of a Concept for Organismic Systems Biology"; M. Bizzarri, A. Giuliani, A. Cucina, et al., "Fractal Analysis in a Systems Biology Approach to Cancer," *Seminars in Cancer Biology* 21 (2011): 175–182.
40. G. Longo and M. Montévil, "From Physics to Biology by Extending Criticality and Symmetry Breakings," *Progress in Biophysics and Molecular Biology* 106 (2011): 340–347; Paul-Antoine Miquel, "Extended Physics as a Theoretical Framework for Systems Biology?," *Progress in Biophysics and Molecular Biolology* 106 (2011) 348–352.

8. The Fruitless Search for Genes in Psychiatry and Psychology

1. N. Risch, R. Herrell, T. Lehner, et al., "Interaction between the Serotonin Transporter Gene (*5-HTTLPR*), Stressful Life Events, and Risk of Depression," *JAMA* 301 (2009): 2462–2471.
2. A. Caspi, K. Sugden, T. E. Moffitt, et al., "Influence of Life Stress on Depression: Moderation by a Polymorphism in the 5-HTT Gene," *Science* 301 (2003): 386–389.
3. C. Ratner, "Genes and Psychology in the News," *New Ideas in Psychology* 22 (2004): 29–47.
4. J. Egland, D. S. Gerhard, D. L. Pauls, et al., "Bipolar Affective Disorders Linked to DNA Markers on Chromosome 11," *Nature* 325 (1987): 783–787; R. Sherrington, J. Brynjolfsson, H. Petursson, et al., "Localization of a Susceptibility Locus for Schizophrenia on Chromosome 5," *Nature* 336 (1988): 164–167.
5. S. V. Faraone, J. W. Smoller, C. N. Pato, et al., "The New Neuropsychiatric Genetics," *American Journal of Medical Genetics,* Part B, *Neuropsychiatric Genetics* 147B (2008): 1–2, at 1.
6. Robert Plomin, J. C. DeFries, G. E. McClearn, et al., *Behavioral Genetics,* 4th ed. (New York: Worth, 2008).
7. R. P. Ebstein and S. Israel, "Molecular Genetics of Personality: How Our Genes Can Bring Us to a Better Understanding of Why We Act the Way We Do," in *Handbook of Behavior Genetics,* ed. Yong-Kyu Kim (New York: Springer, 2009), 239–250, at 240. See also K. J. H. Verweij, B. P. Zietsch, S. Medland, et al., "A Genome-Wide Association Study of Cloninger's Tem-

perament Scales: Implications for the Evolutionary Genetics of Personality," *Biological Psychology* 85 (2010): 306–317.

8. I. J. Deary, L. Penke, and W. Johnson, "The Neuroscience of Human Intelligence Differences," *Nature Reviews Neuroscience* 11 (2010): 201–211, at 205.
9. I. J. Deary, "Intelligence," *Annual Review of Psychology* 63 (2012): 453–482; J. Joseph, "The Crumbling Pillars of Behavioral Genetics," *GeneWatch* 24 (2011): 4–7.
10. Risch et al., "Interaction," 2463. The term "pathognomonic" in this case refers to a biological marker characteristic or indicative of a particular disease.
11. A. L. Collins, Y. Kim, P. Sklar, et al., "Hypothesis-Driven Candidate Genes for Schizophrenia Compared to Genome-Wide Association Results," *Psychological Medicine* 42 (2012): 607–616, at 614.
12. J. Latham and A. Wilson, "The Great DNA Data Deficit: Are Genes for Disease a Mirage?," *Bioscience Research Project* (2010), http://www.bioscienceresource.org/commentaries/article.php?id=46.
13. E. S. Gershon, N. Alliey-Rodriguez, and C. Liu, "After GWAS: Searching for Genetic Risk for Schizophrenia and Bipolar Disorder," *American Journal of Psychiatry* 168 (2011): 253–256; E. S. Lander, "Initial Impact of the Sequencing of the Human Genome," *Nature* 470 (2011): 187–197; B. Maher, "The Case of the Missing Heritability," *Nature* 456 (2008): 18–21; T. A. Manolio, F. S. Collins, N. J. Cox, et al., "Finding the Missing Heritability of Complex Diseases," *Nature* 461 (2009): 747–753.
14. Maher, "Case of the Missing Heritability," 18.
15. Gershon, Alliey-Rodriguez, and Liu, "After GWAS"; O. Zuk, E. Hechter, S. R. Sunyaev, and E. S. Lander, "The Mystery of Missing Heritability: Genetic Interactions Create Phantom Heritability," *PNAS* 109 (2012): 1193–1198.
16. E. E. Eichler, J. Flint, G. Gibson, et al., "Missing Heritability and Strategies for Finding the Underlying Causes of Complex Disease," *Nature Reviews Genetics* 11 (2010): 446–450, at 446.
17. Manolio et al., "Finding the Missing Heritability."
18. Ibid., 748.
19. C. Chaufan, "How Much Can a Large Population Study on Genes, Environments, Their Interactions and Common Diseases Contribute to the Health of the American People?," *Social Science and Medicine* 65 (2007): 1730–1741.
20. Plomin et al., *Behavioral Genetics,* 70.
21. Ibid., 151.
22. Jay Joseph, *The Gene Illusion: Genetic Research in Psychiatry and Psychology under the Microscope* (New York: Algora, 2004); Jay Joseph, *The Missing Gene: Psychiatry, Heredity, and the Fruitless Search for Genes* (New York: Algora, 2006); K. Kendler, "Overview: A Current Perspective on Twin Studies of Schizophrenia," *American Journal of Psychiatry* 140 (1983): 1413–1425.
23. Although many twin researchers believe that treatment similarity based on physical appearance should be counted as a genetic effect and therefore does

not violate the equal environment assumption of the twin method, the physical resemblance of monozygotic twin pairs is the result of the splitting of a fertilized egg. Thus, although this physical resemblance is based on the pair's identical genotype, it is not an inherited characteristic because the same physical resemblance will be present regardless of who their biological parents are, or in whose womb they develop. See J. Joseph, "The Equal Environment Assumption of the Classical Twin Method: A Critical Analysis," *Journal of Mind and Behavior* 19 (1998): 325–358.

24. E. Charney, "Behavior Genetics and Post Genomics," *Behavioral and Brain Sciences* (forthcoming).
25. T. J. Bouchard Jr. and M. McGue, "Genetic and Environmental Influences on Human Psychological Differences," *Journal of Neurobiology* 54 (2003): 4–45; Kendler, "Overview"; David Rowe, *The Limits of Family Influence: Genes, Experience, and Behavior* (New York: Guilford, 1994).
26. For a recent example of the "trait relevant" argument in defense of the EEA, see K. Smith, J. R. Alford, P. K. Hatemi, et al., "Biology, Ideology, and Epistemology: How Do We Know Political Attitudes Are Inherited and Why Should We Care?," *American Journal of Political Science* 56 (2012): 17–33.
27. M. J. Lyons, K. Kendler, A. Provet, et al., "The Genetics of Schizophrenia," in *Genetic Issues in Psychosocial Epidemiology,* ed. M. Tsuang, K. Kendler, and M. Lyons (New Brunswick, NJ: Rutgers University Press, 1991), 119–153; J. R. Alford, C. L. Funk, and J. R. Hibbing, "Beyond Liberals and Conservatives to Political Genotypes and Phenotypes," *Perspectives on Politics* 6 (2008): 321–328.
28. Kendler, "Overview"; S. Scarr and L. Carter-Saltzman, "Twin Method: Defense of a Critical Assumption," *Behavior Genetics* 9 (1979): 527–542; J. H. Fowler, L. A. Baker, and C. T. Dawes, "Genetic Variation in Political Participation," *American Political Science Review* 102 (2008): 233–248.
29. D. A. Hay, M. McStephen, and F. Levy, "Introduction to the Genetic Analysis of Attentional Disorders," in *Attention, Genes, and ADHD,* ed. F. Levy and D. Hay (East Sussex, UK: Brunner-Routledge, 2001), 7–34, at 12.
30. N. L. Segal and W. Johnson, "Twin Studies of General Mental Ability," in Kim, *Handbook of Behavior Genetics,* 81–89, at 82.
31. S. O. Lilienfeld, S. J. Lynn, and J. M. Lohr, "Science and Pseudoscience in Clinical Psychology: Initial Thoughts, Reflections, and Considerations," in *Science and Pseudoscience in Clinical Psychology,* ed. S. Lilienfeld, S. Lynn, and J. Lohr (New York: Guilford, 2003), 1–14.
32. Joseph, *Missing Gene;* A. Pam, S. S. Kemker, C. A. Ross, et al., "The 'Equal Environment Assumption' in MZ-DZ Comparisons: An Untenable Premise of Psychiatric Genetics?," *Acta Geneticae Medicae et Gemellologiae* 45 (1996): 349–360.
33. A. S. Reber, *The Penguin Dictionary of Psychology* (London: Penguin, 1985), 123.
34. J. Joseph, "Genetic Research in Psychiatry and Psychology: A Critical Overview," in *Handbook of Developmental Science, Behavior, and Genetics,* ed.

K. Hood, C. Tucker Halpern, G. Greenberg, et al. (Malden, MA: Wiley-Blackwell, 2010), 557–625; J. Joseph, "The 'Missing Heritability' of Psychiatric Disorders: Elusive Genes or Non-Existent Genes?," *Applied Developmental Science* 16 (2012): 65–83. One of the first attempts to use the argument that "twins create their own environment" in support of the twin method was by James Shields in 1954, who wrote, "In so far as binovular [DZ] twins are treated differently from one another and more differently than uniovular [MZ] twins, this is likely to be due, not so much to causes outside the twins as to innate differences in the needs of the binovular twins themselves, manifested by different patterns of behaviour." J. Shields, "Personality Differences and Neurotic Traits in Normal Twin Schoolchildren," *Eugenics Review* 45 (1954): 213–246, at 240. Other examples of researchers invoking this argument since the 1950s include F. J. Kallmann, "The Uses of Genetics in Psychiatry," *Journal of Mental Science* 104 (1958): 542–552; S. V. Vandenberg, "Contributions of Twin Research to Psychology," *Psychological Bulletin* 66 (1966): 327–352; J. C. Loehlin and R. C. Nichols, *Heredity, Environment, and Personality* (Austin: University of Texas Press, 1976); K. S. Kendler, "The Genetics of Schizophrenia: A Current Perspective," in *Psychopharmacology: The Third Generation of Progress,* ed. H. Meltzer (New York: Raven Press, 1987), 705–713; and Rowe, *Limits of Family Influence.*

35. Joseph, *Gene Illusion.*
36. Ibid.; Joseph, *Missing Gene.*
37. D. Jackson, "A Critique of the Literature on the Genetics of Schizophrenia," in *The Etiology of Schizophrenia,* ed. D. Jackson (New York: Basic Books, 1960), 33–87.
38. I. I. Gottesman, *Schizophrenia Genesis* (New York: Freeman, 1991); R. C. Lewontin, S. Rose, and L. J. Kamin, *Not in Our Genes: Biology, Ideology, and Human Nature* (New York: Pantheon, 1984).
39. Jackson, "Critique of the Literature"; Joseph, *Gene Illusion;* C. Ratner, *Vygotsky's Sociohistorical Psychology and Its Contemporary Applications* (New York: Plenum, 1991).
40. J. Joseph, "The Genetics of Political Attitudes and Behavior: Claims and Refutations," *Ethical Human Psychology and Psychiatry* 12 (2010): 200–217.
41. Wilhelm Ostwald, quoted in L. Hogben, *Nature and Nurture* (London: George Allen and Unwin, 1933), 121.
42. T. J. Bouchard Jr., D. T. Lykken, M. McGue, et al., "Sources of Human Psychological Differences: The Minnesota Study of Twins Reared Apart," *Science* 250 (1990): 223–228.
43. S. L. Farber, *Identical Twins Reared Apart: A Reanalysis* (New York: Basic Books, 1981); J. Joseph, "Separated Twins and the Genetics of Personality Differences: A Critique," *American Journal of Psychology* 114 (2001): 1–30; Joseph, *Gene Illusion;* L. Kamin, *The Science and Politics of I.Q.* (Potomac, MD: Erlbaum, 1974); L. J. Kamin and A. S. Goldberger, "Twin Studies in Behavioral Research: A Skeptical View," *Theoretical Population Biology* 61 (2002): 83–95; H. F. Taylor, *The IQ Game: A Methodological Inquiry into*

the Heredity-Environment Controversy (New Brunswick, NJ: Rutgers University Press, 1980).

44. Joseph, *Gene Illusion.*
45. Farber, *Identical Twins Reared Apart.*
46. N. Juel-Nielsen, *Individual and Environment: Monozygotic Twins Reared Apart,* rev. ed. (New York: International Universities Press, 1980), 75; emphasis in original.
47. S. Pinker, *The Blank Slate: The Modern Denial of Human Nature* (New York: Viking, 2002), 33; J. R. Harris, *The Nurture Assumption: Why Children Turn Out the Way They Do* (New York: Touchstone, 1998), 33.
48. R. J. Rose, "Separated Twins: Data and Their Limits," *Science* 215 (1982): 959–960, at 960.
49. Joseph, "Genetic Research in Psychiatry and Psychology."
50. M. McGue and T. J. Bouchard Jr., "Adjustment of Twin Data for the Effects of Age and Sex," *Behavior Genetics* 14 (1984): 325–343.
51. Taking a different approach with a pair sharing only age, sex, and prenatal environment, we could imagine a hypothetical pair of separated-at-birth Arab American female MZAs born in Berkeley, California. One twin is raised in a liberal, nonreligious Berkeley family, while the other is sent to Saudi Arabia and is raised there in a conservative Islamic Saudi Arabian family, and they have no contact with each other and do not know of each other's existence. In Saudi Arabia women experience "gender apartheid" in a strict Islamic society, are not allowed to interact with men much of the time, and are required to be covered in black clothing from head to toe when they are in public. If these twins are reunited for the first time at age 40, how much would we expect them to have in common?
52. Joseph, *Gene Illusion;* Joseph, "Genetic Research in Psychiatry and Psychology."
53. For example, see Bouchard, Lykken, and McGue, "Sources of Human Psychological Differences," 226, table 4.
54. T. J. Bouchard Jr., foreword to *Entwined Lives,* by N. Segal (New York: Dutton, 1999), ix–x, at ix.
55. S. S. Kety, D. Rosenthal, P. H. Wender, et al., "The Types and Prevalence of Mental Illness in the Biological and Adoptive Families of Adopted Schizophrenics," in *The Transmission of Schizophrenia,* ed. D. Rosenthal and S. Kety (New York: Pergamon, 1968): 345–362; S. S. Kety, P. H. Wender, B. Jacobsen, et al., "Mental Illness in the Biological and Adoptive Relatives of Schizophrenic Adoptees: Replication of the Copenhagen Study to the Rest of Denmark," *Archives of General Psychiatry* 51 (1994): 442–455.
56. Joseph, *Gene Illusion;* Joseph, *Missing Gene;* T. Lidz, "Commentary on a Critical Review of Recent Adoption, Twin, and Family Studies of Schizophrenia: Behavioral Genetics Perspective," *Schizophrenia Bulletin* 2 (1976): 402–412; T. Lidz and S. Blatt, "Critique of the Danish-American Studies of the Biological and Adoptive Relatives of Adoptees Who Became Schizophrenic," *American Journal of Psychiatry* 140 (1983): 426–435; T. Lidz, S. Blatt, and B. Cook, "Critique of the Danish-American Studies of the

Adopted-Away Offspring of Schizophrenic Parents," *American Journal of Psychiatry* 138 (1981): 1063–1068; Lewontin et al., *Not in Our Genes*, 218–219.

57. P. Tienari, A. Sorri, I. Lahti, et al., "Genetic and Psychosocial Factors in Schizophrenia: The Finnish Adoptive Family Study," *Schizophrenia Bulletin* 13 (1987): 477–484.
58. M. Stoolmiller, "Implications of the Restricted Range of Family Environments for Estimates of Heritability and Nonshared Environment in Behavior-Genetic Adoption Studies," *Psychological Bulletin* 125 (1999): 392–409; Plomin et al., *Behavioral Genetics;* M. Rutter, *Genes and Behavior: Nature-Nurture Interplay Explained* (Malden, MA: Blackwell, 2006).
59. P. Tienari, I. Lahti, A. Sorri, et al., "The Finnish Adoptive Family Study of Schizophrenia: Possible Joint Effects of Genetic Vulnerability and Family Interaction," in *Understanding Major Mental Disorder: The Contribution of Family Interaction Research,* ed. K. Halweg and M. Goldstein (New York: Family Process Press, 1987), 33–54.
60. Ibid.; Ratner, *Vygotsky's Sociohistorical Psychology.*
61. Joseph, *Gene Illusion,* 228–234, 265–268.
62. Joseph, *Gene Illusion;* E. Hemminki, A. Rasimus, and E. Forssas, "Sterilization in Finland: From Eugenics to Contraception," *Social Science and Medicine* 45 (1997): 1875–1884; M. Hietala, "From Race Hygiene to Sterilization: The Eugenics Movement in Finland," in *Eugenics and the Welfare State: Sterilization Policy in Denmark, Sweden, Norway, and Finland,* ed. G. Broberg and N. Roll-Hansen (East Lansing: Michigan State University Press, 1996), 195–258.
63. Kamin, *Science and Politics of I.Q.;* L. J. Kamin, in *The Intelligence Controversy,* by H. J. Eysenck versus L. J. Kamin (New York: John Wiley and Sons, 1981); H. Munsinger, "The Adopted Child's IQ: A Critical Review," *Psychological Bulletin* 82 (1975): 623–659.
64. Joseph, *Gene Illusion;* J. Joseph and S. Baldwin, "Four Editorial Proposals to Improve Social Sciences Research and Publication," *International Journal of Risk and Safety in Medicine* 13 (2000): 117–127.
65. Joseph, *Gene Illusion;* Joseph, *Missing Gene;* Joseph, "Genetic Research in Psychiatry and Psychology."
66. Joseph, "Genetic Research in Psychiatry and Psychology."
67. Latham and Wilson, "The Great DNA Data Deficit."
68. R. Plomin, M. J. Owen, and P. McGuffin, "The Genetic Basis of Complex Behaviors," *Science* 264 (1994): 1733–1739, at 1737.
69. Collins et al., "Hypothesis-Driven Candidate Genes for Schizophrenia"; Deary, "Intelligence"; Gershon, Alliey-Rodriguez, and Liu, "After GWAS"; Joseph, "Genetic Research in Psychiatry and Psychology" and "Crumbling Pillars of Behavioral Genetics"; Risch et al., "Interaction"; Segal and Johnson, "Twin Studies of General Mental Ability"; Verweij et al., "Genome-Wide Association Study of Cloninger's Temperament Scales."
70. H. Akil, S. Brenner, E. Kandel, et al., "The Future of Psychiatric Research: Genomes and Neural Circuits," *Science* 327 (2010): 1580–1581, at 1580.

71. See Ratner, *Vygotsky's Sociohistorical Psychology;* C. Ratner, *Macro Cultural Psychology: A Political Philosophy of Mind* (New York: Oxford University Press, 2012); and C. Ratner, "A Cultural-Psychological Analysis of Emotions," *Culture and Psychology* 6 (2000): 5–39.
72. C. Ratner and S. El-Badwi, "A Cultural Psychological Theory of Mental Illness, Supported by Research in Saudi Arabia," *Journal of Social Distress and The Homeless* 20 (2011): 217–274, http://www.sonic.net/~cr2/cult%20psy%20mental%20illness.pdf.
73. E. Turkheimer, "Three Laws of Behavior Genetics and What They Mean," *Current Directions in Psychological Science* 9 (2000): 160–164, at 160.
74. For example, see S. E. Hyman, "Introduction to the Complex Genetics of Mental Disorders," *Biological Psychiatry* 45 (1999): 518–521; and J. B. Potash and J. R. DePaulo Jr., "Searching High and Low: A Review of the Genetics of Bipolar Disorder," *Bipolar Disorders* 2 (2000): 8–26.
75. Turkheimer, "Three Laws of Behavior Genetics," 163.

9. Assessing Genes as Causes of Human Disease in a Multicausal World

1. Michael Scriven, "The Logic of Cause," *Theory and Decision* 2 (1971): 52, 64–65.
2. H. L. A. Hart and A. M. Honoré, *Causation in the Law,* 2nd ed. (Oxford: Clarendon Press, 1985), 114.
3. Michael Scriven, "Causation as Explanation," *Nous* 9 (1975): 3–16; Kenneth Rothman, *Epidemiology* (New York: Oxford University Press, 2002), 9–16; Hart and Honoré, *Causation in the Law;* Joel Feinberg, "Sua Culpa," in *Doing and Deserving,* by Joel Feinberg (Princeton, NJ: Princeton University Press, 1970), 187–221.
4. Feinberg, "Sua Culpa," 203–204.
5. Ibid., 189.
6. On childhood leukemia, see M. F. Greaves, "Aetiology of Acute Leukaemia," *Lancet* 349 (1997): 344–349; and Cliona M. McHale and Martyn T. Smith, "Prenatal Origin of Chromosomal Translocations in Acute Childhood Leukemia: Implications and Future Directions," *American Journal of Hematology* 75 (2004): 254–257; on solutions to better preventing the developmental origins of disease from toxicants and similar problems, see Carl F. Cranor, *Legally Poisoned: How the Law Puts Us at Risk from Toxicants* (Cambridge, MA: Harvard University Press, 2011).
7. *Anderson v. Minneapolis, St. Paul & S. St. M.R.R. Co.,* 146 Minn. 432, 179 N.W. 45 (1920).
8. Jerrold J. Heindel, "Animal Models for Probing the Developmental Basis of Disease and Dysfunction Paradigm," *Basic and Clinical Pharmacology and Toxicology* 102 (2008): 76–81, at 79–80.
9. See PubMed Health, "Phenylketonuria," http://www.ncbi.nlm.nih.gov/pubmedhealth/PMH0002150/.
10. Rothman, *Epidemiology,* 11.
11. Andreas Kortenkamp, "Breast Cancer and Exposure to Hormonally Active Chemicals: An Appraisal of the Scientific Evidence" (background paper pub-

lished by the Health and Environment Alliance and CHEM Trust), 4, 5, http://www.chemtrust.org.uk.

12. Zoltan Gregus and Curtis D. Klaassen, "Mechanisms of Toxicity," in *Casarett and Doull's Toxicology,* 6th ed., ed. Curtis D. Klaassen (New York: McGraw-Hill, 2001), 35–81, at 75–76.
13. Kortenkamp, "Breast Cancer and Exposure to Hormonally Active Chemicals," 4.
14. Ibid., 5.
15. See Helen Weiss, Nancy Potischman, Louise Brinton, et al., "Prenatal and Perinatal Risk Factors for Breast Cancer in Young Women," *Epidemiology* 89 (1997): 181–187, on some risk factors for cancers; and Kortenkamp, "Breast Cancer and Exposure to Hormonally Active Chemicals," 6, and A. S. Robbins and C. A. Clarke, "Regional Changes in Hormone Theory and Breast Cancer Incidence in 2003 in California from 2001 to 2004," *Journal of Clinical Oncology* 25 (2007): 3437–3439, on other evidence.
16. National Cancer Institute, "BRCA1 and BRCA2: Cancer Risk and Genetic Testing," http://www.cancer.gov/cancertopics/factsheet/Risk/BRCA.
17. Heindel, "Animal Models for Probing the Developmental Basis of Disease and Dysfunction Paradigm," 76.
18. P. A. Baird, "Genetics and Health Care: A Paradigm Shift," *Perspectives in Biology and Medicine* 33 (1990): 203–213, at 203, 205, 206, 207.
19. See Theodore Friedman, "Progress toward Human Gene Therapy," *Science* 244 (1989): 1275–1281, on the possibilities of gene therapy; and Theodore Friedman, presentation at the University of California Humanities Research Institute, Irvine, April 19, 1991, on the deliberate choice to use language in a particular manner.
20. Ruth Hubbard, "Genes as Causes," in *The Politics of Women's Biology,* by Ruth Hubbard (New Brunswick, NJ: Rutgers University Press, 1990), 76.
21. Ruth Hubbard and Elijah Wald, *Exploding the Gene Myth: How Genetic Information Is Produced and Manipulated by Scientists, Physicians, Employers, Insurance Companies, Educators, and Law Enforcers* (Boston: Beacon Press, 1993), 73.
22. Ruth Hubbard, "The Mismeasure of the Gene," *Rethinking Marxism* 15 (2003): 515–522. (This article is reprinted as Chapter 1 of the present volume.)
23. Hubbard and Wald, *Exploding the Gene Myth,* 64.
24. Ibid.
25. Ian Hacking, "Philosophers of Experiment," *Proceedings of the Biennial Meeting of the Philosophy of Science Association* 2 (1988): 147–156, at 152.
26. Amy Lavoi, "The Genes in Your Congeniality: Researchers Identify Genetic Influence in Social Networks," Harvard Faculty of Arts and Sciences, January 26, 2009, http://www.fas.harvard.edu/home/news-and-notices/news/press-releases/genes-01262009.shtml; Alex, "The Genes for Violence: 'My Genes Made Me Do It,'" Neatorama, July 2, 2010, http://www.neatorama.com/2010/07/02/the-genes-for-violence-my-genes-made-me-do-it/; David Epstein, "Sports Genes: Who Has the Speed Gene, and Who Doesn't? How Much of Performance Is Genetic? How Did Early Humans Become Ath-

letes? And Can the Perfect Athlete Be Genetically Engineered?," http://sportsillustrated.cnn.com/vault/article/magazine/MAG1169440/index/index.htm; Benjamin Weiser, "Court Rejects Judge's Assertion of a Child Pornography Gene," *New York Times,* January 28, 2011.

10. Autism

1. American Psychiatric Association, *Diagnostic and Statistical Manual of Mental Disorders,* 4th ed., Text Revision (Washington, DC: American Psychiatric Association, 2000), 132.
2. A. T. Cavagnaro, *Autism Spectrum Disorders: Changes in the California Caseload; An Update: June 1987–June 2007* (Sacramento: Department of Developmental Services, California Health and Human Services Agency, 2007).
3. M. King and P. Bearman, "Diagnostic Change and the Increased Prevalence of Autism," *International Journal of Epidemiology* 38 (2009): 1224–1234; I. Hertz-Picciotto and L. Delwiche, "The Rise in Autism and the Role of Age at Diagnosis," *Epidemiology* 20 (2009): 84–90.
4. R. R. Grinker, *Unstrange Minds: Remapping the World of Autism* (New York: Basic Books, 2008).
5. C. Stoltenberg, "Autism Spectrum Disorders: Is Anything Left for the Environment?," *Epidemiology* 22 (2011): 489–490; P. H. Patterson, "Maternal Infection and Immune Involvement in Autism," *Trends in Molecular Medicine* 17 (2011): 389–394; E. M. Roberts, P. B. English, J. K. Grether, et al., "Maternal Residence near Agricultural Pesticide Applications and Autism Spectrum Disorders among Children in the California Central Valley," *Environmental Health Perspectives* 115 (2007): 1482–1489; B. Eskenazi, L. G. Rosas, A. R. Marks, et al., "Pesticide Toxicity and the Developing Brain," *Basic and Clinical Pharmacology and Toxicology* 102 (2008): 228–236.
6. I. Hertz-Picciotto, A. Bergman, B. Fangstrom, et al., "Polybrominated Diphenyl Ethers in Relation to Autism and Developmental Delay: A Case-Control Study," *Environmental Health* 10 (2011): 1–11; M. M. Wiest, J. B. German, D. J. Harvey, et al., "Plasma Fatty Acid Profiles in Autism: A Case-Control Study," *Prostaglandins, Leukotrienes, and Essential Fatty Acids* 80 (2009): 221–227; D. K. Kinney, D. H. Barch, B. Chayka, et al., "Environmental Risk Factors for Autism: Do They Help Cause de Novo Genetic Mutations That Contribute to the Disorder?," *Medical Hypotheses* 74 (2010): 102–106.
7. C. Betancur, "Etiological Heterogeneity in Autism Spectrum Disorders: More than 100 Genetic and Genomic Disorders and Still Counting," *Brain Research* 1380 (2011): 42–77.
8. M. Bucan, B. S. Abrahams, K. Wang, et al., "Genome-Wide Analyses of Exonic Copy Number Variants in a Family-Based Study Point to Novel Autism Susceptibility Genes," *PLoS Genetics* 5 (2009): e1000536; S. Girirajan and E. E. Eichler, "Phenotypic Variability and Genetic Susceptibility to Genomic Disorders," *Human Molecular Genetics* 19 (2010): R176–R187.

9. J. Hallmayer, S. Cleveland, A. Torres, et al., "Genetic Heritability and Shared Environmental Factors among Twin Pairs with Autism," *Archives of General Psychiatry* 68 (2011): 1095–1102.
10. D. B. Campbell, J. S. Sutcliffe, P. J. Ebert, et al. "A Genetic Variant That Disrupts MET Transcription Is Associated with Autism," *Proceedings of the National Academy of Sciences of the United States of America* 103 (2006): 16834–16839.
11. E. Redcay and E. Courchesne, "When Is the Brain Enlarged in Autism? A Meta-analysis of All Brain Size Reports," *Biological Psychiatry* 58 (2005): 1–9.
12. M. R. Herbert, "Large Brains in Autism: The Challenge of Pervasive Abnormality," *Neuroscientist* 11 (2005): 417–440.
13. J. E. Lainhart, "Increased Rate of Head Growth during Infancy in Autism," *JAMA* 290 (2003): 393–394; K. D. Mraz, J. Green, T. Dumont-Mathieu, et al., "Correlates of Head Circumference Growth in Infants Later Diagnosed with Autism Spectrum Disorders," *Journal of Child Neurology* 22 (2007): 700–713; K. D. Mraz, J. Dixon, T. Dumont-Mathieu, et al., "Accelerated Head and Body Growth in Infants Later Diagnosed with Autism Spectrum Disorders: A Comparative Study of Optimal Outcome Children," *Journal of Child Neurology* 24 (2009): 833–845.
14. E. Courchesne, C. M. Karns, H. R. Davis, et al., "Unusual Brain Growth Patterns in Early Life in Patients with Autistic Disorder: An MRI Study," *Neurology* 57 (2001): 245–254; M. R. Herbert, D. A. Ziegler, N. Makris, et al., "Localization of White Matter Volume Increase in Autism and Developmental Language Disorder," *Annals of Neurology* 55 (2004): 530–540.
15. E. Werner and G. Dawson, "Validation of the Phenomenon of Autistic Regression Using Home Videotapes," *Archives of General Psychiatry* 62 (2005): 889–895.
16. M. L. Bauman, "Medical Comorbidities in Autism: Challenges to Diagnosis and Treatment," *Neurotherapeutics* 7 (2010): 320–327; M. R. Herbert, "A Whole-Body Systems Approach to ASD," in *The Neuropsychology of Autism,* ed. D. A. Fein (New York: Oxford University Press, 2011), 499–510.
17. T. Buie, D. B. Campbell, G. J. Fuchs, et al., "Evaluation, Diagnosis, and Treatment of Gastrointestinal Disorders in Individuals with ASDs: A Consensus Report," *Pediatrics* 125, suppl. 1 (2010): S1–S18; T. Buie, G. J. Fuchs III, G. T. Furuta, et al., "Recommendations for Evaluation and Treatment of Common Gastrointestinal Problems in Children with ASDs," *Pediatrics* 125, suppl. 1 (2010): S19–S29.
18. A. J. Whitehouse, M. Maybery, J. A. Wray, et al., "No Association between Early Gastrointestinal Problems and Autistic-Like Traits in the General Population," *Developmental Medicine and Child Neurology* 53 (2011): 457–462; L. W. Wang, D. J. Tancredit, and D. Thomas, "The Prevalence of Gastrointestinal Problems in Children across the United States with Autism Spectrum Disorders from Families with Multiple Affected Members," *Journal of Developmental and Behavioral Pediatrics* 32 (2011): 351–360; D. B. Campbell,

T. M. Buie, H. Winter, et al., "Distinct Genetic Risk Based on Association of MET in Families with Co-occurring Autism and Gastrointestinal Conditions," *Pediatrics* 123 (2009): 1018–1024.

19. M. E. Edelson, "Are the Majority of Children with Autism Mentally Retarded? A Systematic Evaluation of the Data," *Focus on Autism and Other Developmental Disabilities* 21 (2006): 66–82.
20. M. Dawson, I. Soulieres, M. A. Gernsbacher, et al., "The Level and Nature of Autistic Intelligence," *Psychological Science* 18 (2007): 657–662.
21. P. Iversen, *Strange Son* (New York: Riverhead Publishers, 2006).
22. L. K. Curran, C. J. Newschaffer, L. C. Lee, et al., "Behaviors Associated with Fever in Children with Autism Spectrum Disorders," *Pediatrics* 120 (2007): e1386–e1392.
23. M. F. Mehler and D. P. Purpura, "Autism, Fever, Epigenetics and the Locus Coeruleus," *Brain Research Reviews* 59 (2009):388–392.
24. M. Helt, E. Kelley, M. Kinsbourne, et al., "Can Children with Autism Recover? If So, How?," *Neuropsychology Review* 18 (2008): 339–366.
25. National Institute of Mental Health, *Identification of Characteristics Associated with Symptom Remission in Autism* (2010), http://clinicaltrials.gov/ct2/show/study/NCT00938054.
26. M. R. Herbert, "Autism: The Centrality of Active Pathophysiology and the Shift from Static to Chronic Dynamic Encephalopathy," in *Autism: Oxidative Stress, Inflammation, and Immune Abnormalities,* ed. A. Chauhan, V. Chauhan, and T. Brown (Boca Raton, FL: CRC Press, 2009), 343–387.
27. S. Dominus, "The Crash and Burn of an Autism Guru," *New York Times Magazine,* April 20, 2011; A. J. Wakefield, J. M. Puleston, S. M. Montgomery, et al., (2002). "Review Article: The Concept of Entero-colonic Encephalopathy, Autism and Opioid Receptor Ligands," *Alimentary Pharmacology and Therapeutics* 16 (2002): 663–674.
28. P. Ashwood and J. Van de Water, "A Review of Autism and the Immune Response," *Clinical and Developmental Immunology* 11 (2004): 165–174.
29. D. L. Vargas, C. Nascimbene, C. Krishnan, et al., "Neuroglial Activation and Neuroinflammation in the Brain of Patients with Autism," *Annals of Neurology* 57 (2005): 67–81.
30. X. Li, A. Chauhan, A. M. Sheikh, et al., "Elevated Immune Response in the Brain of Autistic Patients," *Journal of Neuroimmunology* 207 (2009): 111–116; J. T. Morgan, G. Chana, C. A. Pardo, et al., "Microglial Activation and Increased Microglial Density Observed in the Dorsolateral Prefrontal Cortex in Autism," *Biological Psychiatry* 68 (2010): 368–376; I. Voineagu, X. Wang, P. Johnston, et al., "Transcriptomic Analysis of Autistic Brain Reveals Convergent Molecular Pathology," *Nature* 474 (2011): 380–384.
31. J. J. Gargus, "Mitochondrial Component of Calcium Signaling Abnormality in Autism," in *Autism: Oxidative Stress, Inflammation, and Immune Abnormalities,* ed. A. Chauhan, V. Chauhan, and T. Brown (Boca Raton, FL: CRC Press, 2009), 207–224; D. Rossignol and R. E. Frye, "Mitochondrial Dysfunction in Autism Spectrum Disorders: A Systematic Review and Meta-analysis," *Molecular Psychiatry* (2009): 1–25, doi: 10.1038/mp. 2010.136.

32. K. B. Wallace and A. A. Starkov, "Mitochondrial Targets of Drug Toxicity," *Annual Review of Pharmacology and Toxicology* 40 (2000): 353–388.
33. S. Melnyk, G. J. Fuchs, E. Schulz, et al., "Metabolic Imbalance Associated with Methylation Dysregulation and Oxidative Damage in Children with Autism," *Journal of Autism and Developmental Disorders* 42 (2012): 367–377; Chauhan, Chauhan, and Brown, *Autism;* K. Bowers, Q. Li, J. Bressler, et al., "Glutathione Pathway Gene Variation and Risk of Autism Spectrum Disorders," *Journal of Neurodevelopmental Disordorders* 3 (2011): 132–143; L. Palmieri and A. M. Persico, "Mitochondrial Dysfunction in Autism Spectrum Disorders: Cause or Effect?," *Biochimica et Biophysica Acta* 1797 (2010): 1130–1137.
34. Melnyk et al., "Metabolic Imbalance"; R. Deth, C. Muratore, J. Benzecry, et al., "How Environmental and Genetic Factors Combine to Cause Autism: A Redox/Methylation Hypothesis," *Neurotoxicology* 29 (2008): 190–201; S. J. James, S. Melnyk, S. Jernigan, et al., "Metabolic Endophenotype and Related Genotypes Are Associated with Oxidative Stress in Children with Autism," *American Journal of Medical Genetics* B, *Neuropsychiatric Genetics* 141B (2006): 947–956.
35. D. F. McFabe, N. E. Cain, F. Boon, et al., "Effects of the Enteric Bacterial Metabolic Product Propionic Acid on Object-Directed Behavior, Social Behavior, Cognition, and Neuroinflammation in Adolescent Rats: Relevance to Autism Spectrum Disorder," *Behavioral Brain Resesarch* 217 (2011): 47–54.
36. J. L. Jankowsky and P. H. Patterson, "The Role of Cytokines and Growth Factors in Seizures and Their Sequelae," *Progress in Neurobiology* 63 (2001): 125–149.
37. J. J. Gargas, "Genetic Calcium Signaling Abnormalities in the Central Nervous System: Seizures, Migraine, and Autism," *Annals of the New York Academy of Sciences* 1151 (2009): 133–156; A. Aubert, R. Costalat, and R. Valabregue, "Modelling of the Coupling between Brain Electrical Activity and Metabolism," *Acta Biotheoretica* 49 (2001): 301–326; O. Kahn and R. Kovacs, "Mitochondria and Neuronal Activity," *American Journal of Physiology—Cell Physiology* 292 (2007): C641–C657; M. P. Mattson, M. Gleichmann, and A. Cheng, "Mitochondria in Neuroplasticity and Neurological Disorders," *Neuron* 60 (2008): 748–766; J. W. Pan, A. Williamson, I. Cavus, et al., "Neurometabolism in Human Epilepsy," *Epilepsia* 49, suppl. 3 (2008): 31–41.
38. J. Ochoa-Reparaz, D. W. Mielcarz, S. Begum-Haque, et al., "Gut, Bugs, and Brain: Role of Commensal Bacteria in the Control of Central Nervous System Disease," *Annals of Neurology* 69 (2011): 240–247; A. Gonzalez, J. Stombaugh, C. Lozupone, et al., "The Mind-Body-Microbial Continuum," *Dialogues in Clinical Neuroscience* 13 (2011): 55–62.
39. R. A. Muller, P. Shih, B. Keehn, et al., "Underconnected, but How? A Survey of Functional Connectivity MRI Studies in Autism Spectrum Disorders," *Cerebral Cortex* 21 (2011): 2233–2243; S. Wass, "Distortions and Disconnections: Disrupted Brain Connectivity in Autism," *Brain and Cognition* 75 (2011): 18–28.

40. A. A. Scott–van Zeeland, B. S. Abrahams, A. I. Alverez-Retuerto, et al., "Altered Functional Connectivity in Frontal Lobe Circuits Is Associated with Variation in the Autism Risk Gene CNTNAP2," *Science Translational Medicine* 2 (2010): 56ra80, doi: 10.1126/scitranslmed.3001344.
41. A. Narayanan, C. A. White, S. Saklayen, et al., "Effect of Propranolol on Functional Connectivity in Autism Spectrum Disorder—A Pilot Study," *Brain Imaging and Behavior* 4 (2010): 189–197.
42. M. E. Hasselmo, C. Linster, M. Patil, et al., "Noradrenergic Suppression of Synaptic Transmission May Influence Cortical Signal-to-Noise Ratio," *Journal of Neurophysiology* 77 (1997): 3326–3339.
43. E. Courchesne, K. Campbell, and S. Solso, "Brain Growth across the Life Span in Autism: Age-Specific Changes in Anatomical Pathology," *Brain Research* 1380 (2011): 138–145.
44. S. R. Dager, S. D. Friedman, H. Petropolous, et al., *Imaging Evidence for Pathological Brain Development in Autism Spectrum Disorders* (Totowa, NJ: Humana Press, 2008).
45. M. P. Anderson, B. S. Hooker, and M. R. Herbert, "Bridging from Cells to Cognition in Autism Pathophysiology: Biological Pathways to Defective Brain Function and Plasticity," *American Journal of Biochemistry and Biotechnology* 4 (2008): 167–176, http://www.scipub.org/fulltext/ajbb/ajbb42167–176.pdf; Herbert, "Autism"; C. A. Pardo and C. G. Eberhart, "The Neurobiology of Autism," *Brain Pathology* 17 (2007): 434–447.
46. R. L. Blaylock and J. Maroon, "Immunoexcitotoxicity as a Central Mechanism in Chronic Traumatic Encephalopathy—A Unifying Hypothesis," *Surgical Neurology International* 2 (2011): 107.
47. M. N. Patel, "Oxidative Stress, Mitochondrial Dysfunction, and Epilepsy," *Free Radical Research* 36 (2002): 1139–1146.
48. J. L. Rubenstein and M. M. Merzenich, "Model of Autism: Increased Ratio of Excitation/Inhibition in Key Neural Systems," *Genes, Brain and Behavior* 2 (2003): 255–267.
49. R. D. Fields, *The Other Brain: From Dementia to Schizophrenia; How New Discoveries about the Brain Are Revolutionizing Medicine and Science* (New York: Simon and Schuster, 2009).
50. M. Aschner, J. W. Allen, H. K. Kimelberg, et al., "Glial Cells in Neurotoxicity Development," *Annual Review of Pharmacology and Toxicology* 39 (1999): 151–173.
51. M. M. Bolton and C. Eroglu, "Look Who Is Weaving the Neural Web: Glial Control of Synapse Formation," *Current Opinion in Neurobiology* 19 (2009): 491–497.
52. A. Vezzani, J. French, T. Bartfai, et al., "The Role of Inflammation in Epilepsy," *Nature Reviews Neurology* 7 (2011): 31–40; G. Seifert, G. Carmignoto, and C. Steinhauser, "Astrocyte Dysfunction in Epilepsy," *Brain Research Reviews* 63 (2010): 212–221; R. T. Johnson, S. M. Breedlove, and C. L. Jordan, "Astrocytes in the Amygdala," *Vitamins & Hormones* 82 (2010): 23–45; G. F. Tian, H. Azmi, T. Takano, et al., "An Astrocytic Basis of Epilepsy," *Nature Medicine* 11 (2005): 973–981.

53. M. K. Belmonte and T. Bourgeron, "Fragile X Syndrome and Autism at the Intersection of Genetic and Neural Networks," *Nature Neuroscience* 9 (2006): 1221–1225.
54. I. Bukelis, F. D. Porter, A. W. Zimmerman, et al., "Smith-Lemli-Opitz Syndrome and Autism Spectrum Disorder," *American Journal of Psychiatry* 164 (2007): 1655–1661.
55. E. Tierney, I. Bukelis, R. E. Thompson, et al., "Abnormalities of Cholesterol Metabolism in Autism Spectrum Disorders," *American Journal of Medical Genetics* B, *Neuropsychiatric Genetics* 141B (2006): 666–668.
56. A. Aneja and E. Tierney, "Autism: The Role of Cholesterol in Treatment," *International Review of Psychiatry* 20 (2008): 165–170.
57. T. Page, "Metabolic Approaches to the Treatment of Autism Spectrum Disorders," *Journal of Autism and Developmental Disorders* 30 (2000): 463–469.
58. C. Gillberg and M. Coleman, *The Biology of the Autistic Syndromes* (Cambridge: Cambridge University Press, 2000).
59. N. Coleman, *Autism: Nondrug Biological Treatments* (New York: Plenum Press, 1989); Gillberg and Coleman, *Biology of the Autistic Syndromes;* T. Page and C. Moseley, "Metabolic Treatment of Hyperuricosuric Autism," *Progress in Neuropsychopharmacoly and Biological Psychiatry* 26 (2002): 397–400.
60. E. Fernell, Y. Watanabe, I. Adolfsson, et al., "Possible Effects of Tetrahydrobiopterin Treatment in Six Children with Autism—Clinical and Positron Emission Tomography Data: A Pilot Study," *Developmental Medicine and Child Neurology* 39 (1997): 313–318.
61. Coleman, *Autism.*
62. Ibid.
63. P. Moretti, T. Sahoo, K. Hyland, et al., "Cerebral Folate Deficiency with Developmental Delay, Autism, and Response to Folinic Acid," *Neurology* 64 (2005): 1088–1090; V. T. Ramaekers and N. Blau, "Cerebral Folate Deficiency," *Developmental Medicine and Child Neurology* 46 (2004): 843–851.
64. Aneja and Tierney, "Autism."
65. M. V. Johnston, "Commentary: Potential Neurobiologic Mechanisms through Which Metabolic Disorders Could Relate to Autism," *Journal of Autism and Developmental Disorders* 30 (2000): 471–473.
66. A. W. Zimmerman, "Commentary: Immunological Treatments for Autism in Search of Reasons for Promising Approaches," *Journal of Autism and Developmental Disorders* 30 (2000): 481–484.
67. S. J. James, "Oxidative Stress and the Metabolic Pathology of Autism," in *Autism: Current Theories and Evidence,* ed. A. W. Zimmerman (New York: Humana Press, 2008).
68. M. C. Dolske, J. Spollen, S. McKay, et al., "A Preliminary Trial of Ascorbic Acid as Supplemental Therapy for Autism," *Progress in Neuropsychopharmacoly and Biological Psychiatry* 17 (1993): 765–774.
69. B. N. Ames, I. Elson-Schwab, and E. A. Silver, "High-Dose Vitamin Therapy Stimulates Variant Enzymes with Decreased Coenzyme Binding Affinity

(Increased K(m)): Relevance to Genetic Disease and Polymorphisms," *American Journal of Clinical Nutrition* 75 (2002): 616–658.

70. M. R. Herbert and K. Weintraub, *The Autism Revolution: Whole Body Strategies for Making Life All It Can Be* (New York: Random House, 2012).
71. M. R. Herbert, "Contributions of the Environment and Environmentally Vulnerable Physiology to Autism Spectrum Disorders," *Current Opinion in Neurology* 23 (2010): 103–110.
72. Herbert, "Autism."
73. M. R. Herbert and M. Anderson, "An Expanding Spectrum of Autism Models: From Fixed Developmental Defects to Reversible Functional Impairments," in *Autism: Current Theories and Evidence,* ed. A. W. Zimmerman (New York: Humana Press, 2008), 429–463.
74. E. Thelen, "Development as a Dynamic System," *Current Directions in Psychological Science* 1 (1992): 189–193, at 191; L. B. Smith and E. Thelen, "Development as a Dynamic System," *Trends in Cognitive Science* 7 (2003): 343–348.
75. E. M. Bonker and V. G. Breen, *I Am in Here: The Journey of a Child with Autism Who Cannot Speak but Finds Her Voice* (Ada, MI: Revell, 2011).
76. Herbert and Weintraub, *Autism Revolution.*

11. The Prospects of Personalized Medicine

1. Francis S. Collins, "Medical and Societal Consequences of the Human Genome Project" (Shattuck Lecture), *New England Journal of Medicine* 341 (1999): 28–37, at 28, 36.
2. When the field emerged in the 1950s, researchers initially used the term "pharmacogenetics." In the 1990s a convergence of interests rebranded the field as "pharmacogenomics." See Adam M. Hedgecoe, "Terminology and the Construction of Scientific Disciplines: The Case of Pharmacogenomics," *Science, Technology and Human Values* 28 (2003): 513–537. Although some authors try to distinguish the two terms, they are essentially interchangeable now. Since "pharmacogenomics" is now in wider use, I use that term throughout.
3. Margaret A. Hamburg and Francis S. Collins, "The Path to Personalized Medicine," *New England Journal of Medicine* 363 (2010): 301–304, at 301, 304. The apostrophes are in the original.
4. Harold Varmus, "Ten Years On—The Human Genome and Medicine," *New England Journal of Medicine* 362 (2010): 2028–2029; Nicholas Wade, "Decades Later, Genetic Map Yields Few Cures," *New York Times,* June 12, 2010; Andrew Pollack, "Awaiting the Genome Payoff," *New York Times,* June 14, 2010.
5. Patrick W. Kleyn and Elliot S. Vessel, "Genetic Variation as a Guide to Drug Development," *Science* 281 (1998): 1820–1821.
6. For early classic reviews of the importance of the environment and of noncompliance, see A. H. Conney and J. J. Burns, "Metabolic Interactions among Environmental Chemicals and Drugs," *Science* 178 (1972): 576–586;

D. W. T. Haynes and D. L. Sackett, *Compliance in Health Care* (Baltimore: Johns Hopkins University Press, 1979).

7. Archibald E. Garrod, "The Incidence of Alkaptonuria: A Study in Chemical Individuality," *Lancet* 160 (1902): 1616–1620, at 1620.
8. For discussions of this early history, see Werner Kalow, *Pharmacogenetics: Heredity and the Response to Drugs* (Philadelphia: W. B. Saunders Company, 1962), 1. For primaquine, see P. E. Carson, C. L. Flanagan, C. E. Ickes, et al., "Enzymatic Deficiency in Primaquine-Sensitive Erythrocytes," *Science* 124 (1956): 484–485. For isoniazid, see H. B. Hughes, J. P. Biehl, A. P. Jones, et al., "Metabolism of Isoniazid in Man as Related to the Occurrence of Peripheral Neuritis," *Annual Review of Tuberculosis* 70 (1954): 266–273; and D. A. Price Evans, K. A. Manley, and V. A. McKusick, "Genetic Control of Isoniazid Metabolism in Man," *British Medical Journal* 2 (1960): 485–491.
9. W. Kalow and N. Staron, "On Distribution and Inheritance of Atypical Forms of Human Serum Cholinesterase, as Indicated by Dibucaine Numbers," *Canadian Journal of Biochemistry and Physiology* 35 (1957): 1305–1320; Kalow, *Pharmacogenetics.* For a discussion, see David S. Jones, "How Personalized Medicine Became Genetic, and Racial: Werner Kalow and the Formations of Pharmacogenetics," *Journal of the History of Medicine and Allied Science* (2011): 1–48, doi: 10.1093/jhmas/jrr046.
10. Arno Motulsky, "Drug Reactions, Enzymes, and Biochemical Genetics," *JAMA* 165 (1957): 835–837; F. Vogel, "Moderne Probleme der Humangenetik," *Ergebnisse der inneren Medizin und Kinderheilkunde* 12 (1959): 52–125; Kalow, *Pharmacogenetics.*
11. For the slow development, see Werner Kalow, "Pharmacogenetics: Past and Future," *Life Sciences* 47 (1990): 1385–1397. On malignant hyperthermia, see B. A. Britt, W. G. Locher, and W. Kalow, "Hereditary Aspects of Malignant Hyperthermia," *Clinical Pharmacology and Therapeutics* 16 (1969): 89–98. For a twin study, see E. S. Vesell and J. G. Page, "Genetic Control of Drug Levels in Man: Phenylbutazone," *Science* 159 (1968): 1479–1480.
12. For a review of drug metabolism, see Richard Weinshilboum, "Inheritance and the Drug Response," *New England Journal of Medicine* 348 (2003): 529–537. For the P450 system and its history, see Bernard B. Brodie, Julius Axelrod, Jack R. Cooper, et al., "Detoxification of Drugs and Other Foreign Compounds by Liver Microsomes," *Science* 121 (1955): 603–604; and A. H. Conney, "Induction of Drug-Metabolizing Enzymes: A Path to the Discovery of Multiple Cytochromes P450," *Annual Review of Pharmacology and Toxicology* 43 (2003): 1–30.
13. The growth of the field can be timed by searching Web of Science for titles that include "pharmacogenetics" or "pharmacogenomics." Between 1961 and 1996 somewhere between 1 and 16 articles appeared each year. This grew to 17 in 1997, 33 in 1998, 47 in 1999, 104 in 2000, 201 by 2003, and 320 by 2008. Search performed October 8, 2011.
14. Collins, "Medical and Societal Consequences," 33. For a discussion of tacrine, see Adam M. Hedgecoe, "Pharmacogenetics as Alien Science: Alzheimer's

Disease, Core Sets, and Expectations," *Social Studies of Science* 26 (2006): 723–752.

15. W. E. Evans and M. V. Relling, "Pharmacogenomics: Translating Functional Genomics into Rational Therapeutics," *Science* 286 (1999): 487–491, quotations at 488, 487, and 488.
16. William E. Evans and Howard L. McLeod, "Pharmacogenomics—Drug Disposition, Drug Targets, Side Effects," *New England Journal of Medicine* 348 (2003): 538–549, quotations at 538, 538, and 546. Although the inherited sequence may be stable over a lifetime, levels of gene expression do vary substantially over a lifetime, often in response to environmental exposures.
17. Weinshilboum, "Inheritance and the Drug Response," 535–536, 529.
18. Liewei Wang, Howard L. McLeod, and Richard M. Weinshilboum, "Genomics and Drug Response," *New England Journal of Medicine* 364 (2011): 1144–1153, at 1144.
19. W. Gregory Feero, Alan E. Guttmacher, and Francis S. Collins, "Genomic Medicine—An Updated Primer," *New England Journal of Medicine* 362 (2010): 2001–2011, at 2001.
20. Evans and Relling, "Pharmacogenomics," 490.
21. Allan D. Roses, "Reducing Pipeline Attrition in Clinical Development via Pharmacogenomics," *Drug Discovery and Development* 6 (August 2003): 15.
22. Wang, McLeod, and Weinshilboum, "Genomics and Drug Response," 1151.
23. Hamburg and Collins, "Path to Personalized Medicine," 302, 303–304, 304.
24. "Table of Pharmacogenomics Biomarkers in Drug Labels," http://www.fda.gov/Drugs/ScienceResearch/ResearchAreas/%20Pharmacogenetics/ucm083378.htm. The March figure is an increase from 81 on August 2011.
25. Varmus, "Ten Years On," 2029, 2028, 2029.
26. The most sustained social science analysis of pharmacogenomics has been provided by Adam Hedgecoe. In a book and a series of articles he has highlighted many aspects of the promised revolution. He has shown how researchers massage the rhetoric of expectations to deflect pharmacogenomic attention away from the usual ethical concerns that have complicated genetic medicine: Adam M. Hedgecoe and Paul Martin, "The Drugs Don't Work: Expectations and the Shaping of Pharmacogenetics," *Social Studies of Science* 33 (2003): 327–364; Hedgecoe, *The Politics of Personalized Medicine: Pharmacogenetics in the Clinic* (Cambridge: Cambridge University Press, 2004). He has traced the shift from pharmacogenetics to pharmacogenomics as an example of researchers' efforts to capitalize on the hype that surrounds the Human Genome Project: Hedgecoe, "Terminology and the Construction of Scientific Disciplines." He has used the example of tacrine for Alzheimer's disease to show how misunderstandings of potential efficacy propagate outside the core set of researchers and contribute to false expectations of pharmacogenomic prospects in the media and general public: Hedgecoe, "Pharmacogenetics as Alien Science." His work, as a whole, helps answer why pharmacogenetics became popular in the late 1990s and early 2000s in the absence of useful clinical technologies. Its early success depended on

promises and expectations of future value. Why did enthusiasm among clinicians not match that of scientists? Myriad "socioethical factors" left clinicians ambivalent about the cost and utility of pharmacogenetics even as funders, regulators, and patients remained enthusiastic. See Hedgecoe, "From Resistance to Usefulness: Sociology and the Clinical Uses of Genetic Tests," *BioSocieties* 3 (2008): 183–194; and Hedgecoe, "Bioethics and the Reinforcement of Socio-technical Expectations," *Social Studies of Science* 40 (2010): 163–186.

27. Hamburg and Collins, "Path to Personalized Medicine," 301.
28. Varmus, "Ten Years On," 2029.
29. Nina P. Paynter, Daniel I. Chasman, Guillaume Paré, et al., "Association between a Literature-Based Genetic Risk Score and Cardiovascular Events in Women," *JAMA* 303 (2010): 631–637; Wade, "Decades Later, Genetic Map Yields Few Cures."
30. For nelfinavir and tacrine, see Evans and McLeod, "Pharmacogenomics." For clopidogrel, see Wang, McLeod, and Weinshilboum, "Genomics and Drug Response."
31. Hamburg and Collins, "Path to Personalized Medicine," 303; Wang, McLeod, and Weinshilboum, "Genomics and Drug Response," 1151.
32. Collins, "Medical and Societal Consequences," 35.
33. Robert B. Diasio, Troy L. Beavers, and John T. Carpenter, "Familial Deficiency of Dihydropyrimidine Dehydrogenase: Biochemical Basis for Familial Pyrimidinemia and Severe 5-Fluorouracil-Induced Toxicity," *Journal of Clinical Investigation* 81 (1988): 47–51; Weinshilboum, "Inheritance and the Drug Response," 533.
34. Eugene Y. Krynetski and William E. Evans, "Pharmacogenetics of Cancer Therapy: Getting Personal," *American Journal of Human Genetics* 63 (1998): 11–16.
35. William E. Evans, Mary V. Relling, John D. Rodman, et al., "Conventional Compared with Individualized Chemotherapy for Childhood Acute Lymphoblastic Leukemia," *New England Journal of Medicine* 338 (1998): 499–505. A similar benefit was not seen when they tested clearance and then individualized the dosages of teniposide or cytarabine.
36. For the promise of gene chips, see Evans and Relling, "Pharmacogenomics," 491. For the results, see Ching-Hon Pui, Mary V. Relling, and James R. Dowling, "Acute Lymphoblastic Leukemia," *New England Journal of Medicine* 350 (2004): 1535–1548.
37. This story has now been told many times. The following narrative was developed from David G. Savage and Karen H. Antman, "Imatinib Mesylate—A New Oral Targeted Therapy," *New England Journal of Medicine* 346 (2002): 683–693; Hamburg and Collins, "Path to Personalized Medicine," 303 (table); Siddhartha Mukherjee, *The Emperor of All Maladies: A Biography of Cancer* (New York: Scribner, 2010), 430–440; and Ultan McDermott, James R. Downing, and Michael R. Stratton, "Genomics and the Continuum of Cancer Care," *New England Journal of Medicine* 346 (2011): 340–350, at 343.
38. Wang, McLeod, and Weinshilboum, "Genomics and Drug Response."

39. Marcia Angell, *The Truth about Drug Companies: How They Deceive Us and What to Do about It* (New York: Random House, 2004), 62–64; Mukherjee, *Emperor of All Maladies*, 430–440.
40. Robin K. Kelley and Alan P. Venook, "Nonadherence to Imatinib during an Economic Downturn," *New England Journal of Medicine* 363 (2010): 596–598.
41. McDermott, Downing, and Stratton, "Genomics and the Continuum of Cancer Care," 349.
42. For the St. Louis cancer genome, see Timothy J. Ley, Elaine R. Mardis, Li Ding, et al., "DNA Sequencing of a Cytogenetically Normal Acute Myeloid Leukemia Genome," *Nature* 456 (2008): 66–72. For the lung cancer, see Erin D. Pleasance, Philip J. Stephens, Sarah O'Meara, et al., "A Small-Cell Lung Cancer Genome with Complex Signatures of Tobacco Exposure," *Nature* 463 (2010): 184–190. For the melanoma, see Erin D. Pleasance, R. Keira Cheetham, Philip J. Stephens, et al., "A Comprehensive Catalogue of Somatic Mutations from a Human Cancer Genome," *Nature* 463 (2010): 191–196, quotation at 195.
43. Guruprasad P. Aithal, Christopher P. Day. Patrick J. L. Kesteven, et al., "Association of Polymorphisms in the Cytochrome P450 CYP2C9 with Warfarin Dose Requirement and Risk of Bleeding Complications," *Lancet* 353 (1999): 717–719; International Warfarin Pharmacogenetics Consortium, "Estimation of Warfarin Dose with Clinical and Pharmacogenetic Data," *New England Journal of Medicine* 360 (2009): 753–764; Robert S. Epstein, Thomas P. Moyer, Ronald E. Aubert, et al., "Warfarin Genotyping Reduces Hospitalization Rates," *Journal of the American College of Cardiology* 55 (2010): 2804–2812; Wang, McLeod, and Weinshilboum, "Genomics and Drug Response," 1145.
44. For warfarin pharmacogenomics, see Aithal et al., "Association of Polymorphisms"; and Grant R. Wilkinson, "Drug Metabolism and Variability among Patients in Drug Response," *New England Journal of Medicine* 352 (2005): 2211–2221, at 2219. For warfarin as an exemplar, see Howard L. McLeod, "Pharmacogenetics: More than Skin Deep," *Nature Genetics* 29 (2001): 247–248; Weinshilboum, "Inheritance and the Drug Response."
45. Roberto Padrini and Mariano Ferrari, "Pharmacogenetics," *New England Journal of Medicine* 248 (2003): 2041. This letter was published in response to Weinshilboum's 2003 review to dampen the enthusiasm that he fostered.
46. International Warfarin Pharmacogenetics Consortium, "Estimation of Warfarin Dose," 754; Wang, McLeod, and Weinshilboum, "Genomics and Drug Response," 1145.
47. International Warfarin Pharmacogenetics Consortium, "Estimation of Warfarin Dose," 760.
48. J. Woodcock, "Assessing the Clinical Utility of Diagnostics Used in Drug Therapy," *Clinical Pharmacology and Therapeutics* 88 (2010): 765–773, especially 770.

49. The 2009 decision: Andrew Pollack, "Gene Test for Warfarin Is Rebuffed," *New York Times,* May 4, 2009. The 2010 revision: Wang, McLeod, and Weinshilboum, "Genomics and Drug Response," 1146. The comparative effectiveness trial: Epstein et al., "Warfarin Genotyping Reduces Hospitalization Rates."
50. Wang, McLeod, and Weinshilboum, "Genomics and Drug Response," 1146–1147.
51. Joseph P. Kitzmiller, David K. Groen, Mitch A. Phelps, et al., "Pharmacogenomic Testing: Relevance in Medical Practice," *Cleveland Clinic Quarterly* 78 (2011): 243–257, at 247–248, 248.
52. Ibid., 248; B. Nhi Beasley, Ellis F. Unger, and Robert Temple, "Anticoagulant Options—Why the FDA Approved a Higher but Not a Lower Dose of Dabigatran," *New England Journal of Medicine* 364 (2011): 1788–1790.
53. For the development of codeine pharmacogenomics, see J. Desmeules, M.-P. Gascon, P. Dayer, et al., "Impact of Environmental and Genetic Factors on Codeine Analgesia," *European Journal of Clinical Pharmacology* 41 (1991): 23–26; and Yvan Gasche, Youssef Daali, Marc Fathi, et al., "Codeine Intoxication Associated with Ultrarapid CYP2D6 Metabolism," *New England Journal of Medicine* 351 (2004): 2827–2831. For the commentary, see Evans and Relling, "Pharmacogenomics," 489.
54. Gasche et al., "Codeine Intoxication," 2827.
55. Evans and Relling, "Pharmacogenomics"; Weinshilboum, "Inheritance and the Drug Response"; Nuffield Council on Bioethics, *Pharmacogenetics: Ethical Issues* (London: Nuffield Council on Bioethics, 2003); Wilkinson, "Drug Metabolism and Variability."
56. For instance, it is not mentioned in Evans and McLeod, "Pharmacogenomics," or Wang, McLeod, and Weinshilboum, "Genomics and Drug Response."
57. For the Toronto infant, see Gideon Koren, James Cairns, David Chitayat, et al., "Pharmacogenetics of Morphine Poisoning in a Breastfed Neonate of a Codeine-Prescribed Mother," *Lancet* 368 (2006): 704. For the posttonsillectomy deaths, see Ulrike M. Stamer, Lan Zhang, and Frank Stüber, "Personalized Therapy in Pain Management: Where Do We Stand?," *Pharmacogenomics* 11 (2010): 843–864, at 845. For the FDA's decision, see Kitzmiller et al., "Pharmacogenomic Testing," 253.
58. Stamer, Zhang, and Stüber, "Personalized Therapy in Pain Management," 859.
59. The following narrative is adapted from Jonathan Kahn, "How a Drug Becomes 'Ethnic': Law, Commerce, and the Production of Racial Categories in Medicine," *Yale Journal of Health Policy, Law, and Ethics* 4 (2004): 1–46; Pamela Sankar and Jonathan Kahn, "BiDil: Race Medicine or Race Marketing?," *Health Affairs* (2005): Web exclusive W5-455–W5-463, http://content.healthaffairs.org/content/early/2005/10/11/hlthaff.w5.455; Jonathan Kahn, "Exploiting Race in Drug Development: BiDil's Interim Model of Pharmacogenomics," *Social Studies of Science* 38 (2008): 737–758; and Gregory M. Dorr and David S. Jones, "Facts and Fictions: BiDil and the Resurgence of

Racial Medicine," *Journal of Law, Medicine and Ethics* 36 (2008): 443–448.

60. "BiDil" (approved FDA labeling information), June 2005, http://www.accessdata.fda.gov/drugsatfda_docs/label/2005/020727lbl.pdf.
61. For problems with the evidence, see G. T. H. Ellison, J. S. Kaufman, R. F. Head, et al., "Flaws in the U.S. Food and Drug Administration's Rationale for Supporting the Development and Approval of BiDil as a Treatment for Heart Failure Only in Black Patients," *Journal of Law, Medicine and Ethics* 36 (2008): 449–457. For Tuskegee in the background, see Susan M. Reverby, "'Special Treatment': BiDil, Tuskegee, and the Logic of Race," *Journal of Law, Medicine and Ethics* 36 (2008): 478–484. For NitroMed's funding, see Keith J. Winstein, "NAACP Presses U.S. on Heart Drug," *Wall Street Journal*, January 25, 2007, A20; Anne Pollock, "Medicating Race: Heart Disease and Durable Preoccupations with Difference" (PhD diss., MIT, 2007), 242–305; and Kahn, "Exploiting Race in Drug Development," 750–751. For patent exploitation, see Kahn, "How a Drug Becomes 'Ethnic'"; and Kahn, "Exploiting Race in Drug Development."
62. Observations by the author at the conference "Race, Pharmaceuticals, and Medical Technology," Center for the Study of Diversity in Science, Technology, and Medicine, Massachusetts Institute of Technology, April 7, 2006. See also Pollock, "Medicating Race," 306–311; and Dorothy E. Roberts, "Is Race-Based Medicine Good for Us? African American Approaches to Race, Biomedicine, and Equality," *Journal of Law, Medicine and Ethics* 36 (2008): 537–545, especially 540–541.
63. For Puckrein's defense, see Gary Puckrein, "BiDil: From Another Vantage Point," *Health Affairs* 25 (2006): W368–W374, at W373, http://content.healthaffairs.org/content/25/5/w368.full.pdf. The strategy of using race in the meantime is described in Richard Tutton, Andrew Smart, Paul A. Martin, et al., "Genotyping the Future: Scientists' Expectations about Race/Ethnicity after BiDil," *Journal of Law, Medicine and Ethics* 36 (2008): 464–470.
64. For the fate of BiDil, see Dorr and Jones, "Facts and Fictions." For NitroMed's layoffs, see David Armstrong, "NitroMed Halts Sale of Drug," *Wall Street Journal*, January 15, 2008.
65. For the search, see "Researchers Identify Gene Variations That May Determine Which Heart Failure Patients Are Likely to Benefit from Treatment with BiDil®," *Business Wire*, March 13, 2006; Keith Ferdinand, "Fixed-Dose Isosorbide Dinitrate Hydralazine: Race-Based Cardiovascular Medicine Benefit or Mirage?," *Journal of Law, Medicine and Ethics* 36 (2008): 458–463, especially 460–461. For NitroMed's hopes, see Jane Kramer, quoted in Turna Ray, "HHS Draft Report Suggests Genetic Test for BiDil; NitroMed Does Not Rule Out Dx," *Pharmacogenomics Reporter*, April 4, 2007, http://www.genomeweb.com/dxpgx/hhs-draft-report-suggests-genetic-test-bidil-nitromed-does-not-rule-out-dx.
66. Dennis M. McNamara, S. William Tam, Michael L. Sabolinski, et al., "Endothelial Nitric Oxide Synthase (NOS3) Polymorphisms in African Americans

with Heart Failure: Results from the A-HeFT Trial," *Journal of Cardiac Failure* 15 (2009): 191–198, quotations at 192, 196.

67. Jones, "How Personalized Medicine Became Genetic, and Racial."
68. Evans and Relling, "Pharmacogenomics," 488. For race in other reviews, see McLeod, "Pharmacogenetics," 247; Evans and McLeod, "Pharmacogenomics," 542, 547; Nuffield Council on Bioethics, *Pharmacogenetics;* Esteban González Burchard, Elad Ziv, Natasha Coyle, et al., "The Importance of Race and Ethnic Background in Biomedical Research and Clinical Practice," *New England Journal of Medicine* 348 (2003): 1170–1175, especially 1173; and Wilkinson, "Drug Metabolism and Variability," 2217. For a popular manifestation of this, see Sally Satel, "I Am a Racial Profiling Doctor: Illness Isn't Colorblind. So Why Is It Taboo for Doctors to Take Note of a Patient's Race?," *New York Times Magazine,* May 5, 2002, 56–58. For a critique, see David S. Jones and Roy H. Perlis, "Pharmacogenetics, Race, and Psychiatry," *Harvard Review of Psychiatry* 14 (March–April 2006): 92–108.
69. For Wood's claim, see Alastair J. J. Wood, "Racial Differences in the Response to Drugs—Pointers to Genetic Differences," *New England Journal of Medicine* 344 (2001): 1393–1395, at 1395. For the contrasting result, see Hong-Guang Zie, Richard B. Kim, Alastair J. J. Wood, et al., "Molecular Basis of Ethnic Differences in Drug Disposition and Balance," *Annual Review of Pharmacology* 41 (2001): 815–850, at 818.
70. Ashwini R. Sehgal, "Overlap between Whites and Blacks in Response to Antihypertensive Drugs," *Hypertension* 43 (March 2004): 566–572, especially 571.
71. For the rise in patents, see Jonathan Kahn, "Race-ing Patents/Patenting Race: An Emerging Political Geography of Intellectual Property in Biotechnology," *Iowa Law Review* 92 (2007): 353–416. For the Nike Air Native, see Richard Lyons, "The Curious Return of 'Race' in 2007," *NatNews,* December 24, 2007, http://groups.yahoo.com/group/NatNews/message/46469; and "Nike N7 Sport Summit," http://niken7.com/n7-event/nike-n7-sport-summit/. For the durable preoccupation, see Pollock, "Medicating Race."
72. Feero, Guttmacher, and Collins, "Genomic Medicine," 2001.
73. This has already happened when Britain's National Health Service has tried to limit access to expensive therapies if the benefit is small or unclear. See Robert Steinbrook, "Saying No Isn't NICE—The Travails of Britain's National Institute for Health and Clinical Excellence," *New England Journal of Medicine* 359 (2008): 1977–1981.
74. McDermott, Downing, and Stratton, "Genomics and the Continuum of Cancer Care," 344.
75. Peter Lipton, preface to Nuffield Council on Bioethics, *Pharmacogenetics,* v.
76. Nuffield Council on Bioethics, *Pharmacogenetics,* xiii.
77. For early recognition of the problem, see D. N. Mohler, D. G. Wallin, and E. G. Dreyfus, "Studies in the Home Treatment of Streptococcal Disease: I. Failure of Patients to Take Penicillin by Mouth as Prescribed," *New England Journal of Medicine* 252 (1955): 1116–1118. For the first influential review, see Haynes and Sackett, *Compliance in Health Care.* For a recent

assessment, see L. Osterberg and T. Blaschke, "Adherence to Medication," *New England Journal of Medicine* 353 (2005): 487–497.

78. Nuffield Council on Bioethics, *Pharmacogenetics,* xx.
79. Soren Holm, "Pharmacogenetics, Race and Global Injustice," *Developing World Bioethics* 8 (2008): 82–88. See also Nuffield Council on Bioethics, *Pharmacogenetics,* xiii.
80. Varmus, "Ten Years On," 2029.

12. The Persistent Influence of Failed Scientific Ideas

1. L. Zenderland, *Measuring Minds: Henry Herbert Goddard and the Origins of American Intelligence Testing* (Cambridge: Cambridge University Press, 1998); H. H. Goddard, *The Kallikak Family: A Study in the Heredity of Feeble-Mindedness* (New York: Macmillan, 1912); R. Dugdale, *The Jukes: A Study in Pauperism, Disease, and Heredity* (New York: G. P. Putnam's Sons, 1877).
2. Zenderland, *Measuring Minds,* 331.
3. H. Garrett and H. Bonner, *General Psychology* (New York: Macmillan, 1961).
4. P. A. Jacobs, M. Brunton, N. M. Melville, et al., "Aggressive Behavior, Mental Sub-normality and the XYY Male," *Nature* 208 (1965): 1351–1352.
5. E. Engel, "The Making of an XYY," *American Journal of Mental Deficiency Research* 77 (1972): 123–127.
6. D. S. Borgaonkar and S. A. Shah, "The XYY Chromosome Male—or Syndrome," *Progress in Medical Genetics* 10 (1974): 135–222.
7. H. A. Witkin, S. A. Mednick, F. Schulsinger, et al., "Criminality in XYY and XXY Men," *Science* 193 (1976): 547–555.
8. P. A. Jacobs, "The William Allan Memorial Award Address: Human Population Cytogenetics; The First Twenty-five Years," *American Journal of Human Genetics* 34 (1982): 689–698.
9. M. J. Gotz, E. C. Johnstone, and S. G. Ratcliffe, "Criminality and Antisocial Behavior in Unselected Men with Sex Chromosome Abnormalities," *Psychological Medicine* 29 (1999): 953–962, at 953, 958.
10. K. Royce, *The XYY Man* (New York: Avon, 1973).
11. J. Q. Wilson and R. J. Herrnstein, *Crime and Human Nature* (New York: Simon and Schuster, 1985).
12. R. Pyeritz, H. Schreier, C. Madansky, et al., "The XYY Male: The Making of a Myth," in *Biology as a Social Weapon,* ed. Ann Arbor Science for the People (Minneapolis: Burgess, 1977), 86–100.
13. Hastings Center, "The XYY Controversy: Researching Violence and Genetics," *Hastings Center Report Special Supplement,* 1980, http://www.jstor.org/stable/3560454.
14. H. G. Brunner, M. Nelen, X. O. Breakefield, et al., "Abnormal Behavior Associated with a Point Mutation in the Structural Gene for Monoamine Oxidase A," *Science* 262 (1993): 578–583; H. G. Brunner, M. R. Nelen, P. van Zandvoort, et al., "X-Linked Borderline Mental Retardation with Prominent

Behavioral Disturbance: Phenotype, Genetic Localization, and Evidence for Disturbed Monoamine Metabolism," *American Journal of Human Genetics* 52 (1993): 1032–1039.

15. V. Morell, "Evidence Found for a Possible 'Aggression Gene,'" *Science* 260 (1993): 1722–1723.
16. G. Cowley and C. Hall, "The Genetics of Bad Behavior," *Newsweek,* November 1, 1993, 57.
17. W. Herbert, "Politics of Biology: How the Nature vs. Nurture Debate Shapes Public Policy and Our View of Ourselves," *U.S. News and World Report,* cover, April 21, 1997.
18. E. Felsenthal, "Man's Genes Made Him Kill, His Lawyers Claim," *Wall Street Journal,* November 1, 1994, B1, B5.
19. S. Blakeslee, "Genetic Questions Are Sending Judges Back to Classroom," *New York Times,* July 9, 1996, C1, C9.
20. A. Caspi, J. McClay, T. E. Moffitt, et al., "Role of Genotype in the Cycle of Violence in Maltreated Children," *Science* 297 (2002): 851–854.
21. C. Morris, A. Shen, K. Pierce, et al., "Deconstructing Violence," *GeneWatch* 20 (2007): 3–10; Z. Prichard, A. MacKinnon, A. F. Jorn, et al., "No Evidence for Interaction between *MAOA* and Childhood Adversity for Antisocial Behavior," *American Journal of Medical Genetics* 147B (2008): 228–232.
22. C. Benbow and J. Stanley, "Sex Differences in Mathematical Ability: Fact or Artifact?," *Science* 210 (1980): 1262–1264.
23. G. B. Kolata, "Math and Sex: Are Girls Born with Less Ability?," *Science* 210 (1980): 1234–1235.
24. "The Gender Factor in Math: A New Study Says Males May Be Naturally Abler than Females," *Time,* December 15, 1980, 57.
25. D. A. Williams and P. King, "Do Males Have a Math Gene?," *Newsweek,* December 15, 1980, 73.
26. B. Beckwith, "He Man and She Woman: Cosmo and Playboy Groove on Genes," *Columbia Journalism Review* 12 (1982): 48–51.
27. J. Beckwith, "Gender and Math Performance: Does Biology Have Implications for Educational Policy?," *Journal of Education* 165 (1983): 158–174.
28. Beckwith, "Gender and Math Performance."
29. L. H. Summers, "Remarks at NBER Conference on Diversifying the Science & Engineering Workforce," Office of the President, Harvard University, January 14, 2005, http://www.president.harvard.edu/speeches/summers_2005/nber.php.
30. D. J. Hemel, "Sociologist Cited by Summers Calls His Talk 'Uninformed,'" *Harvard Crimson,* January 14, 2005, http://www.thecrimson.com/article.aspx?ref=505363.
31. J. S. Hyde and J. E. Mertz, "Gender, Culture, and Mathematics Performance," *Proceedings of the National Academy of Sciences of the United States of America* 106 (2009): 8801–8807; R. Monastersky, "Primed for Numbers," *Chronicle of Higher Education,* March 4, 2006, A1; L. Brody and C. Millis, "Talent Search Research: What Have We Learned?," *High Ability Students* 16 (2005): 97–111.

32. L. Guiso, F. Monte, P. Sapienza, et al., "Diversity, Culture, Gender, and Math," *Science* 320 (2008): 1164–1165; B. A. Nosek, F. L. Smyth, N. Sriram, et al., "National Differences in Gender-Science Stereotypes Predict National Sex Differences in Science and Math Achievement," *Proceedings of the National Academy of Sciences of the United States of America* 106 (2009): 10593–10597.
33. P. D. Evans, S. L. Gilbert, N. Mekel-Bobrov, et al., "Microcephalin, a Gene Regulating Brain Size, Continues to Evolve Adaptively in Humans," *Science* 309 (2005): 1717–1720.
34. M. Balter, "Evolution: Are Human Brains Still Evolving? Brain Genes Show Signs of Selection," *Science* 309 (2005): 1662–1663.
35. N. Wade, "Brain May Still Be Evolving, Studies Hint," *New York Times,* September 9, 2000, http://query.nytimes.com/gst/fullpage.html?res=9B01E1DE1331F93AA3575AC0A9639C8B63&sec=health; R. Kotulak, "Two Evolving Genes May Allow Humans to Become Smarter," *Baltimore Sun,* September 9, 2005, Telegraph Section, 3A.
36. J. Derbyshire, "The Spectre of Difference," *National Review Online,* November 7, 2005, http://www.johnderbyshire.com/Opinions/HumanSciences/specterofdifference.htm.
37. M. Inman, "Human Brains Enjoy Ongoing Evolution," *New Scientist,* September 9, 2005, http://www.newscientist.com/article/dn7974-human-brains-enjoy-ongoing-evolution.html.
38. N. Timpson, J. Heron, G. D. Smith, et al., "Comment on Papers by Evans et al. and Mekel-Bobrov et al. on Evidence for Positive Selection of MCPH1 and ASPM," *Science* 317 (2007): 1036.
39. N. Mekel-Bobrov, D. Posthuma, S. Gilbert, et al., "The Ongoing Adaptive Evolution of *ASPM* and *Microcephalin* Is Not Explained by Increased Intelligence," *Human Molecular Genetics* 16 (2007): 600–608; J. P. Rushton, P. A. Vernon, and T. A. Bons, "No Evidence That Polymorphisms of Brain Regulator Genes *Microcephalin* and *ASPM* Are Associated with General Mental Ability, Head Circumference or Altruism," *Biology Letters* 3 (2007): 157–160.
40. M. Balter, "Bruce Lahn Profile: Links between Brain Genes, Evolution, and Cognition Challenged," *Science* 314 (2006): 1872; A. Regalado, "Head Examined: Scientist's Study of Brain Genes Sparks a Backlash," *Wall Street Journal,* June 16, 2006, http://online.wsj.com/article/SB115040765329081636.html.
41. S. J. Gould, "A Positive Conclusion," in *The Mismeasure of Man,* 2nd ed. (New York: W. W. Norton, 1996), 351–353.
42. C. Dean, "Groups Call for Scientists to Engage the Body Politic," *New York Times,* August 8, 2011, D1.

13. Map Your Own Genes!

1. Theodosius Dobzhansky, "Biology, Molecular and Organismic," *American Zoologist* 4 (1964): 449.

2. See John P. Jackson Jr. and Nadine M. Weidman, *Race, Racism, and Science: Social Impact and Interaction* (Santa Barbara, CA: ABC-CLIO, 2004); M. Susan Lindee, *Suffering Made Real: American Science and the Survivors at Hiroshima* (Chicago: University of Chicago Press, 1994).
3. See Mike Fortun, *Promising Genomics: Iceland and deCODE Genetics in a World of Speculation* (Berkeley: University of California Press, 2008).
4. These categories are common, but Sandra Soo-Jin Lee is exploring how they work for 23andMe in her ongoing study of race and distributive justice in pharmacogenomics research. See also Barbara A. Koenig, Sandra Soo-Jin Lee, and Sarah S. Richardson, eds., *Revisiting Race in a Genomic Age* (New Brunswick, NJ: Rutgers University Press, 2008).
5. http://www.gtldna.com/ancestral-origins-dna-ancestry.html.
6. A flurry of scholarly and medical analysis of DTC genetic testing has appeared over the past few years, including work by a few social scientists and other scholars who have begun to explore this phenomenon in persuasive ways. See, for example, Alison Harvey, "Genetic Risks and Healthy Choices: Creating Citizen-Consumers of Genetic Services through Empowerment and Facilitation," *Sociology of Health and Illness* 32 (2010): 365–381; Paula Sakko, Matthew Reed, Nicky Britten, et al., "Negotiating the Boundary between Medicine and Consumer Culture: Online Marketing of Nutrigenetic Tests," *Social Science and Medicine* 70 (2010): 744–753; T. Caulfield, N. M. Ries, P. N. Ray, et al., "Direct-to-Consumer Genetic Testing: Good, Bad or Benign?," *Clinical Genetics* 77 (2010): 101–105; Amy L. McGuire, Christina M. Diaz, Tao Wang, et al., "Social Networkers' Attitudes toward Direct-to-Consumer Personal Genome Testing," *American Journal of Bioethics* 9 (2008): 3–10; and J. P. Evans, "Recreational Genomics: What's in It for You?," *Genetics in Medicine* 10 (2008): 709–710.
7. Their respective websites are 23andMe, https://www.23andme.com; Navigenics, http://www.navigenics.com; and deCODEme, http://decodediagnostics.com/. See Wayne Hall and Coral Gartner, "Direct-to-Consumer Genome-Wide Scans: Astrologicogenomics or Simple Scams?," *American Journal of Bioethics,* 9 (2009): 54–56; first published online, June 1, 2009, http://dx.doi.org/10.1080/15265160902894021.
8. United States Federal Trade Commission, "At-Home Genetic Tests: A Healthy Dose of Skepticism May Be the Best Prescription," 2006, http://www.ftc.gov/bcp/edu/pubs/consumer/health/hea02.shtm. On state responses, see R. Langreth, "California Orders Stop to Gene Testing," *Forbes,* June 14, 2008, http://www.forbes.com/2008/06/14/stop-gene-testing-biz-healthcare-cz_rl_0614genetest.html; and R. Langreth and M. Herper, "States Crack Down on Online Gene Tests," *Forbes,* April 17, 2008, http://www.forbes.com/2008/04/17/genes-regulation-testing-biz-cx_mh_bl_0418genes.html.
9. The announcement of the Walgreens plan on May 11 was quickly followed by a retraction on May 13, a sequence of events that captured the uncertainty, ambiguity, and market stakes surrounding DTC testing. For

a legal analysis of these events, see Dan Vorhaus, "FDA Puts the Brakes on Pathway-Walgreens Pairing: What's Next for DTC?," May 13, 2010, http://www.genomicslawreport.com/index.php/2010/05/13/fda-puts-the-brakes-on-pathway-walgreens-pairing-whats-next-for-dtc/.

10. See Waxman's announcement and at http://energycommerce.house.gov/index.php?option=com_content&view=article&id=2009:committee-investigates-personal-genetic-testing-kits&catid=122:media-advisories&Itemid=55.
11. A description, the agenda, and full webcasts for both days of the conference, which was held in a Marriott hotel in Hyattsville, MD, are available at http://www.fda.gov/MedicalDevices/NewsEvents/WorkshopsConferences/ucm212830.htm.
12. Vorhaus's law and genomics blog, http://www.genomicslawreport.com/index.php/category/badges/fda-ldt-regulation/, has coverage and links describing the many remaining uncertainties and the mixed stakes in enhanced regulation of these tests. The FDA's plans would lead to more comprehensive oversight not only of genomics DTC testing but also of tests used in transplant medicine, all urine and alcohol testing, blood-clotting protein tests, and radiology devices of some kinds.
13. For a critical discussion of this argument, which the authors do not accept, see Christopher F. C. Jordens, Ian H. Kerridge, and Gabrielle N. Samuel, "'Direct-to-Consumer Personal Genome Testing: The Problem Is Not Ignorance—It Is Market Failure," *American Journal of Bioethics* 9 (2009): 13–15.
14. A. E. Guttmacher, M. E. Porteous, and J. D. McInney, "Educating Health Care Professionals about Genetics and Genomics," *Nature Reviews Genetics* 8 (2007): 151; D. J. Hunter, M. J. Khoury, and J. M. Drazen, "Letting the Genome out of the Bottle—Will We Get Our Wish?," *New England Journal of Medicine* 358 (2008): 105–107.
15. Jordens, Kerridge, and Samuel, "Direct-to-Consumer Personal Genome Testing," 13. See also W. Burke and B. M. Psaty, "Personalized Medicine in the Era of Genomics," *JAMA* 298 (2007): 1682–1684.
16. The GTR website is now operational at http://www.ncbi.nlm.nih.gov/gtr/. The press release, "NIH Announces Genetic Testing Registry," is on the NIH website, http://www.nih.gov/news/health/mar2010/od-18.htm.
17. See http://www.ncbi.nlm.nih.gov/gtr/.
18. This remarkable institution has never been the focus of a serious historical study by a scholar in the history of science, and although there might be issues of access and records, Coriell is certainly one possible point in the web at which all post-1945 biological science intersects. See the description of its collections at http://ccr.coriell.org/.
19. Adriana Petryna, *Life Exposed: Biological Citizens after Chernobyl* (Princeton, NJ: Princeton University Press, 2002).
20. The calculation of the numbers of geneticists and astronauts is from J. Bobe, "The Personal Genome: Genomics as a Medical Tool and Lifestyle Choice" (2009), http://thepersonalgenome.com/2007/12/shortage-of-geneticists-in-the-united-states/. Bobe notes that "for perspective, there are nearly as many

professional astronauts in the world as there are board certified geneticists that see patients in the United States. These 509 geneticists are not distributed evenly across the United States. Four states have no physician-geneticists at all. California has the most, with 84."

21. Misha Angrist, "We Are the Genes We've Been Waiting For: Rational Responses to the Gathering Storm of Personal Genomics," *American Journal of Bioethics* Volume 9, Number 6–7, (2009): 30–31, at 31. See also Angrist, "Personal Genomics: Access Denied?," *Technology Review* 111 (2008): 98–99.
22. For the ACCP position statement, see Barbara Ameer and Norberto Krivoy, "Direct-to-Consumer/Patient Advertising of Genetic Testing: A Position Statement of the American College of Clinical Pharmacology," *Clinical Pharmacology* 49 (2009): 886–888, at 888. See also J. Kaye, "The Regulation of Direct-to-Consumer Genetic Tests," *Human Molecular Genetics* 17 (2008): R180–R183.
23. Jennifer Reardon, personal communication, May 2010.
24. Dorothy Nelkin and Susan Lindee, *The DNA Mystique: The Gene as a Cultural Icon* (New York: W. H. Freeman, 1995).
25. Ameer and Krivoy, "Direct-to-Consumer/Patient Advertising of Genetic Testing," 886.
26. One of the more helpful analyses of cultural conceptions of high and low in American culture is Lawrence Levine, *Highbrow/Lowbrow: The Emergence of Cultural Hierarchy in America* (Cambridge, MA: Harvard University Press, 1988).
27. ETC Group, "Direct-to-Consumer DNA Testing and the Myth of Personalized Medicine: Spit Kits, SNP Chips and Human Genomics," http://www.scribd.com/doc/8309349/DirecttoConsumer-DNA-Testing-and-the-Myth-of-Personalized-Medicine-Spit-Kits-SNP-Chips-and-Human-Genomics.
28. Ancestral Origins DNA, http://www.gtldna.com/ancestral-origins-dna-ancestry.html.
29. ISOGG, http://www.isogg.org/.
30. Joanna Radin, "Consuming Identity: Direct-to-Consumer Marketing, Genetic Genealogy, and the Genographic Project" (master's thesis, University of Pennsylvania, August 2007).
31. Gísli Pálsson, "Decode Me! Anthropology and Personal Genomics," in "The Biological Anthropology of Living Human Populations: World Histories, National Styles, and International Networks," Special Issue S5, *Current Anthropology* 53 (2012): S185–S195, at S185.
32. The term "imagined community" was Benedict Anderson's way of understanding the modern state; see *Imagined Communities: Reflections on the Origin and Spread of Nationalism* (New York: Verso, 1991).
33. Eric Wolf's "people without history" are described in his book *Europe and the People without History* (Berkeley: University of California Press, 1982).
34. A PDF explaining the importance of ancestry testing to African Americans, by virture of the history of slavery, is on the website of Kittles's company,

Ancestry by DNA, at http://www.africanancestry.com/cmsdocuments/Ancestry_Tracing_Tips.pdf.

35. The CEPH samples are described at http://www.cephb.fr/en/hgdp/diversity.php.
36. Ancestry by DNA, http://www.ancestrybydna.com/.
37. Ibid.
38. Genographic Project, https://genographic.nationalgeographic.com/genographic/participate.html.
39. Nelkin and Lindee, *DNA Mystique,* 12.
40. National Geographic, Migration Stories, "I Wish My Dad Was Alive to See the Results: He Would Have Been Stunned," http://migration-stories.nationalgeographic.com/story/89.
41. "Number of Ancestors in a Given Generation," http://dgmweb.net/Ancillary/OnE/NumberAncestors.html.
42. The technical term is "pedigree collapse." In its simplest form it refers to the fact that the many individuals in a hypothetical family tree are not distinct. A single individual may occupy multiple places in a family tree; ancestor trees are not binary but highly variable in their structure. When first cousins produce offspring together, a not-uncommon event in human history, the number of great-grandparents collapses (four instead of eight).
43. ISOGG, posted on January 1, 2007, http://www.isogg.org/.
44. Family Tree DNA (a DTC DNA-testing company), http://www.familytreedna.com/testimonials.aspx.
45. Ibid.
46. Ibid.
47. National Geographic, Migration Stories, "Dave's Adventure," http://migration-stories.nationalgeographic.com/story/39/.
48. C. Thauvin-Robinet, A. Munck, F. Huet, et al., "The Very Low Penetrance of Cystic Fibrosis for the R117H Mutation: A Reappraisal for Genetic Counseling and Newborn Screening," *Journal of Medical Genetics* 46 (June 2009): 752–758, at 752.
49. Ian Hacking, "Making Up People," *London Review of Books* 28 (2006): 23–26, at 23. The quote is from Nietzsche's *Gay Science.*
50. Hannah Arendt, *The Human Condition,* 2nd ed., intro. Margaret Canovan (Chicago: University of Chicago Press, 1998), 47, 150, 231, 233.
51. See her *Ordinary Genomes: Science, Citizenship, and Genetic Identities* (Durham, NC: Duke University Press, 2009).
52. Patent Insights, Inc., *Genetic Testing in Medical Diagnostics: A Glimpse into the Future by an Analysis of U.S. Patenting Trends,* January 1, 2002. The price of this report is $7,000.
53. Frost and Sullivan, *U.S. Genetic Testing Markets,* October 1, 2001. The price of this report is $3,450.
54. BCC Research, *Diagnostics and Therapeutics for Genetic Diseases,* April 1, 2006. The price of this report is $4,850.
55. Fuji-Keizai USA, Inc., *US DNA-Based Diagnostic and Test Market,* January 7, 2008. The price of this report is $1,800. J. S. Bertino Jr., H. E. Greenberg,

and M. D. Reed, "American College of Clinical Pharmacology Position Statement on the Use of Microdosing in the Drug Development Process," *Journal of Clinical Pharmacology* 47 (2007): 418–422.

56. Nelkin and Lindee, *DNA Mystique,* 197.

14. Creating a "Better Baby"

1. Centers for Disease Control and Prevention, American Society for Reproductive Medicine, and Society for Assisted Reproductive Technologies, *Assisted Reproductive Technology Success Rates: National Summary and Fertility Clinic Reports* (Atlanta: U.S. Department of Health and Human Services, 2011); Maurizio Macaluso, Tracie Wright-Schnapp, Anjani Chandra, et al., "A Public Health Focus on Infertility Prevention, Detection, and Management," *Fertility and Sterility* 93, no. 1 (2010): 6–16; Gillian R. Bentley and Nicholas Mascie-Taylor, *Infertility in the Modern World: Present and Future Prospects* (Cambridge: Cambridge University Press, 2000).
2. American Society for Reproductive Medicine (ASRM), *Oversight of Assisted Reproductive Technology* (Birmingham, AL: ASRM, 2010), 1–11; Nanette R. Elster, "Assisted Reproductive Technologies: Contracts, Consents, and Controversies," *American Journal of Family Law* 18, no. 4 (2005): 193–199; Faye Ginsburg and Rayna Rapp, "The Politics of Reproduction," *Annual Review of Anthropology* 20 (1991): 311–343; Hedva Eyal, "Old Patterns, New Ideas," *GeneWatch* 24, no. 3–4 (2011): 33–38; Abby Lippman, "Willful Ignorance," *GeneWatch* 24, no. 3–4 (2011): 35–36; Debora L. Spar, *The Baby Business: How Money, Science and Politics Drive the Commerce of Conception* (Boston: Harvard Business School Press, 2006); Tina Stevens, Pat Jennings, and Diane Beeson, "Finding the Active Voice," *GeneWatch* 24, no. 3–4 (2011): 23–24.
3. American Society for Reproductive Medicine (ASRM), *Patient Fact Sheet: Risks of In Vitro Fertilization (IVF)* (Birmingham, AL: ASRM, 2007); Jolande A. Land and Johannes L. H. Evers, "Risks and Complications in Assisted Reproduction Techniques: Report of an ESHRE Consensus Meeting," *Human Reproduction* 18, no. 2 (2003): 455–457.
4. Elizabeth Ettore, *Reproductive Genetics, Gender and the Body* (London: Routledge, 2002); Jana Sawicki, "Disciplining Mothers: Feminism and the New Reproductive Technologies," in *Feminist Theory and the Body: A Reader*, ed. Janet Price and Margrit Shildrick (Edinburgh: Edinburgh University Press, 1999), 190–202; Shirley Shalev and Dafna Lemish, "'Dynamic Infertility': The Contribution of News Coverage of Reproductive Technologies to Gender Politics," *Feminist Media Studies* 12, no. 3 (2012).
5. Barbara Katz Rothman, "Of Maps and Imaginations: Sociology Confronts the Genome," *Social Problems* 42, no. 1 (1995): 1–10; Sawicki, "Disciplining Mothers"; Shalev and Lemish, "'Dynamic Infertility.'"
6. American Society for Reproductive Medicine (ASRM), *Third Party Reproduction: A Guide for Patients* (Birmingham, AL: ASRM, 2006), 1–18; Eric Blyth and Ruth Landau, "Introduction," in *Third Party Assisted Conception*

across Cultures, ed. Eric Blyth and Ruth Landau (London: Jessica Kingley, 2004), 7–20; Eric Blyth and Jean Benward, "The United States of America: Regulation, Technology and the Marketplace," in Blyth and Landau, *Third Party Assisted Conception across Cultures*, 246–265.

7. Lori B. Andrews, "Surrogate Motherhood: The Challenge for Feminists," in *Surrogate Motherhood: Politics and Privacy*, ed. Larry Gostin (Bloomington: Indiana University Press, 1990), 167–182; Martha A. Field, *Surrogate Motherhood: The Legal and Human Issues* (Cambridge, MA: Harvard University Press, 1990); Diederika Pretorius, *Surrogate Motherhood: A Worldwide View of the Issues* (Springfield, IL: Charles C. Thomas, 1994); Helena Ragoné, *Surrogate Motherhood: Conception in the Heart* (Boulder, CO: Westview, 1994), 51–86; Kelly D. Weisberg, *The Birth of Surrogacy in Israel* (Gainesville: University Press of Florida, 2005).
8. Ethics Committee of the American Society for Reproductive Medicine, "Access to Fertility Treatment by Gays, Lesbians, and Unmarried Persons," *Fertility and Sterility* 92, no. 4 (2009): 1190–1193.
9. See, for example, "Ultra Modern Family: Dad+Dad+Baby," *ABC News*, September 29, 2010, http://abcnews.go.com/Nightline/growing-generations-surrogacy-agency-gay-families/story?id=11749014#.T3tGSb9rPF8; and "Four Parents and a Baby," *Independent*, June 27, 1993, http://www.independent.co.uk/arts-entertainment/four-parents-and-a-baby-the-boy-is-turning-two-now-having-him-seemed-such-a-simple-idea-at-first-a-way-to-make-both-couples-happy-two-gay-men-mixing-their-sperm-to-make-a-child-they-would-share-with-its-two-lesbian-mothers-then-nature-found-a-way-to-assert-the-old-mum-and-dad-routine-now-its-very-complicated-indeed-1494250.html.
10. Art L. Caplan and Pasquale Patrizio, "Are You Ever Too Old to Have a Baby? The Ethical Challenges of Older Women Using Infertility Services," *Seminars in Reproductive Medicine* 28, no. 4 (2010): 281–286.
11. Ibid.; American Society for Reproductive Medicine (ASRM), *Age and Fertility: A Guide for Partients* (Birmingham, AL: ASRM, 2003): 1–17; Karen M. Benzies, "Advanced Maternal Age: Are Decisions about the Timing of Childbearing a Failure to Understand the Risks?," *Canadian Medical Association Journal* 178, no. 2 (2008): 183–184; Carrie Friese, Gay Becker, and Robert D. Nachtigall, "Older Motherhood and the Changing Life Course in the Era of Assisted Reproductive Technologies," *Journal of Aging Studies* 22 (2008): 65–73; Sylvia Ann Hewlett, *Creating a Life: Professional Women and the Quest for Children* (New York: Talk Miramax Books, 2002); Hamisu M. Salihu, Nicole M. Shumpert, Martha Slay, et al., "Childbearing beyond Maternal Age 50 and Fetal Outcomes in the United States," *Obstetrics and Gynecology* 102, no. 5 (2003): 1006–1014.
12. Ethics Committee of the American Society for Reproductive Medicine, "Posthumous Reproduction," *Fertility and Sterility* 82, no. 1 (2004): 260–262; Rajesh Bardale and P. G. Dixit, "Birth after Death: Questions about Posthumous Sperm Retrieval," *Indian Journal of Medical Ethics* 3, no. 4 (2006): 122–123; Charles P. Kindregan and Maureen McBrien, "Posthumous Reproduction," *Family Law Quarterly* 39, no. 3 (2005): 579–597.

13. Timothy Caulfield, "Human Cloning Laws, Human Dignity and the Poverty of the Policy Making Dialogue," *BMC Medical Ethics* 4, no. 3 (2003): 1–7; Judith A. Johnson and Erin D. Williams, *CRS Report for Congress: Human Cloning* (Washington, DC: Congressional Research Service, 2006); William P. Cheshire, Edmund D. Pellegrino, Linda K. Bevington, et al., "Stem Cell Research: Why Medicine Should Reject Human Cloning," *Mayo Clinic Proceedings* 78, no. 8 (2003): 1010–1018.
14. American Society for Reproductive Medicine (ASRM), *Patient Fact Sheet: Genetic Screening for Birth Defects* (Birmingham, AL: ASRM, 2005); Ingrid Lobo and Kira Zhaurova Kira, "Birth Defects: Causes and Statistics," *Nature Education* 1 (2008), http://www.nature.com/scitable/topicpage/birth-defects-causes-and-statistics-863.
15. Yael Hashiloni-Dolev, "Between Mothers, Fetuses and Society: Reproductive Genetics in the Israeli-Jewish Context," *Nashim* 12 (2006): 129–150; Ruth Hubbard, "Childbearing in the Age of Biotechnology," *Gene Watch* 14, no. 4 (2001): 7–9; Rayana Rapp, *Testing Women, Testing the Fetus: The Social Impact of Amniocentesis in America* (New York: Routledge, 1999).
16. See, for example, *SEQureDx Technology* by Sequenom Center for Molecular Medicine, http://www.sequenomcmm.com/home/health-care-professionals/fetal-rhd-genotyping/.
17. Marcy Darnovsky, "One Step Closer to Designer Babies: New Noninvasive Prenatal Genetic Testing Could Change Human Pregnancy Forever," *Science Progress*, April 22, 2011, http://scienceprogress.org/2011/04/one-step-closer-to-designer-babies/; Lori Haymon, "The Fast and the Furious," *GeneWatch* 24, no. 3–4 (2011): 16–20; Lori Haymon, *Non-Invasive Prenatal Genetic Diagnosis (NIPD)* (Cambridge, MA: Council for Responsible Genetics, 2011), http://www.councilforresponsiblegenetics.org/pageDocuments/E3RTQAOVMU.pdf; Kat Zambon, "Case Studies Illustrate the Dilemmas of Genetic Testing," American Association for the Advancement of Science, April 29, 2011, http://www.aaas.org/news/releases/2011/0429genetic_testing.shtml.
18. Sue Hall, Martin Bobrow, and Theresa M. Marteau, "Psychological Consequences of Parents of False Negative Results on Prenatal Screening for Down's Syndrome: Retrospective Interview Study," *British Medical Journal* 320 (2000): 407–412.
19. Susannah Baruch, David Kaufman, and Kathy Hudson, "Genetic Testing of Embryos: Practices and Perspectives of U.S. IVF Clinics," *Fertility and Sterility* 89, no. 5 (2006): 1053–1058.
20. Brian J. Zikmund-Fisher, Angela Fagerlin, Kristie Keeton, et al., "Does Labeling Prenatal Screening Test Results as Negative or Positive Affect Women's Responses?," *American Journal of Obstetrics and Gynecology* 197, no. 5 (2007): 528.e1–528.e6.
21. Darnovsky, "One Step Closer to Designer Babies"; Haymon, *Non-Invasive Prenatal Genetic Diagnosis*.
22. See, for example, *The Universal Carrier Test* by Counsyl, http://www.counsyl.com.
23. Baruch, Kaufman, and Hudson, "Genetic Testing of Embryos."

24. American Society for Reproductive Medicine, *Patient Fact Sheet: Genetic Screening for Birth Defects;* Hashiloni-Dolev, "Between Mothers, Fetuses and Society"; Barry Starr, "The Myth of Genetic Improvement," *GeneWatch* 22, no. 6 (2009): 14–16.
25. Georges Canguilhem, *The Normal and the Pathological* (New York: Zone Books, 1991); Peter Conrad and Rochelle Kern, "The Social Production of Disease and Illness," in *The Sociology of Health and Illness*, ed. Peter Conard and Rochelle Kern (New York: St. Marin's Press, 1990), 9–11.
26. Ethics Committee of the American Society for Reproductive Medicine, "Preconception Gender Selection for Nonmedical Reasons," *Fertility and Sterility* 75, no. 5 (2001): 861–864; Marcy Darnovsky, "High-Tech Sex Selection," *GeneWatch* 17, no. 1 (2003); Tarun Jain, Stacy A. Missmer, Raina S. Gupta, et al., "Preimplantation Sex Selection Demand and Preferences in an Infertility Population," *Fertility and Sterility* 83, no. 3 (2005): 649–658.
27. Baruch, Kaufman, and Hudson, "Genetic Testing of Embryos."
28. David Heyd, "Male or Female, We Will Create Them: The Ethics of Sex Selection for Non-Medical Reasons," *Ethical Perspectives* 10, no. 3–4 (2003): 204–214.
29. Laxmi Murthy, "Sex Selection: Getting Down to Business," *Infochange*, February 2003, http://infochangeindia.org/population/features/sex-selection-getting-down-to-business.html; Wei Xing Zhu, Li Lu, and Therese Hesketh, "China's Excess Males, Sex Selective Abortion, and One Child Policy: Analysis of Data from 2005 National Intercensus Survey," *British Medical Journal* 338 (2009): 920–936.
30. "160 Million and Counting," *New York Times*, June 26, 2011, http://www.nytimes.com/2011/06/27/opinion/27douthat.html; "Tens of Millions of 'Missing' Girls," *CNN Opinion*, September 5, 2010, "http://articles.cnn.com/2010-09-05/opinion/wudunn.women.oppression_1_baby-girls-sheryl-wudunn-girls-in-many-countries?_s=PM:OPINION; and Mara Hvistendahl, "Where Have All the Girls Gone?," *Monterey Institute of International Studies*, June 27, 2011, http://www.foreignpolicy.com/articles/2011/06/27/where_have_all_the_girls_gone?page=full.
31. Christophe Z. Guilmoto, "The Sex Ratio Transition in Asia," *Population and Development Review* 35, no. 3 (2009): 519–549.
32. Stephanie A. Devaney, Glenn E. Palomaki, Joan A. Scott, et al., "Noninvasive Fetal Sex Determination Using Cell-Free Fetal DNA," *Journal of the American Medical Association* 306, no. 6 (2011): 627–636; Haymon, "The Fast and the Furious."
33. Heyd, "Male or Female, We Will Create Them."
34. Ettore, *Reproductive Genetics, Gender and the Body;* Susan Gal and Gail Kligman, "Introduction," in *Reproducing Gender: Politics, Publics, and Everyday Life after Socialism*, ed. Susan Gal and Gail Kligman (Princeton, NJ: Princeton University Press, 2000), 3–20.
35. See, for example, "Smoking Out the Smoking Gene," *Economist*, April 3, 2008, http://www.economist.com/research/articlesBySubject/PrinterFriendly

.cfm?story_id=10952815; "Studies Find Genetic Link to Smoking," *New York Times*, April 3, 2008, http://query.nytimes.com/gst/fullpage.html?res=980CE0DF143DF930A35757C0A96E9C8B63; and "Can't Quit Smoking? Blame Your Genes," *MSNBC*, April 2, 2008, http://www.msnbc.msn.com/id/23919596/ns/health-addictions.

36. See, for example, "Having Trouble Squeezing into Your Jeans? Blame It on Your Genes," *Scientific American*, April 12, 2007, http://www.scientificamerican.com/article.cfm?id=having-trouble-squeezing-into-jeans-blame-genes; and "Clear Obesity Gene Link 'Found,'" *BBC News*, April 12, 2007, http://news.bbc.co.uk/2/hi/6547891.stm.
37. See, for example, "Bad Driver? Blame Your Genes," *Reuters*, October 29, 2009, http://www.reuters.com/article/2009/10/29/us-genes-driving-idUSTRE59S0M720091029; and "Are You a Bad Driver? Now You Can Blame It on Your Genes," *Mail Online*, October 29, 2009, http://www.dailymail.co.uk/motoring/article-1223620/Are-bad-driver-Now-blame-genes-.html.
38. See, for example, "Ability to Navigate May Be Linked to Genes, Researcher Says," *Science Codex*, February 1, 2010, http://www.sciencecodex.com/ability_to_navigate_may_be_linked_to_genes_researcher_says; "Ability to Navigate May Be Linked to Genes," *Innovations Report*, February 2, 2010, http://www.innovations-report.com/html/reports/life_sciences/ability_navigate_linked_genes_147775.html; and "Ability to Navigate May Be Linked to Genes," *Science Daily*, February 2, 2010, http://www.sciencedaily.com/releases/2010/02/100201171920.
39. See, for example, "Bullying Behavior: Blame It on Bad Genes?," *Science Daily*, March 10, 1999, http://www.sciencedaily.com/releases/1999/03/990310053751; and "Bullies Are Born and Not Made," *Independent*, March 10, 1999, http://www.independent.co.uk/news/bullies-are-born-and-not-made-1079488.
40. See, for example, "Hate Broccoli? Spinach? Blame Your Genes," *Los Angeles Times*, February 19, 2007, http://articles.latimes.com/2007/feb/19/health/he-eat19; and "Don't Like Your Veggies? Blame Your Genes," *Science Line*, April 1, 2011, http://scienceline.org/2011/04/dont-like-your-veggies-blame-your-genes.
41. See, for example, "Bad-Tempered Women 'Can Blame It on Genes,'" *Telegraph*, March 9, 2007, http://www.telegraph.co.uk/news/uknews/1545010/Bad-tempered-women-can-blame-it-on-genes; "'Angry Gene' Could Help Spur Hostility," *Washington Post*, March 9, 2007, http://www.washingtonpost.com/wp-dyn/content/article/2007/03/09/AR2007030901449.html; and "Anger and Aggression in Women: Blame It on Genetics," *Science Daily*, March 10, 2007, http://www.sciencedaily.com/releases/2007/03/070309103136.htm.
42. See, for example, "Study Suggests Gene Linked to Credit Card Debt," *Bio News*, May 10, 2010, http://www.bionews.org.uk/page_59621.asp; "Millions of Useless Purchases Explained at Last," *Newsweek*, November 11, 2009, http://www.newsweek.com/2009/11/10/millions-of-useless-purchases

-explained-at-last.html; and "Born into Debt: Gene Linked to Credit-Card Balances," *Scientific American*, August 12, 2010, http://www.scientificamerican .com/article.cfm?id=born-into-debt.

43. Nicholas Agar, "Liberal Eugenics," *Public Affairs Quarterly* 12, no. 2 (1998): 137–155; Jeremy Rifkin, "Ultimate Therapy: Commercial Eugenics in the 21st Century," *Harvard International Review* 27, no. 1 (2005): 44–45.
44. Council for Responsible Genetics, "Special Topic: Genetic Reductionism," *GeneWatch* 24, no. 3–4 (2011): 39.
45. Mohan J. Dutta-Bergman, "Primary Sources of Health Information: Comparison in the Domain of Health Attitudes, Health Cognitions, and Health Behaviors," *Health Communication* 16 (2004): 393–409; Uwe Flick, "Introduction: Social Representations in Knowledge and Language as Approaches to a Psychology of the Social," in *The Psychology of the Social,* ed. Uwe Flick (Cambridge: Cambridge University Press, 1998): 1–14; Nurit Guttman, "Ethics in Health Communication Interventions," in *Handbook of Health Communication,* ed. Teresa L. Thompson et al. (Mahwah, NJ: Lawrence Erlbaum Associates, 2003), 651–679; Susan E. Morgan, Tyler R. Harrison, Lisa Chewning, et al., "Entertainment (Mis)Education: The Framing of Organ Donation in Entertainment Television," *Health Communication* 22, no. 2 (2007): 143–151; Phyllis Tilson Piotrow, D. Lawrence Kincaid, Jose G. Rimon II, et al., *Health Communication: Lessons from Family Planning and Reproductive Health* (Westport, CT: Praeger, 1997), 1–17.
46. "Too Many One-Night Stands? Blame Your Genes," *Time*, December 2, 2010, http://healthland.time.com/2010/12/02/too-many-one-night-stands -blame-your-genes/.
47. "Genes May Be to Blame for Infidelity," *BBC News*, June 7, 2004, http:// news.bbc.co.uk/2/hi/health/3783031.stm.
48. "Could Monogamy Gene Combat Infidelity?," *ABC News*, July 23, 2005, http://abcnews.go.com/GMA/OnCall/story?id=970035&page=1.
49. "Women's Infidelity Is All in the Genes," *Times Online*, June 6, 2004, http:// www.timesonline.co.uk/tol/news/uk/health/article442311.ece?token–ull& offset=12&page=2.
50. "BU Researchers Connect Gene to Infidelity," *FOX 40 WICZ*, December 2, 2010, http://www.wicz.com/news2005/printarticle.asp?a=16646.
51. "Infidelity Gene? Genetic Link to Relationship Difficulties Found," *Science Daily*, September 2, 2008, http://www.sciencedaily.com/releases/2008/09 /080902161213.htm.
52. "The Urge to Infidelity . . . It's in Her Genes," *Guardian*, November 25, 2004, http://www.guardian.co.uk/science/2004/nov/25/science.research1.
53. "Infidelity Might Be in the Genes," *Business Week*, December 3, 2010, http:// www.businessweek.com/lifestyle/content/healthday/646957.html.
54. "The Love-Cheat Gene: One in Four Born to Be Unfaithful, Claim Scientists," *Mail Online*, December 3, 2010, http://www.dailymail.co.uk/sciencetech /article-1334932/The-love-cheat-gene-One-born-unfaithful-claim-scientists .html.

55. "Women's Infidelity Is All in the Genes," *Times Online,* June 6, 2004, http://www.timesonline.co.uk/tol/news/uk/health/article442311.ece?token=null&offset=12&page=2.
56. "Can't Quit Smoking? Blame Your Genes" (see note 35).
57. "If You Smoke Too Much 'Blame Your Genes', Say Experts," *BBC News*, April 26, 2010, http://news.bbc.co.uk/2/hi/health/8643803.stm.
58. "Having Trouble Squeezing into Your Jeans?" (see note 36).
59. "Bad Driver? Blame Your Genes" (see note 37).
60. "Early Birds, Night Owls: Blame Your Genes," *National Geographic*, January 28, 2008, http://news.nationalgeographic.com/news/2008/01/080126-sleep-genes.html.
61. "Like to Sleep Around? Blame Your Genes," *CBS News*, December 2, 2010, http://www.cbsnews.com/8301-501465_162-20024414-501465.html.
62. "Blame Your Genes for That Desperate Craving for Coffee, Research Suggests," *Vancouver Sun*, April 16, 2011, http://www.vancouversun.com/health/Blame+your+genes+that+desperate+craving+coffee+research+suggests/4627996/story.html.
63. "Overeating? Blame Your Genes," *Washington Post*, October 16, 2008, http://www.washingtonpost.com/wp-dyn/content/article/2008/10/16/AR2008101602109.html.
64. "Binge Drinking? Blame It on Your Genes," *Mid Day*, February 3, 2011, http://www.mid-day.com/lifestyle/2011/mar/020311-Binge-drinking-Blame-it-on-your-genes.htm.
65. "Lost It All in the Stock Market? Blame Your Genes," *Discover Magazine*, February 11, 2009, http://blogs.discovermagazine.com/80beats/2009/02/11/lost-it-all-in-the-stock-market-blame-your-genes.
66. "Don't Want to Exercise? Blame Your Genes," *Live Science*, June 17, 2010, http://www.livescience.com/6602-exercise-blame-genes.html.
67. "Feeling Lonely? Genes Might Be at Fault," *CNN*, December 10, 2008, http://www.cnn.com/2008/HEALTH/12/08/loneliness.psychology.
68. "Cluttered Home? Blame Your Genes," *ABC Science*, October 30, 2009, http://www.abc.net.au/science/articles/2009/10/30/2728509.htm.
69. "Speaking in Tones? Blame It on Your Genes," *Times Online*, May 29, 2007, http://www.timesonline.co.uk/tol/news/science/article1851794.ece.
70. "Bad-Tempered Women 'Can Blame It on Genes'" (see note 41).
71. "Hate Broccoli? Spinach? Blame Your Genes" (see note 40).
72. "If You Don't Eat Greens, Blame It on Your Genes," *Telegraph*, October 23, 2007, http://www.telegraph.co.uk/news/uknews/1567084/If-you-dont-eat-greens-blame-it-on-your-genes.html.
73. Kenneth Weiss, "Genetics without Ideology," *GeneWatch* 24, no. 3–4 (2011): 43–46.
74. See, for example, donor profiles at the Sperm Bank of California, http://www.thespermbankofca.org/pages/page.php?pageid=4#catalog.
75. See, for example, screening and testing at Fairfax Cryobank, http://www.fairfaxcryobank.com/donorscreen.shtml; Fairfax Cryobank, http://www.fair

faxcryobank.com/geneticdisease.shtml; and California Cryobank, http://www.cryobank.com/Services/Genetic-Counseling/Donor-Screening.

76. See, for example, Fairfax Cryobank, http://www.fairfaxcryobank.com/geneticdisease.shtml; and Xytex Cryo International Sperm Bank, http://www.xytex.com/sperm-donor-bank-patient/sperm-donor-bank-patient-screening.cfm.
77. In *Donovan v. Idant Laboratories,* a New York sperm bank was sued for failing to detect fragile X defect in a sperm donation that resulted in the birth of a mentally retarded girl from Pennsylvania. The initial ruling from 2009 was in favor of the plaintiff, viewing the sale of spern as a product that is subject to liability, yet the rulling was reversed and the case was entirly dismissed. A third Circuit Courst of Appeals in 2010 upheld the District ruling that genetic defects in sperm cannot form the basis for product liability suits. For further reading see, for example, Jennifer M. Vagle, "Putting the 'Product' in Reproduction: The Viability of Products Liability Action for Genetically Defective Sperm," *Pepperdine Law Review* 38, no. 4(5) (2011): 1175–1236.
78. The litigation of Myriad Genetics regarding the validity of patents on two human genes associated with breast and ovarian cancer has not been finally resolved (see *Association for Molecular Pathology v. U.S. Patent and Trademark Office*). A New York District Court ruling from 2010 invalidated certain Myriad patents, yet a notice of appeal was filed by the defendants, and in March 2012 it was reported that the U.S. Supreme Court ordered a lower court to revisit the patentability of human genes. See, for example, "Myriad's Human-Gene Patent Rehearing Ordered by High Court," *Bloomberg*, March 27, 2012, http://www.bloomberg.com/news/2012-03-27/illumina-myriad-egis-apple-pinterest-intellectual-property.html. For further reading, see for example Robert Cook-Deegan, "The Overstated Case," *GeneWatch* 23, no. 5 (2010): 20–22; and Debra Greenfield, "Freedon of Genes," *GeneWatch* 23, no. 5 (2010): 36–38.
79. See, for example, Fairfax Cryobank, http://donorsearch.fairfaxcryobank.com; Lisa C. Ikemoto, "Match, Mate, Trait," *GeneWatch* 24, no. 3–4 (2011): 4–6.
80. Starr, "The Myth of Genetic Improvement."
81. For example, see reports regarding several European production companies that considered reality television shows like *Make Me a Mom* and *I Want Your Child and Nothing Else*, in which a single woman is challenged to choose the ultimate genetic profile for her child from hundreds of competing potential sperm donors. See, for example, "Dutch TV Show Seeks Sperm Donor," *BBC*, August 25, 2005, http://news.bbc.co.uk/2/hi/entertainment/4183324.stm; "'Sperm Idol' Reality TV Show?," *Bio News*, July 26, 2004, http://www.bionews.org.uk/page_12051.asp; "Procreation May Become a New Reality Show Reward," *Reality TV World*, July 28, 2004, http://www.realitytvworld.com/news/procreation-may-become-new-reality-show-reward-2783.php; "Reality TV Search for Sperm Donor," *Bio News*, August 25, 2005, http://www.bionews.org.uk/page_12481.asp; "Sperm Donor TV Show

Seems Inconceivable," *MSN*, January 17, 2005, http://today.msnbc.msn.com/id/6807612/ns/today-entertainment/t/sperm-donor-tv-show-seems-inconceivable/#; and "Big Sperm Race Is Staged on German Reality TV," *Guardian*, January 29, 2005, http://www.guardian.co.uk/media/2005/jan/30/realitytv.germany.

82. Hewlett, *Creating a Life;* Ikemoto, "Match, Mate, Trait"; Kathleen Sloan, "Abuses of Women's Human Rights in Third Party Reproduction," *GeneWatch* 24, no. 3–4 (2011): 20–23; Spar, *The Baby Business.*
83. It was widely reported that *Playboy* photographer Ron Harris announced online auctions of human eggs from beautiful models to the highest bidder at http://ronsangels.com/Egg_Donations (these auctions were later suspected to be a scam). See, for example, "On Web, Models Auction Their Eggs to Bidders for Beautiful Children," *New York Times*, October 23, 1999, http://www.nytimes.com/1999/10/23/us/on-web-models-auction-their-eggs-to-bidders-for-beautiful-children.html?pagewanted=all&src=pm; "World: America's Anger Greets Online Egg Auction," *BBC News*, October 24, 1999, http://news.bbc.co.uk/2/hi/americas/483783.stm; "Model Eggs for Sale," *Bio News*, October 25, 1999, http://www.bionews.org.uk/page_10521.asp; and "'Ron's Angels' Internet Egg Auction Is a Scam," *Bio News*, November 1, 1999, http://www.bionews.org.uk/page_10528.asp.
84. See, for example, "Number of Egg Donors Doubles as Economy Tanks: As the Economy Worsens, More Men and Women Are Hoping to Cash In by Selling Their Sperm and Eggs," *Fox News*, January 25, 2009, http://www.foxnews.com/story/0,2933,482728,00.html; and "Stalled Economy Fertile Ground for Baby Business: Bay State Agencies See Rise in Numbers of Applicants," *Boston Herald*, January 25, 2009, http://www.bostonherald.com/news/hard_times/view.bg?articleid=1147570.
85. Practice Committee of the American Society for Reproductive Medicine, "Ovarian Hyperstimulation Syndrome," *Fertility and Sterility* 80, no. 5 (2003): 1309–1314; Diane Beeson and Abby Lippman, "Egg Harvesting for Stem Cell Research: Medical Risks and Ethical Problems," *Reproductive BioMedicine Online* 13, no. 4 (2006): 573–579; Antonio Girolami, Raffaella Scandellari, Fabiana Tezza, et al., "Arterial Thromobosis in Young Women after Ovarian Stimulation: Case Report and Review of the Literature," *Journal of Thrombosis and Thrombolysis* 24 (2007): 169–174; Linda Giudice, Eileen Santa, and Robert Pool, *Assessing the Medical Risks of Human Oocyte Donation for Stem Cell Research: Workshop Report* (Washington, DC: National Academies Press, Institute of Medicine, and National Research Council, 2007); Wendy Kramer, Jennifer Schneider, and Natalie Schultz, "US Oocyte Donors: A Retrospective Study of Medical and Psychosocial Issues," *Human Reproduction* 24, no. 12 (2009): 3144–3149; Jennifer Schneider, "Fatal Colon Cancer in a Young Egg Donor: A Physician Mother's Call for Follow-Up Research on the Long-Term Risks of Ovarian Stimulation," *Fertility and Sterility* 90, no. 5 (2008): 2016.e1–2016.e5.
86. See, for example, the *East-West Plan* at Tammuz International Surrogacy, http://www.tammuz.com/main.php?lang=eng&action rackseastwest.

87. Center for Social Research, *Surrogate Motherhood—Ethics or Commercial* (2012), 1–88, http://www.womenleadership.in/Csr/SurrogacyReport.pdf; Vrinda Marwah, "Commercial Surrogacy in India," *GeneWatch* 24, no. 3–4 (2011): 10–13; Sloan, "Abuses of Women's Human Rights in Third Party Reproduction"; Marsha Darling, "Commercial Surrogacy and the Reproductive 'Freedom'," *GeneWatch* 24, no. 3–4 (2011): 27–30.
88. See, for example, Fairfax Cryobank, http://www.fairfaxcryobank.com/FaceMatchFAQs.shtml.
89. See, for example, Morth Thing, http://www.morphthing.com/blog/138-What-Will-My-Baby-Look-Like; and Make Me Babies, http://www.makemebabies.com/.
90. See, for example, California Cryobank, http://www.cryobank.com/Donor-Search/Look-A-Likes/; and "Ben Affleck Tops Celebrity Look-A-Like Sperm Donors List," *Telegraph*, October 7, 2011, http://www.telegraph.co.uk/news/celebritynews/6884489/Ben-Affleck-tops-celebrity-look-a-like-sperm-donors-list.html.
91. David Plotz, *The Genius Factory: The Curious History of the Nobel Prize Sperm Bank* (New York: Random House, 2005).
92. Margrit Shildrick and Janet Price, "Splitting the Difference: Adventures in the Anatomy and Embodiment of Women," in *Stirring It: Challenges for Feminism*, ed. Gabriel Griffin, Marianne Hester, Shirin Rai, et al. (Bristol, PA: Taylor and Francis, 1994), 156–179; Janice G. Raymond, *Women as Wombs: Reproductive Technologies and the Battle over Women's Freedom* (San Francisco: HarperSanFrancisco, 1993), 29–75; Barbara Katz Rothman, *Recreating Motherhood: Ideology and Technology in Patriarchal Society* (New York: W. W. Norton and Company, 1989).
93. Weiss, "Genetics without Ideology."
94. Agar, "Liberal Eugenics"; Rifkin, "Ultimate Therapy."
95. Sandra Lee Bartky, "Foucault, Femininity, and the Modernization of Patriarchal Power," in *Feminism and Foucault*, ed. Diamond Irene (Boston: Northeastern University Press, 1988), 61–86; Naomi Wolf, *The Beauty Myth: How Images of Beauty Are Used against Women* (New York: Doubleday, 1991).
96. Michel Foucault, *The History of Sexuality*, vol. 1: *An Introduction* (New York: Vintage, 1978); Shalev and Lemish, "'Dynamic Infertility.'"
97. Agar, "Liberal Eugenics."
98. Merle Spriggs, "Lesbian Couple Create a Child Who Is Deaf like Them," *Journal of Medical Ethics* 28 (2002): 283.
99. "Sperm Bank to Redheads: We Don't Want Your Semen," *CBS News*, September 19, 2011, http://www.cbsnews.com/8301-504763_162-20108310-10391704.html; "World's Biggest Sperm Bank, Cryos, Tells Redheads: We Don't Want Your Semen," *New York Daily News*, September 18, 2011, http://articles.nydailynews.com/2011-09-18/entertainment/30196092_1_redheads-sperm-bank-cryos-international; and "World's Biggest Sperm Bank Shows Redheads to the Door," *ABC News*, September 19, 2011, http://abcnews.go.com/blogs/health/2011/09/19/worlds-biggest-sperm-bank-shows-redheads-to-the-door/.

100. Ethics Committee of the American Society for Reproductive Medicine, "Informing Offspring of Their Conception by Gamete Donation," *Fertility and Sterility* 81, no. 3 (2004): 527–531; Ethics Committee of the American Society for Reproductive Medicine, "Interests, Obligations, and Rights of the Donor in Gamete Donation," *Fertility and Sterility* 91, no. 1 (2009): 22–27; Erica Haimes, "Gamete Donation and the Social Management of Genetic Origins," in *Changing Human Reproduction: Social Science Perspectives*, ed. Meg Stacey (London: Sage, 1992), 119–147.
101. Tabitha Freeman, Vasanti Jadva, Wendy Kramer, et al., "Gamete Donation: Parents' Experiences of Searching for Their Child's Donor Siblings and Donor," *Human Reproduction* 24, no. 3 (2009): 505–516; Vasanti Jadva, Tabitha Freeman, Wendy Kramer, et al., "Sperm and Oocyte Donors' Experiences of Anonymous Donation and Subsequent Contact with Their Donor Offspring," *Human Reproduction* 26, no. 3 (2011): 638–645; and see, for example, Donor Sibling Registry, http://www.donorsibilingregistry.com.
102. Njabulo S. Ndebele, "Maintaining Domination through Language," *Academic Development* 1, no. 1 (1995): 1–5; Ikemoto, "Match, Mate, Trait."
103. Aila Collins and Judith Rodin, "The New Reproductive Technologies: What Have We Learned?," in *Women and New Reproductive Technologies: Medical, Psychological, Legal, and Ethical Dilemmas*, ed. Judith Rodin and Aila Collins (Hillsdale, NJ: Lawrence Erlbaum Associates, 1991), 153–161; Sarah Franklin, "Making Sense of Missed Conceptions: Anthropological Perspectives on Unexplained Infertility," in Stacey, *Changing Human Reproduction,* 75–91; Erica Haimes, "Gamete Donation and the Social Management of Genetic Origins"; Helena Ragoné, "Chasing the Blood Tie," in *Situated Lives*, ed. Louise Lamphere, Helena Ragoné, and Patricia Zavella (New York: Routledge, 1997), 110–127.
104. Rothman, "Of Maps and Imaginations"; Roger Gosden, *Designing Babies: The Brave New World of Reproductive Technology* (New York: W. H. Freeman, 1999); Carmel Shalev, *Birth Power* (New Haven, CT: Yale University Press, 1989).
105. Rose M. Kreider and Renee Ellis, "Living Arrangements of Children: 2009," in *Current Population Reports* (Washington, DC: U.S. Census Bureau, 2011), 1–25.
106. Susan M. Kahn, *Reproducing Jews* (Durham, NC: Duke University Press, 2000); Ragoné, *Surrogate Motherhood*; Elly Teman," Bonding with the Field: On Researching Surrogate Motherhood Arrangements in Israel," in *Dispatches from the Field: Neophyte Ethnographers in a Changing World*, ed. Andrew Gardner and David M. Hoffman (Long Grove, IL: Waveland Press, 2006), 179–194.
107. See, for example, 23 and Me, https://www.23andme.com; DeCodeMe, http://www.decodeme.com/; and Navigenics, http://www.navigenics.com/.
108. Jeremy Gruber, "DTC Genetic Testing: Consumer Privacy Concerns," *GeneWatch* 23, no. 4 (2010): 18–20; Sheldon Krimsky, "Not What the Doctor Ordered," *GeneWatch* 23, no. 4 (2010): 11–12; Jordan P. Lerner-Ellis, J. David Ellis, and Robert Green, "Direct-to-Consumer Genetic Testing:

What's the Prognosis?," *GeneWatch* 23, no. 4 (2010): 6–8; Patricia A. Roche and George J. Annas, "DNA Testing, Banking, and Genetic Privacy," *New England Journal of Medicine* 355, no. 6 (2006): 545–546.

109. Troy Duster, "Ancestry Testing and DNA: Uses, Limits and Caveat Emptor," *GeneWatch* 22, no. 3–4 (2009): 16–18; Osagie Obasogie, "The Color of Our Genes," *GeneWatch* 22, no. 3–4 (2009): 25–27; Jessica Bardill, "DNA and Tribal Citizenship," *GeneWatch* 23, no. 3 (2010): 8–10.

110. Stuart Hall, "Introduction: Who Needs 'Identity'?," in *Questions of Cultural Identity*, ed. Stuart Hall and Paul Du Gay (London: Sage, 1996), 1–17; Kathyryn Woodward, "Concepts of Identity and Difference," in *Identity and Difference*, ed. Kathyryn Woodward (London: Sage in association with the Open University, 1997), 7–61.

15. Forensic DNA Evidence

1. Jay Aronson, *Genetic Witness: Science, Law, and Controversy in the Making of DNA Profiling* (New Brunswick, NJ: Rutgers University Press, 2007); Jonathan Koehler, "Error and Exaggeration in the Presentation of DNA Evidence," *Jurimetrics* 34 (1993): 21–39.
2. Michael Lynch, Simon Cole, Ruth McNally, et al., *Truth Machine: The Contentious History of DNA Fingerprinting* (Chicago: University of Chicago Press, 2008).
3. Ibid.
4. Sheldon Krimsky and Tania Simoncelli, *Genetic Justice: DNA Data Banks, Criminal Investigations, and Civil Liberties* (New York: Columbia University Press, 2011).
5. David Lazer, ed., *DNA and the Criminal Justice System: The Technology of Justice* (Cambridge, MA: MIT Press, 2004).
6. Committee on Identifying the Needs of the Forensic Sciences Community, National Research Council, *Strengthening Forensic Science in the United States: A Path Forward* (Washington, DC: National Academies Press, 2009).
7. Koehler, "Error and Exaggeration," 24.
8. Committee on Identifying the Needs of the Forensic Sciences Community, National Research Council, *Strengthening Forensic Science in the United States,* 121.
9. Ibid., 133.
10. See Katherine Roeder, "DNA Fingerprinting: A Review of the Controversy," *Statistical Science* 9 (1994): 222–247 and accompanying commentary.
11. Opinion in the Reference by the Scottish Criminal Cases Review Commission in the Case of Brian Kelly, Appeals Court, High Court of Justiciary, August 6, 2004, http://www.scotcourts.gov.uk/opinions/XC458.html.
12. This account is based on the author's review of a corrective action report produced by the Washington State Patrol laboratory. The report was included in documents obtained by the *Seattle Post-Intelligencer* as part of an investigation of errors at the laboratory. The author assisted the newspaper in this investigation.

13. California and Florida: William Thompson, "Tarnish on the 'Gold Standard': Understanding Recent Problems in Forensic DNA Testing," *Champion* 30 (2006): 10–16; New Jersey: M. Spoto, "Murder, Rape Charges Dropped Due to Botched DNA Evidence," *New Jersey Star Ledger,* February 7, 2006; K. Sudol, "Murder Charges Dismissed Because of Unreliable DNA Evidence," *Asbury Park Press,* February 7, 2006; New Zealand: Michael Strutt, "Legally Scientific? A Brief History of DNA Evidence in the Criminal Justice System," June 9, 2001, http://www.justiceaction.org.au/actNow/Campaigns/DNA/pdf_files/02_Legal.pdf; Australia: Amanda Banks, "DNA Lab Admits Rape Case Bungle," *Australian,* March 16, 2006; Graeme Johnson, State Coroner, *Inquest into the Death of Jaidyn Raymond Leskie,* Coroner's Case No. 007/98 (July 31, 2006), http://www.bioforensics.com/articles/Leskie_decision.pdf; William Thompson, "Victoria State Coroner's Inquest into the Death of Jaidyn Leskie: Report" (December 3, 2003), http://www.bioforensics.com/articles/Thompsonreport.pdf; William Thompson, "Victoria State Coroner's Inquest into the Death of Jaidyn Leskie: Supplemental Report" (January 29, 2004) (available from the author).
14. Mark Butler and Anthony Dowsley, "DNA Fiasco with 7000 Crimes to Be Re-examined by Police," *Herald Sun* (Melbourne, Australia), August 7, 2008; Anthony Dowsley, Mark Butler, and Nick Higginbottom, "DNA Samples 'Left Exposed,'" *Herald Sun* (Melbourne, Australia), August 8, 2008; Milanda Rout, "Doubts on DNA Evidence Let Convicted Rapist Walk Free," *Australian,* December 8, 2009.
15. Fran Yeoman, "The Phantom of Heilbronn, the Tainted DNA and an Eight-Year Goose Chase," *London Times Online,* March 26, 2009, http://www.timesonline.co.uk/tol/news/world/europe/article5983279; "A Very Embarrassing History," *SpeigelOnline,* March 26, 2009, http://www.spiegel.de/panorama/justiz/0,1518,druck-615547,00.html.
16. Jackie Valley, "Metro to Probe Old Cases after DNA Mix-up Led to Wrongful Conviction," *Las Vegas Sun,* July 7, 2011, http://www.lasvegassun.com/news/2011/jul/07/dna-lab-switch-led-wrongful-conviction-man-who-ser/.
17. Accounts of the Pennsylvania and California errors (based on a review of laboratory records) are included in William Thompson, Franco Taroni, and Colin Aitken, "How the Probability of a False Positive Affects the Value of DNA Evidence," *Journal of Forensic Sciences* 48 (2003): 47–54; and in Thompson, "Tarnish on the 'Gold Standard.'" The other Las Vegas case is described in Glen Puit, "DNA Evidence: Officials Admit Error, Dismiss Case (LV Lab Put Wrong Name on Sample)," *Las Vegas Review Journal,* April 18, 2002.
18. K. Danks, "DNA Bungles See Wrongful Charges, Jailing," *Daily Telegraph,* January 25, 2007, http://www.news.com.au/dailytelegraph/story/0,22049,21117550-5001028,00.html.
19. A. McDonald, "DNA Evidence Claim Clouds Convictions," *Australian,* July 8, 2006.
20. "DNA Blunder Let Night Stalker Continue," *Independent,* March 24, 2011, http://www.independent.co.uk/news/uk/crime/dna-blunder-let-night-stalker-continue-2252048.html.

21. M. S. Enkoji, "DNA Lapse Puts Scrutiny on Lab Work." *Sacramento Bee,* September 14, 2006.
22. Robert A. Jarzen, "Technical Problem Review—Analyst Casework (Criminalist Mark Eastman)" (unpublished report, Office of the District Attorney, Sacramento County, August 14, 2006; available from the author).
23. John M. Butler, *Forensic DNA Typing: Biology, Technology, and Genetics of STR Markers,* 2nd ed. (Burlington, MA: Elsevier Academic Press, 2005).
24. William Thompson, "Painting the Target around the Matching DNA Profile: The Texas Sharpshooter Fallacy in Forensic DNA Interpretation," *Law, Probability and Risk* 8 (2009): 257–276.
25. Itiel E. Dror and Greg Hampikian, "Subjectivity and Bias in Forensic DNA Mixture Interpretation," *Science and Justice* 51 (2011): 204–208.
26. Brandon Garrett, *Convicting the Innocent: Where Criminal Prosecutions Go Wrong* (Cambridge, MA: Harvard University Press, 2011), 98, 102, 157; Thompson, Taroni, and Aitken, "How the Probability of a False Positive Affects the Value of DNA Evidence."
27. A detailed account of the DNA-testing problems in Sutton's case can be found in William Thompson, "Beyond Bad Apples: Analyzing the Role of Forensic Science in Wrongful Convictions," *Southwestern Law Review* 37 (2008): 971–994. On Gilbert Alejandro, see Garrett, *Convicting the Innocent,* 61, 70, 101; and Brandon Garrett and Peter Neufeld, "Invalid Forensic Science Testimony and Wrongful Conviction," *Virginia Law Review* 95 (2009): 1–94, at 64.
28. During the four and a half years he spent in prison, Josiah Sutton made a number of unsuccessful requests for retesting. His case was rejected by the Innocence Project, which at that time did not consider cases in which DNA testing had already been conducted by the state. A retest was conducted only after problems in the initial DNA test were highlighted in a television exposé about misconduct in the Houston Police Department Crime Laboratory. Thompson, "Beyond Bad Apples."
29. The first report was L. Brenner and B. Pfleeger, "Investigation of the Sexual Assault of Danah H. Philadelphia, PA: Philadelphia Police Department DNA Identification, Laboratory; 1999 Sept. 24; Lab No.: 97-70826." The second report was L. Brenner and B. Pfleeger, "Amended Report: Investigation of the Sexual Assault of Danah H. Philadelphia, PA: Philadelphia Police Department DNA Identification Laboratory; 2000 Feb. 7; Lab No.: 97-70826." Copies of these unpublished reports are available from the author. This case is discussed in Thompson, Taroni, and Aitken, "How the Probability of a False Positive Affects the Value of DNA Evidence." A sample that was described as a "seminal stain" in the first report was relabeled a "bloodstain" in the second report, where it was correctly identified as matching the profile of the female victim.
30. Jonathan Koehler, "The Random Match Probability in DNA Evidence: Irrelevant and Prejudicial?," *Jurimetrics* 35 (1995): 201–219; William Thompson, "Accepting Lower Standards: The National Research Council's Second Report on Forensic DNA Evidence," *Jurimetrics* 37 (1997): 405–424; Lau-

rence Mueller, "The Use of DNA Typing in Forensic Science," *Accountability in Research* 3 (1993): 1–13; Roeder, "DNA Fingerprinting."

31. DNA Advisory Board Quality Assurance Standards for Forensic DNA Testing Laboratories, *Standard 14.1.1,* July 1998, http://www.fbi.gov/about-us/lab/forensic-science-communications/fsc/july2000/codis2a.htm.
32. These records are discussed in more detail in Thompson, "Tarnish on the 'Gold Standard.'"
33. Ibid.
34. Maura Dolan and Jason Felch, "DNA: Genes as Evidence; The Danger of DNA: It Isn't Perfect," *Los Angeles Times,* December 26, 2008, A1. The author of this chapter assisted the *Los Angeles Times* in reviewing the laboratory files and was quoted in the *Times* article.
35. Peter Jamison, "SFPD Concealed DNA Sample Switch at Crime Lab," *San Francisco Weekly,* December 15, 2010; Jaxon Van Derbeken, "San Francisco Police Crime Lab Accused of Cover-Up," *San Francisco Chronicle,* December 5, 2010.
36. Maine State Police: Sharon Mack, "DNA Mix-up Results in Mistrial," *Bangor Daily News,* December 18, 2009; U.S. Army: Marisa Taylor and Michael Doyle, "Army Slow to Act as Crime-Lab Worker Falsified, Botched Tests," McClatchy Newspapers, March 20, 2011, http://www.mcclatchydc.com/2011/03/20/110551/army-slow-to-act-as-crime-lab.html; North Carolina State Bureau of Investigation: "SBI Culture Resists Change," *News Observer* (Raleigh, NC), December 12, 2010. The most detailed account of the Houston Police Department Crime Laboratory scandal is Michael R. Bromwich, "Final Report of the Independent Investigator for the Houston Police Department Crime Laboratory and Property Room," July 13, 2007, http://hpdlabinvestigation.org/. Bromwich was hired by the City of Houston to conduct a detailed investigation of the laboratory.
37. For a detailed review, see Paul Giannelli, "Wrongful Conviction and Forensic Science: The Need to Regulate Crime Labs," *North Carolina Law Review* 163 (2007): 165–236.
38. See Bromwich, "Final Report of the Independent Investigator."
39. Rosanna Ruiz and Robert Crowe, "HPD Again Shuts Down Crime Lab's DNA Unit: Move Follows Resignation of Division's Leader in Cheating Probe," *Houston Chronicle,* January 26, 2008, http://www.chron.com/CDA/archives/archive.mpl?id=2008_4501987.
40. Garrett, *Convicting the Innocent,* 215–221.
41. William Thompson and Rachel Dioso-Villa, "Turning a Blind Eye to Misleading Scientific Testimony: Failure of Procedural Safeguards in a Capital Case," *Albany Law Journal of Science and Technology* 18 (2008): 151–204.
42. William Thompson, "Painting the Target," 269–271. The *Virginian-Pilot* and the *Richmond Times-Dispatch* have published a series of news article and editorials about DNA-testing problems in the Virginia State Division of Forensic Sciences. See, e.g., "Confusion over DNA a Threat to Justice,"

Virginian-Pilot, August 29, 2005; Frank Green, "Study Will Assess Whether Errors in Washington Case Are 'Endemic to the System,'" *Richmond Times-Dispatch,* June 14, 2005; and "Alarming Indifference from Crime Lab Boss," *Virginian-Pilot,* May 10, 2005.

43. See "SBI Culture Resists Change;" also Phoebe Zerwick, "State Crime Lab Is Faulted: Lawyers' Group Calls for Probe, Cites DNA Errors in Three Cases," *Winston-Salem Journal,* July 20, 2005.
44. "SBI Culture Resists Change."
45. Associated Press, "Worker in Army Lab May Have Falsified DNA Test Result," August 27, 2005; Taylor and Doyle, "Army Slow to Act."
46. The problems in the FBI laboratory are described in U.S. Department of Justice, Office of the Inspector General, *The FBI DNA Laboratory: A Review of Protocol and Practice Vulnerabilities* (May 2004), http://www.usdoj.gov/oig/special/0405/index.htm. The firing at Orchid-Cellmark is mentioned in Laura Cadiz, "Md.-Based DNA Lab Fires Analyst over Falsified Tests," *Baltimore Sun,* November 18, 2004; details are provided in Thompson, "Tarnish on the 'Gold Standard,'" which also discusses the Office of the Chief Medical Examiner case. The fraudulent work of the notorious Fred Zain in Bexar County, Texas, is discussed in Garrett and Neufeld, "Invalid Forensic Science," 63–64. The problems with the work of analyst Pamela Fish in Chicago are discussed in Giannelli, "Wrongful Conviction and Forensic Science," 185–187 (describing problems in the Chicago lab).
47. I elaborate on this issue and provide extensive citations and examples in William Thompson, "What Role Should Investigative Facts Play in the Evaluation of Scientific Evidence?," *Australian Journal of Forensic Sciences* 43 (2011): 117–128.
48. See Bromwich, "Final Report of the Independent Investigator."
49. Thompson, "Tarnish on the 'Gold Standard'" (elaborating on these points).
50. Taylor and Doyle, "Army Slow to Act."
51. Giannelli, "Wrongful Convictions and Forensic Science," discusses the misconduct of Zain, Gilchrist, and Fish.
52. B. Weir, "The Rarity of DNA Profiles," *Annals of Applied Statistics* 1, no. 2 (2007): 358–370. In a structured population, particular genetic characteristics or sets of characteristics are more common in some subgroups than others.
53. John M. Butler, *Forensic DNA Typing.* The loci examined are those selected by the FBI for CODIS, the national DNA database. Some of the newer test kits also examine two additional STR loci.
54. In general, as the number of alleles in a DNA profile decreases, the probability that a randomly chosen person will, by coincidence, happen to match that profile increases. Because the alleles vary greatly in their rarity, however, it is possible for a profile containing a few rare alleles to be rarer overall than a profile containing a larger number of more common alleles. Consequently, in discussing the likelihood of a coincidental match, it is more helpful to focus on the estimated frequency of the profile than the number of loci or alleles encompassed in the profile.

55. I computed these profile frequencies (and the match probabilities for relatives presented later in this discussion) using Genostat, a free software program available at Forensic Bioinformatics, http://www.bioforensics.com. Genostat generates profile frequencies for a variety of published databases. For the examples presented here, I used the FBI's database of U.S. Caucasians.
56. Yun S. Song, Anand Patil, Erin Murphy, et al., "Average Probability That a 'Cold Hit' in a DNA Database Search Results in an Erroneous Attribution," *Journal of Forensic Sciences* 54 (2009): 22–26.
57. David J. Balding, *Weight-of-Evidence for Forensic DNA Profiles* (London: John Wiley and Sons, 2005), 32, 148.
58. If genetic profile G from a given individual occurs with probability P_G, then the probability of finding at least one additional individual who has the profile in a population of N unrelated individuals is $1-(1-P_G)^N$. An approximate estimate of this probability is the simpler expression NP_G. National Research Council, *The Evaluation of Forensic DNA Evidence* (Washington, DC: National Academy Press, 1996), 137.
59. British Home Office, *The National DNA Database Annual Report, 2005–2006,* September 2006, 35, http://www.cellmarkforensics.co.uk/assets/docs/DNA-report2005-06.pdf.
60. BBC Panorama, "Give Us Your DNA," September 24, 2007, http://news.bbc.co.uk/go/pr/fr/-/1/hi/programmes/panorama/7010687.stm; Balding, *Weight-of-Evidence for Forensic DNA Profiles,* 32, 148.
61. A. Sweeney and F. Main, "Botched Case Forces State to Change DNA Reports," *Chicago Sun Times,* November 8, 2004.
62. Elizabeth Gibson, "10-Year-Old Case: Retest of DNA Clears Defendant of Charges," *Columbus Dispatch,* June 3, 2010, http://www.dispatch.com/content/stories/local/2010/06/03/retest-of-dna-clears-defendant-of-charges.html.
63. Bromwich, "Final Report of the Independent Investigator"; Thompson, "Beyond Bad Apples."
64. James Curran and John Buckleton, "Inclusion Probabilities and Dropout," *Journal of Forensic Sciences* 55 (2010): 1171–1173; Thompson, "Painting the Target."
65. Koehler, "Error and Exaggeration," 21, 27, 32; Dawn McQuiston-Surrett and Michael J. Saks, "Communicating Opinion Evidence in the Forensic Identification Sciences: Accuracy and Impact," *Hastings Law Journal* 59 (2008): 1159, 1178–1179; Robert Aronson and Jacqueline McMurtrie, "The Use and Misuse of High-Tech Evidence by Prosecutors: Ethical and Evidentiary Issues," *Fordham Law Review* 76 (2007): 1453, 1478–1479; David Balding and Peter Donnelly, "The Prosecutor's Fallacy and DNA Evidence," *Criminal Law Review* (October 1994): 711.
66. *McDaniels v. Brown,* 130 S. Ct. 665, 675 (2010); "Brief of 20 Scholars of Forensic Evidence as Amici Curiae Supporting Respondents," in *McDaniels v. Brown,* 130 S. Ct. 665 (2010). The scholars' brief was reprinted, with brief commentary, in *Criminal Law Bulletin* 46 (2010): 709–757.
67. John M. Butler, *Forensic DNA Typing*; Balding, *Weight-of-Evidence for Forensic DNA Profiles.*

68. David Kaye, *The Double Helix and the Law of Evidence* (Cambridge, MA: Harvard University Press, 2010); Aronson, *Genetic Witness.*
69. The history of litigation on the admissibility of DNA statistics is traced by Kaye, *Double Helix,* 79–114; Aronson, *Genetic Witness,* 56–88; and William Thompson, "The National Research Council's Plan to Strengthen Forensic Science: Does the Path Forward Run through the Courts?," *Jurimetrics* 50 (2009): 35–51, at 41–43.
70. Weir, "Rarity of DNA Profiles."
71. Weir's "Rarity of DNA Profiles" was critiqued in Laurence Mueller, "Can Simple Population Genetic Models Reconcile Partial Match Frequencies Observed in Large Forensic Databases?," *Journal of Genetics* 87 (2008): 101–108. The quotes are from Jason Felch and Maura Dolan, "How Reliable Is DNA in Identifying Suspects?," *Los Angeles Times,* July 20, 2008, A1.
72. 42 U.S.C. Section 14132.
73. Felch and Dolan, "How Reliable Is DNA in Identifying Suspects?" The FBI's resistance is also discussed in Mueller, "Can Simple Population Genetic Models," and David Kaye, "Trawling DNA Databases for Partial Matches: What Is the FBI Afraid Of?," *Cornell Journal of Law and Public Policy* 19 (2009): 145–171.
74. D. Krane, V. Bahn, D. Balding, et al., "Time for DNA Disclosure," *Science* 326 (2009): 1631–1632.
75. Balding, *Weight-of-Evidence for Forensic DNA Profiles,* 14–15.
76. Paul Ekblom, "Can We Make Crime Prevention Adaptive by Learning from Other Ecological Struggles?," *Studies on Crime and Crime Prevention* 8 (1998): 27–51; Ekblom, "How to Police the Future: Scanning for Scientific and Technological Innovations Which Generate Potential Threats and Opportunities in Crime, Policing and Crime Reduction," in *Crime Science: New Approaches to Preventing and Detecting Crime,* ed. M. Smith and N. Tilley (Cullompton, UK: Willan, 2005); William J. Chambliss, *Boxman: A Professional Thief's Journey* (1984; reprint, Lincoln, NE: iUniverse, 2004).
77. Richard Willing, "Criminals Try to Outwit DNA," *USA Today,* August 28, 2000, http://www.fairfaxidlab.com/html/news.html; A. Dutton, "More Rape Cases Go Unsolved: Police and Experts Put Part of the Blame on Crime Show, Which Can Provide Clues on Covering One's Tracks," *Newsday,* September 19, 2006; Simon A. Cole and Rachel Dioso-Villa, "CSI and Its Effects: Media, Juries, and the Burden of Proof," *New England Law Review* 41 (2007): 435–470; "CSI TV Shows Are Training Murderers to Get Away with Their Crimes," *New Criminologist,* March 16, 2006.
78. Harry G. Levine, Jon Gettman, Craig Reinarman, et al., "Drug Arrests and DNA: Building Jim Crow's Database" (paper produced for the Council for Responsible Genetics and its national conference, "Forensic DNA Databases and Race: Issues, Abuses and Actions," held June 19–20, 2008, at New York University), http://www.councilforresponsiblegenetics.org/pagedocuments/0rrxbggaei.pdf.
79. Rebecca Dent, "The Detection and Characterization of Forensic DNA Profiles Manipulated by the Addition of PCR Amplicon" (master's thesis, Centre for Forensic Science, University of Western Australia, 2006).

80. For a critique of the NRC's comments on "culture" and a cogent discussion of what the term "scientific culture" might mean in the context of forensic science, see Simon Cole, "Acculturating Forensic Science: What Is 'Scientific Culture', and How Can Forensic Science Adopt It?," *Fordham Urban Law Journal* 36 (2010): 435–472.
81. Under the 2004 Justice for All Act, agencies applying for federal Coverdell grants to support forensic DNA testing must certify that a government entity exists and an appropriate process is in place to conduct independent investigations into allegations of serious negligence or misconduct substantially affecting the integrity of forensic results. However, a review of the program by the inspector general of the U.S. Department of Justice found that many state and local crime laboratories that had received grants did not meet this standard. For many labs, no qualified entity existed to examine allegations of misconduct, and no process was in place to ensure that allegations were referred to qualified entities. U.S. Department of Justice, Office of the Inspector General, Evaluations and Inspections Division, "Review of the Office of Justice Programs' Paul Coverdell Forensic Science Improvement Grants Program" (January 2008), http://www.justice.gov/oig/reports/OJP/e0801/final.pdf.

16. Nurturing Nature

1. Ruth Hubbard, "The Theory and Practice of Genetic Reductionism—From Mendel's Laws to Genetic Engineering," in *Against Biological Determinism, and Towards a Liberatory Biology,* ed. C. M. Barker, L. Birke, A. D. Muir, et al. (London: Allison and Busby, 1982), 62–78; M. W. Ho and P. T. Saunders, "Adaptation and Natural Selection: Mechanism and Teleology," in *Against Biological Determinism,* 85–102.
2. R. Hubbard and E. Wald, *Exploding the Gene Myth: How Genetic Information Is Produced and Manipulated by Scientists, Physicians, Employers, Insurance Companies, Educators, and Law Enforcers,* with a new afterword (Boston: Beacon Press, 1997); "Exploding the Gene Myth," a conversation between Ruth Hubbard and Frank R. Aqueno, 1997, http://gender.eserver.org/exploding-the-gene-myth.html.
3. Ho and Saunders, "Adaptation and Natural Selection"; M.-W. Ho and P. T. Saunders, "Beyond Neo-Darwinism—An Epigenetic Approach to Evolution," *Journal of Theoretical Biology* 78 (1979): 673–691; M.-W. Ho, *Genetic Engineering: Dream or Nightmare? Turning the Tide on the Brave New World of Bad Science and Big Business,* 2nd ed., rev. and updated (New York: Continuum, 2000). My definition of science is from Ho, *Genetic Engineeering,* 6–7.
4. Ibid.; M.-W. Ho, "Living with the Fluid Genome," ISIS/TWN, London/Penang, 2003, http://www.i-sis.org.uk/fluidGenome.php.
5. "Quantum Jazz Biology," interview with Mae-Wan Ho by David Riley, Rollin McCraty, and Suzanne Schneider, *Science in Society* 47 (forthcoming).
6. Nicholas Wade, "A Decade Later, Genetic Map Yields Few New Cures," *New York Times,* June 12, 2010.

7. M.-W. Ho, "Human Genome Map Spells Death of Genetic Determinism," *ISIS News* 7/8, February 2001, http://www.i-sis.org.uk/isisnews/i-sisnews7-1.php.
8. N. P. Paynter, D. I. Chasman, G. Paré, et al., "Association between a Literature-Based Genetic Risk Score and Cardiovascular Events in Women," *Journal of the American Medical Association* 303 (2010): 631–637.
9. Wade, "Decade Later, Genetic Map Yields Few New Cures."
10. Ho and Saunders, "Adaptation and Natural Selection."
11. M.-W. Ho, "Epigenetic Toxicology," *Science in Society* 41 (2009): 13–15, http://www.i-sis.org.uk/epigeneticToxicology.php.
12. M.-W. Ho, "Epigenetic Inheritance through Sperm Cells: The Lamarckian Dimension in Evolution," *Science in Society* 42 (2009): 40–42, http://www.i-sis.org.uk/epigeneticInheritance.php.
13. I. Sciamanna, P. Vitullo, A. Curatolo, et al., "Retrotransposons, Reverse Transcriptase and the Genesis of New Genetic Information," *Gene* 448 (2009): 180–186.
14. Ho, "Living with the Fluid Genome."
15. Ho, "GM Is Dangerous and Futile."
16. ENCODE Project Consortium, "Identification and Analysis of Functional Elements in 1% of the Human Genome by the ENCODE Pilot Project," *Nature* 2007 (2007): 799–816.
17. M.-W. Ho, "GM Is Dangerous and Futile," *Science in Society* 40 (2008): 4–8, http://www.i-sis.org.uk/isisnews/sis40.php.
18. Ho, "Living with the Fluid Genome."
19. A. Caspi, J. McClay, T. E. Moffitt, et al., "Role of the Genotype in the Cycle of Violence in Maltreated Children," *Science* 297 (2002): 851–854.
20. E. Balaban and R. Lewontin, "Criminal Genes," *GeneWatch* 20 (March/April 2007): 10–15, http://www.councilforresponsiblegenetics.org/genewatch/GeneWatchPage.aspx?pageId=275&archive=yes.
21. M.-W. Ho, "From Genomics to Epigenomics," *Science in Society* 41 (2009): 10–12, http://www.i-sis.org.uk/isisnews/sis41.php.
22. C. H. Waddington, *The Strategy of the Genes* (London: Allen and Unwin, 1957).
23. Ho and Saunders, "Beyond Neo-Darwinism."
24. "Epigenetics," in *Oxford Dictionary of Biochemistry and Molecular Biology,* ed. Richard Cammack et al. (New York: Oxford University Press, 2006), 221.
25. Ho, "Epigenetic Inheritance through Sperm Cells"; M.-W. Ho, "How Development Directs Evolution: Epigenetics and Generative Dynamics" (invited lecture for "Evolution and the Future," Belgrade, Serbia, October 2009), forthcoming in proceedings, http://www.i-sis.org.uk/howDevelopmentDirectsEvolution.php.
26. H.-W. Ho, "Epigenetic Inheritance, "What Genes Remember," *Science in Society* 41 (2009): 4–5, http://www.i-sis.org.uk/isisnews/sis41.php.
27. Ho, "GM Is Dangerous and Futile."
28. C. Lindqvist, A. D. Janczak, D. Nätt, et al., "Transmission of Stress-Induced Learning Impairment and Associated Brain Gene Expression from Parents to

Offspring in Chickens," *PLOS One* 4 (2009): 1–7, e364; M. Weinstock, "The Potential Influence of Maternal Stress Hormones on Development and Mental Health of the Offspring," *Brain and Behavioural Immunology* 19 (2005): 296–308.

29. M. S. Martin-Gronert and S. E. Ozanne, "Maternal Nutrition during Pregnancy and Health of the Offspring," *Biochemical Society Transactions* 34 (2006): 779–792.
30. F. A. Champagne and J. P. Curley, "Epigenetic Mechanisms Mediating the Long-Term Effects of Maternal Care on Development," *Neuroscience and Biobehavioral Reviews* 33 (2009): 593–600.
31. Martin-Gronert and Ozanne, "Maternal Nutrition during Pregnancy."
32. Institute of Corrections, "Relationship between Childhood Experiences and Adult Criminal Behaviour," Corrections Library, October 31, 2006, http://www.nationalinstituteofcorrections.gov/Library/015873.
33. D. Chapman and K. Scott, "The Impact of Maternal Intergenerational Risk Factors on Adverse Development Outcomes," *Developmental Reviews* 21 (2001): 305–325; B. Egeland, D. Jacobvitz, and K. Papatola, *Child Abuse and Neglect: Biosocial Dimensions* (New York: Aldine, 1987).
34. F. A. Champagne, "Epigenetic Mechanisms and the Transgenerational Effects of Maternal Care," *Frontiers in Neuroendocrinology* 29 (2008): 386–397.
35. Champagne and Curley, "Epigenetic Mechanisms Mediating the Long-term Effects of Maternal Care on Development"; Institute of Corrections, "Relationship between Childhood Experiences and Adult Criminal Behaviour"; Chapman and Scott, "Impact of Maternal Intergenerational Risk Factors on Adverse Development Outcomes"; Egeland, Jacobvitz, and Papatola, *Child Abuse and Neglect;* Champagne, "Epigenetic Mechanisms and the Transgenerational Effects of Maternal Care"; M. J. Meaney, "Maternal Care Gene Expression and the Transmission of Individual Differences in Stress Reactivity across Generations," *Annual Review of Neuroscience* 24 (2001): 1161–1192; M.-W. Ho, "Caring Mothers Reduce Stress for Life," *Science in Society* 24 (2004): 11+55, http://www.i-sis.org.uk/isisnews/sis24.php; M.-W. Ho, "Caring Mothers Strike Fatal Blow against Genetic Determinism," *Science in Society* 41 (2009): 6–9, http://www.i-sis.org.uk/isisnews/sis41.php.
36. Champagne, "Epigenetic Mechanisms and the Transgenerational Effects of Maternal Care."
37. F. A. Champagne, I. C. Weaver, J. Diorio, et al., "Maternal Care Associated with Methylation of the Estrogen Receptor-Alpha1b Promoter and Estrogen Receptor-Alpha Expression in the Medial Preoptic Area of Female Offspring," *Endocrinology* 147 (2006): 2909–2915.
38. See Champagne, "Epigenetic Mechanisms and the Transgenerational Effects of Maternal Care."
39. L. A. Smit-Rigter, D. L. Champagne, and J. A. van Hooft, "Lifelong Impact of Variations in Maternal Care on Dendritic Structure and Function of Cortical Layer 2/3 Pyramidal Neurons in Rat Offspring," *PLoS One* 4 (2009): 1–7, e5167, doi:10.1371/journal.pone.0005167.

40. C. G. Weaver, M. J. Meaney, and M. Szyf, "Maternal Care Effects on the Hippocampal Transcriptome and Anxiety-Mediated Behaviours in the Offspring That Are Reversible in Adulthood," *Proceedings of the National Academy of Sciences of the United States of America* 103 (2006): 3480–3485.
41. J. A. Arai, S. Li, D. M. Harley, et al., "Transgenerational Rescue of a Genetic Defect in Long-Term Potentiation and Memory Formation by Juvenile Enrichment," *Journal of Neuroscience* 29 (2009): 1496–1502.
42. P. O. McGowan, A. D. Sasaki, A. C. D'Alessio, et al., "Epigenetic Regulation of the Glucocorticoid Receptor in Human Brain Associates with Childhood Abuse," *Nature Neuroscience* 12 (2009): 342–348.
43. A. R. Tyrka, L. H. Price, H.-T. Kao, et al., "Childhood Maltreatment and Telomere Shortening: Preliminary Support for an Effect of Early Stress on Cellular Aging," *Biological Psychiatry* 67 (2010): 531–534.
44. Champagne, "Epigenetic Mechanisms and the Transgenerational Effects of Maternal Care."
45. P. O. McGowan, M. J. Meaney, and M. Szyf, "Diet and the Epigenetic (Re) programming of Phenotypic Differences in Behaviour," *Brain Research* 1237 (2008): 12–24.
46. T. B. Franklin and I. M. Mansuy, "The Prevalence of Epigenetic Mechanisms in the Regulation of Cognitive Functions and Behaviour," *Current Opinion in Neurobiology* 10 (2010): 1–9.
47. McGowan, Meaney, and Szyf, "Diet and the Epigenetic (Re)programming of Phenotypic Differences in Behaviour."

Conclusion

Epigraph sources: "President Clinton Announces the Completion of the First Survey of the Human Genome: Hails Public and Private Efforts Leading to This Historic Achievement," White House Office of the Press Secretary, June 25, 2000, U.S. Department of Energy Genome Programs, http://genomics.energy.gov; Francis Collins, "A Genome Story: 10th Anniversary Commentary," *Scientific American Blog*, June 25, 2010, http://blogs.scientificamerican.com.

1. Leon Jaroff, "Happy Birthday, Double Helix," *Time*, March 15, 1993.
2. Michael R. Alvarez, "Genes and Social Networks: New Research Links Genes to Friendship Networks," *Psychology Today*, February 14, 2011; Charlene Laino, "New Alzheimer's Gene Found," *WebMD Health News*, April 14, 2010, http://www.webmd.com; Robert Cooke, "Search Narrowing for Alzheimer's Gene," *Newsday*, February 20, 1987; Mark Wheeler, "Obesity Gene, Carried by More than a Third of the U.S. Population, Leads to Brain Tissue Loss," *UCLA Newsroom*, April 19, 2010; Jane Kirby, "Thin Parents Pass on 'Skinny Genes' to Children," *Independent*, October 4, 2011.
3. Kenna R. Mills Shaw, Katie Van Horne, Hubert Zhang, et al., "Essay Contest Reveals Misconceptions of High School Students in Genetics Content," *Genetics* 178 (2008): 1157–1168, at 1166.

4. M. J. Dougherty, C. Pleasants, L. Solow, et al., "A Comprehensive Analysis of High School Genetics Standards: Are States Keeping Pace with Modern Genetics?," *CBE-Life Sciences Education* 10 (2011): 318–327.
5. NIH Curriculum Supplements for High School, "Human Genetic Variation," 1999, http://science.education.nih.gov/customers.nsf/HSGenetic.htm.
6. Sheldon Krimsky and Kathleen Sloan, eds., *Race and the Genetic Revolution: Science, Myth, and Culture* (New York: Columbia University Press, 2011); Perez Hilton, "Hypocrisy Must Be Genetic for the Palins," *Perez Hilton Blog,* April 7, 2011, http://perezhilton.com.
7. Camilla Burnett, Sara Valentini, Filipe Cabreiro, et al., "Absence of Effects of Sir2 Overexpression on Lifespan in *C. Elegans* and *Drosophila,*" *Nature* 477 (2011): 482–485.
8. J. Ferlay, H. R. Shin, F. Bray, et al., GLOBOCAN, *Cancer Incidence and Mortality Worldwide: IARC CancerBase, 10* (Lyon, France: International Agency for Research on Cancer, 2008), http://globocan.iarc.fr/.
9. Fabio Pammolli and Massimo Riccaboni, *Innovations and Industrial Leadership: Lessons from Pharmaceuticals* (Baltimore: Center for Transatlantic Relations, Johns Hopkins University, 2008).
10. Fabio Pammolli, Laura Magazzini, and Massimo Riccaboni, "The Productivity Crisis in Pharmaceutical R&D," *Nature Reviews Drug Discovery* 10 (2011): 428–438.
11. Evaluation of Genomic Applications in Practice and Prevention (EGAPP), "EGAPP Working Group Recommendations," February 9, 2012, http://www.egappreviews.org/default.htm.
12. Bourne Partners, "Biotechnology Market Overview," last updated July 1, 2011, http://bournepartners.wordpress.com/2011/07/01/biotechnology-market-overview-june-2011/.
13. Bill Clinton, "President Clinton's State of the Union Address" (speech delivered to Congress, Washington, DC, January 27, 1998).
14. Jennifer Reineke Pohlhaus and Robert M. Cook-Deegan, "Genomics Research: World Survey of Public Funding," *BMC Genomics* 9 (2008): 8, 15.
15. Biotechnology Industry Organization, "Unleashing the Promise of Biotechnology," June 28, 2011, http://www.bio.org/unleashing-promise-biotechnology.
16. White House Science and Technology Policy Office, "Request for Information: Building a 21st Century Bioeconomy," October 11, 2011, http://www.federalregister.gov/agencies/science-and-technology-policy-office.
17. Biotechnology Industry Organization, "BIO Applauds the Intent to Create a National Bioeconomy Blueprint," press release, October 13, 2011, http://www.bio.org/media/press-release/bio-applauds-intent-create-national-bioeconomy-blueprint.
18. Sheldon Krimsky, *Science in the Private Interest: Has the Lure of Profits Corrupted the Virtue of Biomedical Research?* (Lanham, MD: Rowman and Littlefield, 2003).
19. Phyllis Gardner, MD, "NIH: Moving Research from the Bench to the Bedside" (testimony before the House Subcommittee on Health, Washington, DC, July 10, 2003).

20. Ibid.
21. Government Accountability Office, "Direct-to-Consumer Genetic Tests: Misleading Test Results Are Further Complicated by Deceptive Marketing and Other Questionable Practices," GAO-10-847T (Washington, DC, July 22, 2010).
22. 23andMe, https://www.23andme.com/health.
23. Navigenics. http://www.navigenics.com/visitor/genetics_and_health.
24. DeCODEme, http://www.decodeme.com.
25. Pathway Genomics, https://www.pathway.com/dna-reports/health-conditions.
26. Representative Stupak, "Direct-to-Consumer Genetic Testing and the Consequences to Public Health" (House Committee on Energy and Commerce Subcommittee on Oversight and Investigations Hearing, Washington, DC, July 22, 2010).
27. Thomas Lord, "Cal's Genetic Testing of Freshmen: Retreat and Declare Victory," *Berkeley Daily Planet*, August 12, 2010.
28. Claire Robinson, "Biotech Critic Denied Tenure," *Science in Society* 21 (2004): 33.
29. W. C. Willett, "Balancing Life-style and Genomics Research for Disease Prevention," *Science* 296, no. 5568 (2002): 695–698.
30. Colin Mathers, Gretchen Stevens, and Maya Mascarenhas, *Global Health Risks: Mortality and Burden of Disease Attributable to Selected Major Risks* (Geneva, Switzerland: World Health Organization, 2009), 31.
31. "NIH Estimates of Funding for Various Research, Condition, and Disease Categories (RCDC) Table," last updated February 13, 2012, http://report.nih.gov/rcdc/categories/Default.aspx.
32. Christopher King, "Hottest Research of 2010," Thomson Reuters Research Analytics Group, last updated March 2011, *Thomson Reuters Science Watch Survey,* http://sciencewatch.com/ana/fea/11maraprFea/.
33. "Tuberculosis Fact Sheet," World Health Organization, November 2010, http://www.who.int/mediacentre/factsheets/fs104/en/.
34. Christian Frech and Nansheng Chen, (2011) "Genome Comparison of Human and Non-Human Malaria Parasites Reveals Species Subset-Specific Genes Potentially Linked to Human Disease," *PLoS Computational Biology* 7(12): e1002320. doi:10.1371/journal.pcbi.1002320.

Selected Readings

Carey, Nessa. *The Epigenetics Revolution: How Modern Biology Is Rewriting Our Understanding of Genetics, Disease, and Inheritance.* New York: Columbia University Press, 2012.

Cranor, Carl F., ed. *Are Genes Us?* New Brunswick, NJ: Rutgers University Press, 1994.

Fagot-Largeault, Anne, Shahid Rahman, and Juan Manuel Torres, eds. *The Influence of Genetics on Contemporary Thinking.* Vol. 6 of *Logic, Epistemology, and the Unity of Science.* Dordrecht: Springer, 2007.

Francis, Richard C. *Epigenetics: The Ultimate Mystery of Inheritance.* New York: W. W. Norton, 2011.

Gould, Stephen Jay. *The Mismeasure of Man.* New York: W. W. Norton, 1981.

Hubbard, Ruth, and Elijah Wald. *Exploding the Gene Myth: How Genetic Information Is Produced and Manipulated by Scientists, Physicians, Employers, Insurance Companies, Educators, and Law Enforcers.* With a new preface. Boston: Beacon Press, 1999.

Jablonka, Eva, and Marion J. Lamb. *Evolution in Four Dimensions: Genetic, Epigenetic, Behavioral, and Symbolic Variation in the History of Life.* Cambridge, MA: MIT Press, 2005.

Keller, Evelyn Fox. *The Century of the Gene.* Cambridge, MA: Harvard University Press, 2000.

Keller, Evelyn Fox. *The Mirage of a Space between Nature and Nurture.* Durham, NC: Duke University Press, 2010.

Lewontin, R. C. *Biology as Ideology: The Doctrine of DNA.* New York: HarperCollins, 1992.

Lewontin, R. C. *The Triple Helix: Gene, Organism, and Environment.* Cambridge, MA: Harvard University Press, 2000.

Lewontin, R. C., Steven Rose, and Leon J. Kamin. *Not in Our Genes: Biology, Ideology, and Human Nature.* New York: Pantheon, 1984.

Moore, David S. *The Dependent Gene: The Fallacy of "Nature vs. Nurture."* New York: Times Books, 2002.
Moss, Lenny. *What Genes Can't Do*. Cambridge, MA: MIT Press, 2003.

Acknowledgments

This book began as a project of the Council for Responsible Genetics under the direction of Sheldon Krimsky. We are grateful for the early work of Dr. Nadine Weidman, who, as a consultant to the council, helped in an initial selection of contributors. We would like to thank Andrew Kimbrell and Andrew Thibedeau for their help in the development of this book. We also wish to acknowledge the substantial support of both Abby Rockefeller and the Cornerstone Campaign.

Michael Fisher, Executive Editor for Science and Medicine at Harvard University Press, provided editorial guidance that helped the editors clarify the book's goals and audience and suggested additional topics for inclusion. Finally, we wish to express our appreciation to the copyeditor, Charles Eberline, and to John Donohue of Westchester Publishing Serivces.

Chapter 1 is reprinted with permission from Routledge, Taylor Francis Group, from *Rethinking Marxism* 15 (2003): 515–522.

Chapter 3 is reprinted with a new conclusion from "Changing the Question to One That Does Make Sense—From Trait to Trait Difference," in *The Mirage of a Space between Nature and Nurture,* by Evelyn Fox Keller, 31–52. © 2010, Duke University Press, www.dukeupress.edu. All rights reserved. Reprinted by permission of the publisher.

Chapter 6 is based on a lecture delivered at the conference "Bioscience and Society: Biodiversity," held on October 1–2, 2009, in Ljubljana, Slovenia.

Chapter 7 was supported by grants from the Bradshaw Foundation, the Parsemus Foundation, the Great Neck Breast Cancer Coalition, the Babylon Breast Cancer Coalition, and the National Institutes of Environmental Health Sciences (grants ES08314 and ES013884).

An earlier version of Chapter 15 was originally prepared for the Council for Responsible Genetics (CRG) and its national conference, "Forensic DNA Databases and Race: Issues, Abuses and Actions," held June 19–20, 2008, at New

York University. An excerpt was published in the council's magazine *Gene-Watch,* November/December 2008. The chapter also draws, in a few instances, on materials the author previously published in "Tarnish on the 'Gold Standard': Understanding Recent Problems in Forensic DNA Testing," *Champion* 30 (January 2006): 10–16.

Contributors

JONATHAN BECKWITH is American Cancer Society Professor at Harvard Medical School, specializing in bacterial genetics. He is active in examining the interaction between genetics and society and is a long-term participant in the Genetics and Society Working Group.

CARL F. CRANOR is Distinguished Professor of Philosophy and faculty member of the Environmental Toxicology Graduate Program at the University of California, Riverside. He has focused on philosophical issues concerning risks, science, and the law.

JEREMY GRUBER is lawyer and president of the Council for Responsible Genetics, a public policy organization that represents the public interest regarding the social, ethical, and environmental implications of developing genetic technologies. His expertise is on issues of genetic privacy and discrimination.

MARTHA R. HERBERT is pediatric neurologist and assistant professor of neurology at Harvard Medical School, specializing in neurodevelopmental disabilities at the Massachusetts General Hospital, where she cofounded and directs the TRANSCEND Research Program, a multimodal and multisystem brain research program.

MAE-WAN HO is director and cofounder of the Institute of Science in Society in the United Kingdom and editor-in-chief and art director of its quarterly magazine *Science in Society*. She advises national and international agencies on science and science policy issues such as genetically modified foods, sustainable agriculture, and renewable energies.

RUTH HUBBARD was professor of biology at Harvard University from 1973 to 1990 and has been professor emerita since 1990. In recent years she has focused on the science and politics of women's health, the critique of genetic reductionism, and the dangers of eugenics.

EVA JABLONKA is professor in the Cohn Institute for the History and Philosophy of Science and Ideas, Tel Aviv University. Her main interest is the understanding of evolution, especially evolution that is driven by nongenetic hereditary variations, and the evolutionary origins of consciousness.

DAVID JONES is A. Bernard Ackerman Professor of the Culture of Medicine at Harvard University. His initial research focused on explanations of health inequalities, and his current research explores the history of decision making in cardiac therapeutics, such as how cardiologists and cardiac surgeons justify and implement new technology.

JAY JOSEPH is a licensed psychologist practicing in the San Francisco Bay Area. His published work in professional journals, books, and book chapters has focused on a critical appraisal of genetic research and theories in psychiatry and psychology.

EVELYN FOX KELLER is professor emerita of history and philosophy of science in the Program in Science, Technology and Society at the Massachusetts Institute of Technology. Her work has focused on the history of molecular genetics and human developmental and evolutionary biology.

SHELDON KRIMSKY is professor of urban and environmental policy and planning and adjunct professor of public health and family medicine at Tufts University. His work focuses on the interactions among science/technology, values/ethics, and public policy in the areas of genetics, biotechnology, environmental risk, chemical health, and ethics in science.

SUSAN LINDEE is a historian and associate dean of the social sciences at the University of Pennsylvania. She studies historical and contemporary questions raised by human and medical genetics, genomic medicine, and the concept of the gene in popular culture.

DAVID S. MOORE is professor of psychology at Pitzer College and Claremont Graduate University, specializing in developmental cognitive neuroscience with a focus on infants' perception of infant-directed speech and on the development of spatial and numerical cognition in infancy.

STUART A. NEWMAN is professor of cell biology and anatomy at New York Medical College, where he directs a research program in developmental biology and evolutionary theory and contributes to several scientific fields, including cell differentiation, theory of biochemical networks and cell pattern formation, protein folding and assembly, and evolutionary developmental biology.

CARL RATNER is director of the Institute for Cultural Research and Education, Trinidad, California. He has developed a cultural approach to psychology, known as macro cultural psychology, which treats psychological phenomena as primarily organized by macrocultural factors.

SHIRLEY SHALEV is a faculty member of the Women, Gender and Health Concentration at Harvard School of Public Health. She conducts her academic work at the Department of Global Health and Social Medicine at Harvard Medical School, where she examines contemporary practices of reproductive technologies and complex interrelations between scientific advances, genetic research, bioethics, mass communication, global commerce, and reproductive rights.

CARLOS SONNENSCHEIN is professor of anatomy and cellular biology at Tufts University School of Medicine. His research interests have centered on the control of cell proliferation by estrogens and androgens, the impact of endocrine disruptors on organogenesis and reproductive function, carcinogenesis during adult life, and the role of stroma/epithelial interactions on rat and human mammary carcinogenesis.

ANA M. SOTO is professor of anatomy and cellular biology at Tufts University School of Medicine and a professor of cancer development at the University of Ulster, U.K. Her research interests have centered on the control of cell proliferation by sex steroids, fetal origins of adult disease, the role of stroma/epithelial interactions on organogenesis and carcinogenesis, and currently on a systems biology approach to investigate morphogenesis.

STEPHEN L. TALBOTT is senior researcher at the Nature Institute in Ghent, New York. His current work relates to genetics, evolution, biotechnology, and the quest for a qualitative science that is faithful to the objective qualities of the world.

WILLIAM C. THOMPSON is professor in the Department of Criminology, Law and Society and the School of Law at the University of California, Irvine. His research focuses on the use and misuse of scientific and statistical evidence in the courtroom and on jurors' reactions to such evidence. He occasionally represents clients in cases involving novel scientific and statistical issues.

Index

Note: Page numbers followed by *f* and *t* indicate figures and tables.